Radiation Dosimetry

Physical and Biological Aspects

RADIATION DOSIMETRY
Physical and Biological Aspects

Radiation Dosimetry

Physical and Biological Aspects

Edited by

Colin G. Orton

Wayne State University School of Medicine
Harper-Grace Hospitals
Detroit, Michigan

Springer Science+Business Media, LLC

Library of Congress Cataloging in Publication Data

Main entry under title:

Radiation dosimetry.

Includes bibliographies and index.
1. Radiation dosimetry. 2. Radiation—Physiological effect. I. Orton, Colin G.
[DNLM: 1. Radiation Dosage. 2. Radiation Effects. 3. Radiation, Ionizing. 4. Dosimetry. WN 650 R1269]
R905.R32 1986 612′.01448 85-30063
ISBN 978-1-4899-0573-4

ISBN 978-1-4899-0573-4 ISBN 978-1-4899-0571-0 (eBook)
DOI 10.1007/978-1-4899-0571-0

Originally published by Plenum Press, New York in 1986
Softcover reprint of the hardcover 1st edition 1986

Contributors

Peter R. Almond, Department of Therapeutic Radiology, James Graham Brown Cancer Center, School of Medicine, University of Louisville, Louisville, Kentucky

Anders Brahme, Department of Radiation Physics, Karolinska Institute, and University of Stockholm, Stockholm, Sweden

Brian Diffey, Regional Medical Physics Department, Dryburn Hospital, Durham, U.K.

Colin G. Orton, Radiation Oncology Center, Harper–Grace Hospitals, Wayne State University School of Medicine, Detroit, Michigan

Harald H. Rossi, Radiological Research Laboratory, Department of Radiology, Cancer Center/Institute of Cancer Research, Columbia University College of Physicians and Surgeons, New York, New York

Hans Svensson, Radiation Physics Department, University of Umeå, Umeå, Sweden

Marco Zaider, Radiological Research Laboratory, Department of Radiology, Cancer Center/Institute of Cancer Research, Columbia University College of Physicians and Surgeons, New York, New York

Preface

Mankind has evolved in a sea of radiation. We have been bombarded constantly by X rays, γ rays, UV rays, and particulate radiations from outer space, and by terrestrial radiations from the ground we walk on, from our building materials, and from our own bodies. Recently, we have become increasingly subjected to man-made radiations, especially from the medical and defense industries. All of these radiations are capable of affecting us biologically, both to our benefit and to our detriment. This book provides a thorough review of the physical and biological dosimetry of these radiations. It is targeted to those health professionals who are concerned with understanding the mechanisms fundamental to the biological action of ionizing radiation or who are involved in the application, measurement, or treatment of the effects of such radiations.

The first chapter, on "Bioeffect Dosimetry in Radiation Therapy," should be of special interest to anyone involved in the treatment of cancer by radiation. It includes a brief review of the history of the manipulation of time–dose parameters in order to improve therapeutic benefit, and an up-to-date analysis of time–dose relationships designed for use in fractionated radiotherapy and brachytherapy. This is followed by two chapters reviewing and comparing national and international protocols for the precise measurement of photon and electron radiations in therapy. These chapters should be invaluable to radiation physicists responsible for treatment machine calibrations. The next chapter, on "Microdosimetry and Its Application to Biological Processes," provides a detailed analysis of the spectrum of ionizing events ultimately responsible for observed biological effects such as cellular lethality, carcinogenesis, and mutations. This is of special concern to those interested in bioeffect dosimetry and epidemiology. Finally, the chapter on "Ultraviolet Radiation Dosimetry and Measurement" presents an overview of the biological effects of UV radiation, physical and biological dosimetry, and methods of protection.

This book should be useful to radiation physicists, biologists, and technologists, radiotherapists, dermatologists, epidemiologists, and all those professionals concerned with the biological effects and uses of radiation.

Colin G. Orton

Detroit, Michigan

Contents

3 Recent Advances in Electron and Photon Dosimetry

Hans Svensson and Anders Brahme 87

4 Microdosimetry and Its Application to Biological Processes 171

Marco Zaider and Harald H. Rossi

Part I. Physics

Chapter 1

Bioeffect Dosimetry in Radiation Therapy

Colin G. Orton

1. INTRODUCTION

Interest in understanding and manipulating time, dose, and fractionation in an attempt to find the "optimal" therapy began on Wednesday, January 29, 1896, less than three months after the discovery of X rays. This was the day that Emil Grubbé initiated the first of several 1-h daily treatments to a Mrs. Rose Lee, who had an advanced carcinoma of the breast. The choice of technique was relatively simple. Grubbé placed the X-ray tube in direct contact with the lesion and treated for the maximum time period he considered reasonable for the comfort of the patient. Since the output of the X-ray tube was so low, it was necessary to deliver multiple daily treatments in order to produce a marked effect. Grubbé had discovered fractionated radiotherapy. It apparently worked to a limited extent, since Grubbé reported marked tumor regression. Little did he realize that he had started a long, tortuous trail in search of the "ideal" treatment technique, a trail which was to take many sudden twists and turns, even on occasions going backwards. We are still on this trail today, although, hopefully, we are progressing mainly in a forward direction.

Another milestone in the development of radiotherapy techniques came about five years later when Robert Abbé, chief surgeon at St. Luke's Hospital in New York, performed the first interstitial radium

Colin G. Orton • Radiation Oncology Center, Harper–Grace Hospitals, Wayne State University School of Medicine, Detroit, Michigan 48201.

implant. As was the case with Grubbé, the technique was inspirational, as of course it had to be. It was also surprisingly close to what we would now consider "good" therapy, considering the limited facilities he had available. The patient was treated for hyperthyroidism by implanting a 100-mg glass encapsulated radium tube into the hypertrophic lobe of the thyroid. It was left there for one day. After removal, the patient went home and remained in good health for at least the next six years, according to Abbé's records.

Unfortunately, these early "successes" were followed by more than a fair share of failure. Oddly enough, these early failures were in part due to improvements in X-ray machine technology. Once X-ray tubes were capable of delivering high enough dose rates to allow a "tumoricidal" dose to be delivered in one sitting, it became fashionable to treat patients with large single doses instead of fractionating them over several sessions. The most commonly used technique was to deliver a dose about 10% higher than that which would produce a brisk erythema. It has since been surmised that this was a dose of about 1100 cGy. It was reported that some skin lesions were cured but no deep tumors. For these deeper cancers, not only was lack of fractionation a problem but so also was penetration. It was not until 1921[1] that the first orthovoltage (deep therapy) treatment was given.

Much experimentation took place in these early years of radiotherapy before fractionation became firmly established in the 1930s. The problem was that many of these "research" activities produced contradictory results. For example, from animal experiments Krönig and Friedrich[2] reported in 1918 that fractionation reduced the effectiveness of a given dose of radiation. This was misinterpreted to mean that it would be best to deliver radiotherapy in large single doses since these would be most effective on the tumor. Earlier, in 1914, Schwarz[3] had theorized that fractionation would be better for curing tumors, since this would allow the greatest number of cancer cells to be caught in mitosis. These conflicting theories led to two opposing concepts of radiotherapy. On the one hand there were those who considered single dose therapy to be ideal,[4] whereas others, most notable being the Fondation Curie in Paris, reverted back to the daily fractionation of the pioneers. There were even some well-known "recipes" which tried to make the most of both worlds. A good example is the Pfahler–Kingery technique of the 1920s,[5,6] in which large initial doses were followed by gradually diminishing fractionated doses over the next 2–3 weeks.

It was not until 1932, when Coutard[7] revealed the results of a careful study of fractionation, that it was finally realized that only by fractionation would the differential response of tumors and normal tissues be adequately exploited. Fortunately this coincided with two other major innovations: (a) the development of a variety of machines capable of producing more deeply penetrating X rays for therapy in the early 1930s and (b) the perfection of the use of ionization chambers to measure radiation accurately in the new international unit, the Roentgen.[8] Modern external beam radiotherapy was born.

The modern era for brachytherapy possibly started a few years earlier. For example, between 1924 and 1929, Geoffrey Keynes, a surgeon at St. Bartholomew's Hospital, London, implanted over 150 patients with radium for primary breast carcinoma. Several fractions were delivered. Indeed, the technique was surprisingly similar to that we would use with Ir-192 in the 1980s. The results were equally impressive and Keynes firmly believed that this technique, which conserved function and appearance of the breast and avoided mutilating surgery, was the treatment of choice for the future. He also believed that this thesis was unlikely to receive rapid acceptance by his surgical colleagues. His insight was impressive.

Numerous other successful applications of radium (and radon) interstitial, intracavitary, and mold therapy were given a firm scientific, quantitative basis during these early years. Edith Quimby, in New York, developed her radium dosage tables,[9] which are still in use today, albeit in a slightly modified form. On the other side of the Atlantic, especially in France and Sweden, and ultimately in Britain with the advent of the Manchester System,[10] the methods of application of radium and radon therapy were formalized, especially in terms of controlling and exploiting the interrelationship between dose, dose rate, and treatment time.

The successes in both external beam treatments and brachytherapy in the early 1930s were impressive—so impressive in fact, that there was little incentive to try to manipulate time and dose in radiotherapy in the hope of achieving better results. In this respect, very few innovative treatment regimes were tried for a period of about 30 years. One of the few notable excursions into the "unknown" was that of Baclesse in the early 1950s, when he attempted to exploit the regenerative capacity of normal tissues by delivering small, daily fractions over very long periods to high doses.[11] Typically, he would give over 50 treatments in about 11 weeks. With this technique it was possible to deliver doses in excess of 9000 cGy without exceeding normal tissue tolerance. Unfortunately, Baclesse discovered that this excessive protraction of his radiotherapy also failed to exceed the tolerance of carcinomas. It was an unsuccessful but valiant experiment.

The 1960s saw a resurgence of time–dose research, possibly spurred on by a surge of interest in the application of radiobiological concepts to radiotherapy. Attempts to make optimal use of the 4 R's of radiotherapy, repair, repopulation, reoxygenation, and redistribution, were rampant. They still are.

We have tried, and still are trying, to exploit repopulation and reoxygenation by the use of split-course radiotherapy.[12] We continue to investigate the optimal use of repair and redistribution by the application of more than one fraction per day.[13] This is being attempted without decreasing the overall treatment time (hyperfractionation) by numerous investigators, such as Backstrom *et al.*[14] and Horiot *et al.*,[15] and more recently cooperative clinical trials groups.[16] Alternatively, multiple daily treatments have been delivered with a significant decrease in overall time (accelerated fractionation),[15,16] as reported, for example, at four fractions per day by Svoboda,[17] or two fractions per day by Chu and Suit,[18] although the latter do in fact allow a rest period during the course of therapy if acute reactions become excessive.

The early results from these hyperfractionation and accelerated fractionation experiments look promising for the treatment of certain cancers in certain sites.[16] It looks as if the ability to control the disease without damaging normal tissues excessively (i.e., the therapeutic ratio) can be increased with these techniques. The same cannot be said for attempts to incorporate large doses per fraction into the treatment schedule, which has been tried in many ways.[13] However, before describing these, it should be stated that, although the results do not look particularly promising, this does not mean that they have no potential utility. Most of these have simply not been adequately studied (controlled clinical trials), and it could well be that they might be effective if applied appropriately. They have also been used primarily for the treatment of far advanced cancers or on patients with a poor prognosis, so long-term tumor control and/or late effects of the radiotherapy have not been demonstrated. Presumably, the rationale for these schedules was to induce rapid reoxygenation of large, advanced cancers. Extensive attempts to use high doses per fraction have been made by Schumacher in West Berlin[19] and by Eichhorn in East Berlin,[20] both for the treatment of bronchial lesions. The courses of therapy tried by Schumacher were as follows:

1. 600 cGy delivered twice per week.
2. 1000 cGy per fraction delivered at increasing intervals from 1 week (first), 2 weeks (second), and then 4 weeks.
3. 3 × 900 cGy fractions delivered at 1 week (first) and 2 week (second) intervals, followed by 8 weekly treatments of 500 cGy.

Eichhorn's experiments were somewhat similar to Schumacher's:

1. A single 800-cGy dose followed by 21 daily doses of 250 cGy.

2. 11 × 550 cGy given every 5 days.
3. Daily doses of 1000, 800, 800 cGy, followed by 17 daily treatments of 250 cGy.

The experiments of Schumacher and Eichhorn were presumably intended to take maximum advantage of reoxygenation. In general, the results were not good and led to a high incidence of late effects such as pneumonitis in the few long-term survivors.

Somewhat more conventional attempts at using high dose fractions in a more regular way have been a little more successful. For example, Habermalz and Fischer[21] demonstrated moderate success in the treatment of malignant melanoma. Typically, these hypofractionation regimes consist of 15 or less fractions, each of dose 300 cGy or more. Although it is often not explicitly stated, it must be presumed that these techniques are used either to decrease the overall treatment time for the convenience of patients or in order to minimize the repair of cancer cells and possibly improve reoxygenation. Recent results have shown, however, that late reactions can be a very significant problem with hypofractionation. This is worsened by the fact that these techniques minimize the occurrence of early reactions.[22–24] Hence, previous experience with conventional radiotherapy, with which early reactions could usually be used as predictors of late effects, was no longer viable when high-dose fractions were used. Consequently, hypofractionation techniques have been used sparingly by the radiotherapy community, especially in the United States, where the fear of malpractice litigation makes it even more desirable to foresee untoward deleterious effects of treatment.

Similar problems with late effects with a somewhat different use of high-dose fractions are being observed by Pierquin, who is treating advanced head and neck cancers with low dose-rate teletherapy.[25] Patients are treated at dose rates of about 2 cGy/min in approximately 8-h sessions each day for, typically, seven successive days. The daily fractions are approximately 1000 cGy each. Early reactions are quite acceptable as also is the tumor regression. Unfortunately, necrosis is significant in long-term survivors. This work is continuing.[26]

Also still continuing are various attempts to optimize *brachytherapy* by the manipulation of dose rate and fractionation. By far the most widely accepted, although probably still far from optimal, is the use of split courses of intracavitary irradiation for the treatment of various gynecological cancers. These intracavitary sessions are usually interspersed with external beam treatments, sometimes with the central region of high dose from the intracavitary sources shielded, sometimes not. The variety of techniques is enormous, although essentially they all derived from the original Stockholm technique,[27] where three intracavitary applications were spaced at 2-week intervals.

An interesting time–dose modification of this technique is rapidly gaining in popularity, namely, the use of high dose-rate fractionated treatments.[28,29] Although initially intended primarily for the treatment of gynecological lesions, these methods are beginning to be applied to the intracavitary treatment of cancers of the esophagus, nasopharynx, bronchus, biliary tract, and rectum.[30] Also, high dose-rate equipment is now being used for interstitial treatments.[31]

Early results of many of the trials of high dose-rate remote afterloading for cervix cancer treatments were not promising. There were several reports of increased occurrence of radiation injuries. The problem was discovered to be in the choice of dose. Owing to lack of any previous experience, many institutions were forced to use time–dose relationships to predict appropriate doses. Some of these relationships have now been shown to be unreliable for these extreme schedules.[32] It appears that this problem has now been solved. The bad news is that hundreds of institutions around the world have adopted these high dose-rate techniques before there has been enough evidence accrued to demonstrate their efficacy. The good news is that, because so many institutions are doing it, it will not take long to determine whether or not these methods are as good as, or maybe even better than, the conventional treatments they are replacing.

Another recent development has been the use of very low dose-rate brachytherapy with I-125 seeds. These are usually used as permanent implants with initial dose rates less than 10 cGy/h. With a half-life of 60 days, this means that the irradiation takes place at low dose rates over very long periods. No previous experience was available when I-125 permanent implants first began to be used, so the choice of activity, and hence total dose, was somewhat of an inspired guess. Recent results have shown that this guess was inspired indeed, since the technique has worked rather well, despite the fact that the "effective" activities of the early I-125 sources may have been overestimated by as much as 30%.[33]

It should be clear by now that developments of many innovative techniques in radiation therapy have been slowed down considerably owing to the lack of a method to predict what dose of radiation would be needed to produce the required effect. Had a reliable time–dose relationship been available, many of the successful fractionation and dose-rate techniques now used routinely would have been developed years sooner. Countless more lives would have been saved and countless painful injuries would have been avoided.

Obviously, this is a problem we have long realized. We have not stood still for the past 90 years of radiotherapy in the hope that this problem would solve itself. Numerous attempts have been made to formulate models to interrelate time, dose, fractionation, and dose rate in radiotherapy. Progress has been slow. It still is. Following is a review of many of the time–dose relationships that have been developed in attempts to answer these needs.

In general, only time–dose relationships for normal tissues will be addressed. This is not because such relationships for tumors would not be useful—they certainly would. Combined with a time–dose model for normal tissues, they could be used to design the "optimal" radiotherapy, i.e., that which would maximize the benefit/risk ratio (therapeutic ratio). Neither is it because time–dose relationships for tumors have not been proposed. Practically all of the normal tissue models that will be reviewed in this chapter have been modified for use with tumors. Rather, the reason that time–dose models for tumors will not be discussed is that they are far too unreliable. Tumors do not act as predictably as normal tissues. No two tumors are alike. Any attempt to model the way in which a tumor will react to variations in time, dose, fractionation, dose rate, or volume is far too speculative to be useful.

2. TIME–DOSE RELATIONSHIPS FOR FRACTIONATED RADIOTHERAPY

2.1. Early Empirical Relationships

Early realization that there was a relationship between fractionation and biological effectiveness was hampered by the lack of accurate dosimetry and the unreliability of X-ray units to perform reproducibly. For example, Fig. 1 shows a typical patient treatment record in the pre–1920 era. Owing to the inability to control X-ray output from day to day, photographic (or other) dosimetry had to be utilized for *each* treatment. In this example the Kienboch system[34] was used. Note that, with this patient, several days passed before the exposure was brought even into the range of the dosimetry system. Not surprisingly, it took about 20 years after the first application of radiation therapy to realize that the biological effectiveness of radiation decreased as the number of fractions and overall treatment time were increased.[2] A few years later, in 1920, Freund[35] proposed a power law relationship between biologically effective dose and time, i.e.,

$$\text{bioeffect dose} \propto T^n$$

although apparently there was little attempt to utilize such a relationship until Strandqvist published his momentous analysis of time–dose data in 1944.[36] Prior to 1944, there were, however, many opportunities to

Name J. G. Very. Medical Attendant Mr Medico

Address City Exchange 8, Hospital Place

Record of X Ray Exposures.

Date	Hardness of Tube	Exposure Min	Sec	Amp	M'a	Filter	Areas	Paper and Value on Scale	Remarks
	[illegible]								
9.3.14	7½	10		5	3½	3mm	1	10+	
	8½	10		5	2	·	2	10+	
12.3.14	9	15		4	·1½	2	3	No paper	
12.3.14	··	"		·	·	2	4	No paper.	
18.3.14	8	·		5	3	2	1	10	
"	9	·		4	1	2	2	8	
25.3.14	9	15		5	2½	3	5	5	
·	9	15		5	2	3	6	4	
30.3.14	9	15		5	2	3	3	4	
"	9	15		5	3	3	4	4	
6.4.14	9	·		4	2	·	1 back	5	
·	9	·		2	1½	·	2 back.	4	Brush not working well
15.4.14	7-8	10		4-4½	2-2½	·	1	6	
	7	10		4½	2½	·	2	5	
	9	15		5	2	Dev. wrong	5	2	Improving
				6	3	·	6	2	
6.5.14	9	15		5	2½	3mm	5	6	
	6½	20		4½	2	·	6	6	
19.6.14	9	15		6	3½		3	8	
·	·	·		·	·		4	8	
Total									

FIG. 349.—Chart of X-ray exposures, to show method used in recording dosage by Kienböck's method.

Figure 1. Typical treatment chart of a patient treated in 1914. Dose measurements on the skin were made daily using the Kienbock quantimeter.[34] Note the wide daily variations in X-ray machine parameters and subsequent skin exposures.

"rediscover" this power law relationship for radiotherapy and, indeed, to determine an appropriate value for the exponent n. For example, Reisner's data[37] showing how isoeffect dose varied with fractionation for skin erythema (Table I) can be closely fitted to a power law equation with $n = 0.36$ between 1 and 12 fractions, although it breaks down as the number of fractions is increased further. Also shown in Table I are the erythema results of Quimby and MacComb,[38] which do not appear to fit a power law equation as well as the Reisner data. Similar results of Witte[39] and Kepp[40] could have been used to preempt Strandqvist.

TABLE I
Some Early Observations on the Effect of Daily Fractionation upon Isoeffect Skin Reaction Doses[a]

Number of fractions (N)	Reisner (4 cm^2)		Quimby and MacComb (70 cm^2)		
	Total dose R	Factor	Total dose (R)	Factor	$N^{0.36}$
1	1000	1.00	525	1.00	1.00
2	1300	1.30	800	1.52	1.28
3	1500	1.50	825	1.57	1.49
4	1600	1.60	—	—	1.65
5	—	—	925	1.76	1.78
7	2100	2.10	—	—	2.01
12	2400	2.40	1200	2.28	2.45
27	2700	2.70	—	—	3.28

[a] The power law with an exponent of 0.36 appears to fit the Reisner[37] data for 1–12 fractions but not the Quimby and MacComb[38] data.

Even closer to "discovering" the Strandqvist-type relationship was Meyer, a radiotherapist from New York, whose name is now almost completely forgotten, but who made several significant observations on practical radiotherapy problems in the 1920s and 1930s. For example, in 1939 Meyer published a graph relating isoeffect dose per fraction to the time between fractions, which he called the treatment interval or dose rhythm (Fig. 2).[41] The overall treatment time was kept constant at 42 days. As shown, he observed that this curve was exponential. Meyer stated that an isoeffect was produced if the time between fractions (T/N) varied as the square of the dose per fraction, d, i.e.,

$$d^2 \propto (T/N)$$

which can be transformed into an equation relating isoeffect total dose, D, to the number of fractions, N, in constant time T, by putting $d = D/N$. This gives

$$D \propto N^{0.5}$$

Meyer had unknowingly "discovered" the Strandqvist-type equation[36] with an exponent of about 0.5.

It was, however, Strandqvist[36] who realized that the relationship between isoeffect dose and overall treatment time was a power law function for both skin erythema *and* control of squamous cell skin cancer. He plotted total dose against the time interval between the first and final treatments in a log–log format and showed that the points fell on parallel straight lines of slopes 0.22 for both skin and skin cancer (Fig. 3).

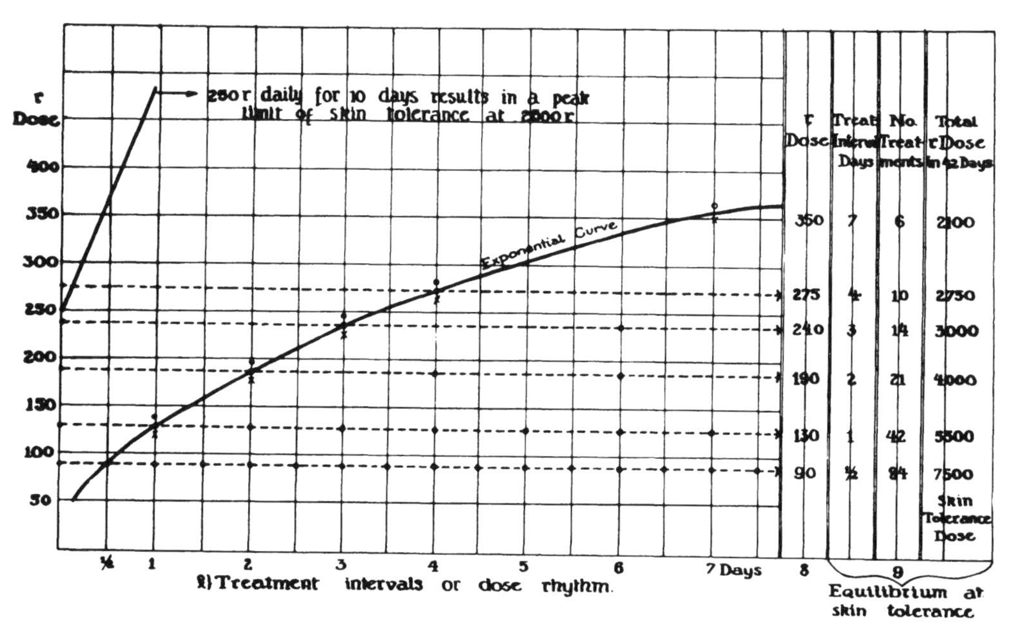

Figure 2. Isoeffect curve of Meyer[41] showing how the "dose" per fraction (in R) needs to be increased as the time between fractions is increased, keeping the overall treatment time constant at 42 days. As shown, Meyer stated that this curve was exponential (see text).

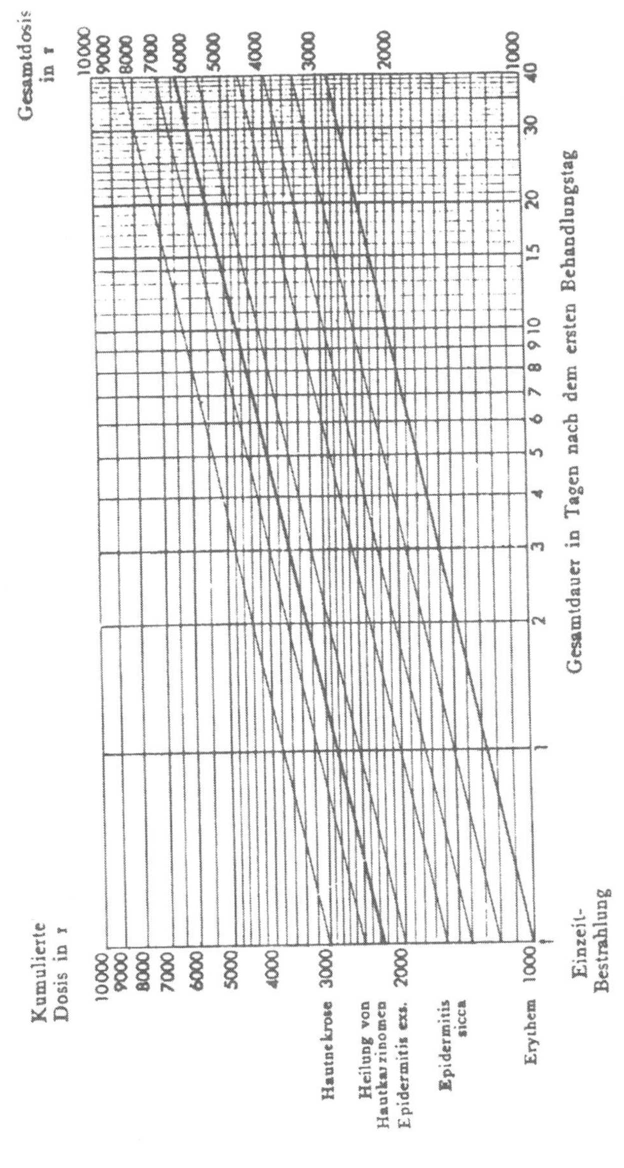

Figure 3. Strandqvist's original isoeffect curves for various levels of skin reaction and for skin cancer control.[36] All lines were parallel, with a slope of 0.22.

Following the lead of Strandqvist, there was a flourish of activity to determine the exponent of the overall treatment time for a variety of tumors and normal tissues such as adenocarcinoma of the uterus, larygeal Ca, brainstem tolerance, cartilage tolerance, bowel tolerance,[42] mycosis fungoides,[42,43] and many others. Values of the exponent ranged from 0.2 to 0.34.

Careful analysis by Cohen[44,45] confirmed Strandqvist's value of 0.22 for the exponent for skin Ca, but showed that a better value for the exponent for both threshold erythema and skin tolerance was 0.33, which became known as the cube-root rule.[46] This latter estimate for skin was later "fine-tuned" using rigorous statistical analysis by Cohen and Kerrich,[47] who determined an average value for the exponent for skin of 0.31 ± 0.026.

It should be noted that in almost all of the above observations used to determine the exponents of T, the number of fractions per week was kept more or less constant, usually between 5 and 7. Hence, there is a linear relationship between T and the number of fractions N. Consequently, if the data had been used to determine the exponent of N instead of T, the same value would have resulted. It was later observed by Fowler and Stern[48] that, on radiobiological grounds, it was probable that part of the dependence of isoeffect dose on time was due to the number of fractions (and hence dose per fraction) and part was due to overall treatment time. They argued that the dose per fraction component was due to Elkind-type repair, and that this was more important than the time component, which was due to repopulation. In a separate study, it was estimated that the effect of repair was about five times the effect of repopulation.[49] Also on radiobiological grounds, Fowler and Stern[48] argued that Strandqvist curves were probably not straight lines but that they could be approximated by straight lines between certain limits of T or N.

In an attempt to find a better fit to clinical observations and experimental animal data, and also to provide an improved fit to the shape of cell survival curves, Fowler and Stern[48,50] proposed a linear–quadratic relationship between number of fractions, N, and dose per fraction, d:

$$\text{const} = N\left(d + \frac{b}{a}d^2\right)$$

They also observed that, for a variety of tissues and tumors, a best fit was obtained when $D_0(b/a) \approx 3$, where D_0 is the equivalent single dose. Using this relationship, they predicted that a/b should vary between about 2 and 13 Gy for most tissues encountered in radiotherapy. This will be shown in the next section to be in remarkably good agreement with average values determined recently for early and late responding normal tissues and for tumors. This linear–quadratic relationship of Fowler and Stern has become the model of choice of many radiobiologists in the 1980s, since it has many features that make it more amenable to radiobiological explanation than power law relationships. Unfortunately, the Fowler and Stern model lay dormant as a "time–dose" relationship for clinical radiotherapy for about two decades. It was "rediscovered" in 1982 by Barendsen[51] and others in response to radiobiological observations of Douglas *et al.*[52] The linear–quadratic model and its application will now be reviewed in detail.

2.2. The Linear–Quadratic Model

Douglas and Fowler and their colleagues at the Gray Laboratory[52] conducted a series of experiments in which they determined isoeffect doses for skin reactions on the legs of mice for a wide range of number of fractions but with constant overall treatment time. They observed that their results, if plotted as reciprocal total dose vs. dose per fraction, fell on a straight line (Fig. 4) of the form

$$1/Nd = \alpha + \beta d$$

where N is the number of fractions, d is the dose per fraction, and α and β are tissue-specific constants. This is clearly the same equation as the

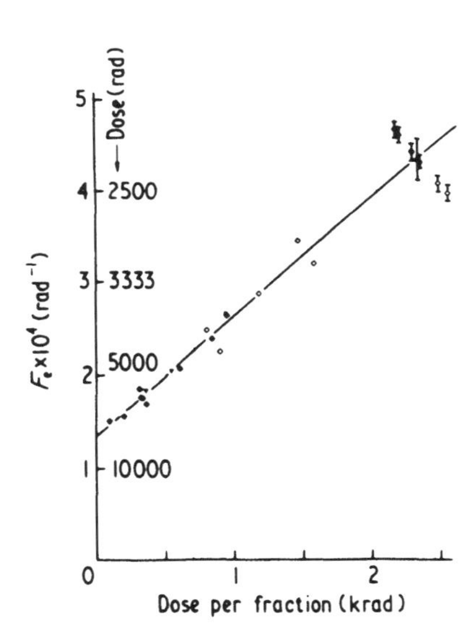

Figure 4. Plot of reciprocal total dose vs. dose per fraction for skin reactions on the legs of mice.[52]

linear–quadratic relationship proposed earlier by Fowler and Stern.[48,50] Rearrangement leads to the form of this equation proposed for use in fractionated radiotherapy by Barendsen[51]:

$$\text{ETD} = Nd[1 + (\beta/\alpha)d]$$

where ETD is the extrapolated total dose, which is the total dose which would produce the biological end point in question if delivered in an infinitely large number of infinitely small doses per fraction. The ETD is thus a single number which represents the biological effectiveness of a course of N fractions of dose d.

A feature of this model that makes it simple to use is the fact that ETD is linearly proportional to N and hence ETDs for different parts of a course of therapy are linearly additive. This is illustrated by the following example.

EXAMPLE. *A patient has been treated with 10 fractions of 3 Gy each in 2 weeks. The treatment is to be completed in a further ten fractions in 4 weeks. What dose per fraction should be used if the overall effect is to be equivalent to that of 30 fractions of 2 Gy each in 6 weeks?*

Since the overall treatment time is constant (6 weeks), it is appropriate to use the ETD equation:

$$\text{ETD}_1 + \text{ETD}_2 = \text{ETD}_{\text{tot}}$$

Hence

$$10 \times 3[1 + (\beta/\alpha)3] + 10 \times d[1 + (\beta/\alpha)d] = 30 \times 2[1 + (\beta/\alpha)2]$$

Rearrangement gives

$$d^2 + (\alpha/\beta)d - 3(\alpha/\beta + 1) = 0 \quad (1)$$

which is a simple quadratic equation which can be solved for d if α/β is known. Values of α/β have been determined for a wide variety of animal and human tissues and cancers.[24,53] Some examples are shown in Table II.

In general, values of α/β fall into two distinct categories: for late reacting tissues, $\alpha/\beta \approx 2.5$ Gy and for early reacting tissues (and tumors) $\alpha/\beta \approx 10$ Gy.[24,53]

In the above example, assume that we are concerned with avoiding injury to an early responding tissue. The substitution of $\alpha/\beta = 10$ Gy in

TABLE II
α/β Values for Multifraction Animal Experiments[a]

Tissue	Specifics	α/β (Gy)
	Early reactions	
Skin	Desquamation	8.6–12.5
Jejunum	Clones	6.0–10.7
Colon	Clones	8–9
	Weight loss	10–12
Testis	Clones	12–13
Callus	50-day	9–10
Mouse lethality	30-day	7–26
	Late reactions	
Spinal cord	Cervical	1.0–2.7
	Lumbar	2.3–4.9
Kidney	Rabbit	1.7–2.0
	Pig	1.7–2.0
	Mouse	1.5–2.4
Lung	LD_{50}	2.0–6.3
	Breathing rate	2.5–5.0
Bladder	Frequency, elasticity	3–7
Tumor bed	45-day	6–7

[a] See Fowler[24] for original references.

Eq. (1) gives

$$d^2 + 10d - 33 = 0$$

which yields the solution

$$d = 2.62\,\text{Gy}$$

The linear–quadratic model is simple to use and, once suitable values of α/β have been determined to acceptable accuracy, it will no doubt become very useful in clinical practice. It does not take into account the influence of overall treatment time, which is known to be of significance for tissues with the capacity for proliferation during treatment, but methods have been proposed for correcting the total dose by a time-dependent parameter of the type presented later in Section 2.4.[51]

2.3. Other Forms of the Fractionation Factor

One of the people responsible for the Fowler and Stern[48,50] model not being applied to clinical radiotherapy for such a long time was Fowler himself, since he and his colleagues introduced two alternative methods to account for fractionation. In 1965, Fowler[54] published a table (Table III) of multiplying factors to be used to correct the total dose for fractionation effects for skin and normal tissues. This table represents "modal" values derived from clinical and animal (pig skin) data. An earlier table by Sambrook was based solely on a hypothetical cell survival curve.[55] Fowler proposed that Table III could be used as a guide for radiotherapy dosage until more reliable data could be obtained. Later these same data were presented in the form of a nomogram (Fig. 5) by Shuttleworth and Fowler[56] in response to the publication of a nomogram by Burns,[57] which the authors criticized since it was based entirely on a hypothetical cell survival curve and not on clinical or animal data.

Like the linear–quadratic model, neither the Fowler tabulation nor the Shuttleworth and Fowler nomogram contained any correction for overall treatment time, although Fowler suggested that the total dose should be increased beyond that inherent in the nominal 5 fractions per week already accounted for in Table III (the data used to derive Table III were primarily for 5 fractions per week, or "daily" treatments). Fowler did not propose a method for making this "time" correction, although he did give an example for pig skin in which a correction of about 25 cGy/day had been observed.[54] Following is a review of some simple time corrections which have been proposed for these models.

TABLE III
Fowler's Table of Multiplying Factors for Total Dose Based on Animal and Modal Clinical Data[54]

Number of fractions (N)	Multiplying factor (F_N)
30	1.00
25	0.96
20	0.91
15	0.85
12	0.80
10	0.76
8	0.72
6	0.67
5	0.62
4	0.58
3	0.52
2	0.46
1	0.33

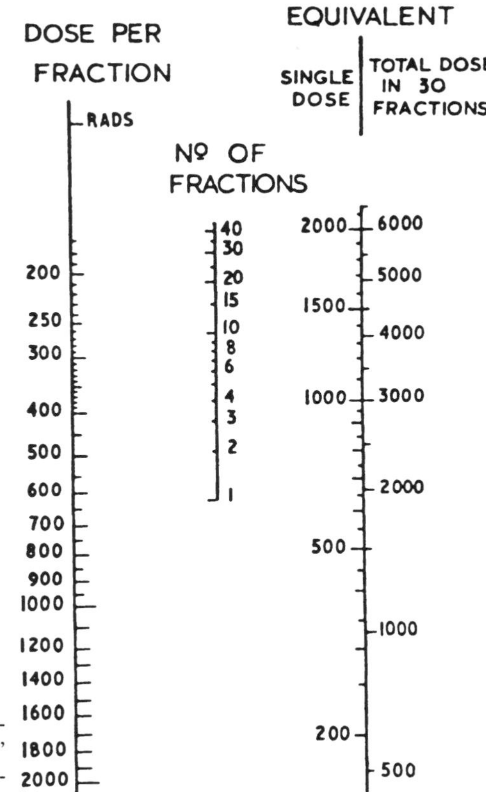

Figure 5. Nomogram for the determination of the "equivalent single dose" or "equivalent 30-fraction dose" for 5-fractions/week treatments.

2.4. Simple Time Corrections

A value of 30 cGy/day was proposed by Cohen,[58] and this was further modified by Liversage,[59] who proposed that the correction should be 30 cGy per day if the change in treatment time is less than 20 days, but 20 cGy per day for changes in time from 20 to 30 days. Mathematically, Liversage proposed a time–dose relationship of the form

$$D_N = (D_1 \times F_N) + C(T - T_F)$$

where D_N is the dose in N fractions equivalent to a single dose D_1, F_N is the Fowler multiplying factor to correct for fractionation (Table III), T_F is the time (in days) that it would take to deliver N fractions at 5 fractions per week, T is the actual overall treatment time (in days), and C is the time correction, which is 30 cGy/day for $(T - T_F) < 20$ days and 20 cGy/day for 20 days $\leqslant (T - T_F) < 30$ days.

Others have noted some evidence from observations on various animal tissues subjected to fractionated radiation that there is little proliferation of significance during the first few weeks of treatment.[60–62] Consequently, the incorporation of a "latent period" or delay time in the time-correction equation has been proposed.[51,63] Even through this may be a perfectly reasonable approach, the resulting equations are mathematically "awkward." It is likely that a more elegant solution to this delayed proliferation problem will be found before it is accepted for use in time–dose relationships.

Probably the main reason why the Liversage model did not gain wide popularity and why the Fowler and Stern linear–quadratic model was not further developed for 20 years was the major impact made by another model, namely, the nominal standard dose (NSD) relationship introduced by Ellis in 1967.[64]

2.5. The Nominal Standard Dose (NSD) Model

Although many investigators before had suggested that the effect of time and number of fractions needed to be considered separately, it was Ellis who first successfully devised a relationship in which the N and T components were separated. As the basis for this theory, Ellis used the slopes of Strandqvist curves for skin tolerance and epidermoid Ca as determined by Cohen.[65] These are reproduced in Fig. 6, and in equation form are represented as follows:

Skin tolerance:

$$\text{Total dose } D = K_1 T^{0.33}$$

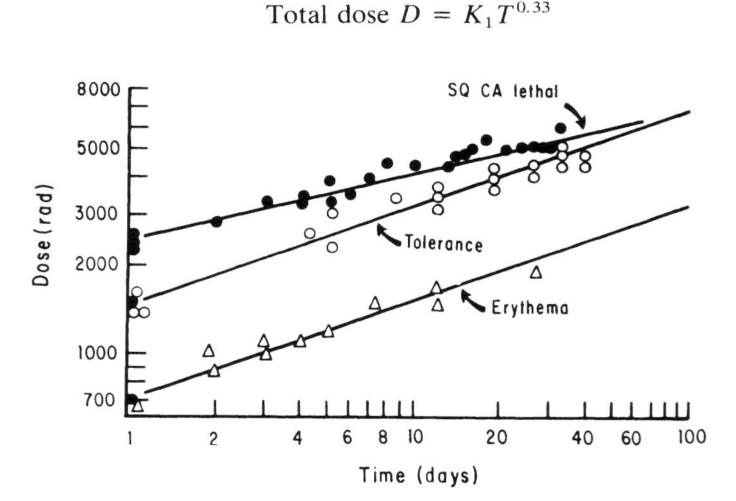

Figure 6. Cohen's plot of Strandqvist curves for skin reactions and skin cancer control.[65]

Epidermoid Ca:

$$\text{Total dose } D = K_2 T^{0.22}$$

These equations, combined with his own clinical observations and the suggestions of Fowler and Stern[48] that the effect of N was greater than the effect of T for normal tissues, led Ellis to propose that, for skin tolerance,

$$D = (\text{NSD}) N^{0.24} T^{0.11}$$

where NSD was the constant of proportionality and represented a "nominal single dose" for skin tolerance. Later, this was renamed nominal *standard* dose due to criticisms of the extrapolation of multifraction data to single doses. Ellis assigned the unit "ret" (rad equivalent therapy) to the NSD. At first sight, it would appear that the exponents of N and T should add up to 0.33, this being the slope of the Strandqvist line for skin tolerance assumed by Ellis. However, this would only be true if there were a linear relationship between N and T for the data shown in Fig. 6. But these data represent "daily" (or 5 fractions per week) treatments. It can be shown that, if the exponent of T is 0.11, the best fit to these data is for an exponent of N of 0.24.[66]

One of the problems that has arisen in the practical application of the NSD equation (and other time–dose relationships) is the inconsistent definition of "overall treatment time" T. Not only has the definition not been consistent from one user to the next, or one model to the next, but it has also frequently been *mathematically* inconsistent. Orton and Ellis[67] proposed that T ought to include the same average time for recovery for the last fraction of a course of treatment as for each previous fraction. Hence, they defined the overall treatment time for a schudule of N fractions as the interval between the first fraction and the $(N + 1)$th fraction. For example, for four successive daily treatments, $T = 4$ days, whereas for 10 treatments given 5 days per week, $T = 14$ days.

The NSD model rapidly gained in application in the years following its introduction[68] (Fig. 7), and has continued to grow in popularity despite continual debate over its accuracy. It is by far the most used time–dose relationship in the history of radiotherapy. Details of its use will now be reviewed.

2.5.1. Use of the NSD Concept

Initially, the NSD model was proposed as a means to determine tolerance regimes of fractionated radiotherapy. Ellis believed that, since

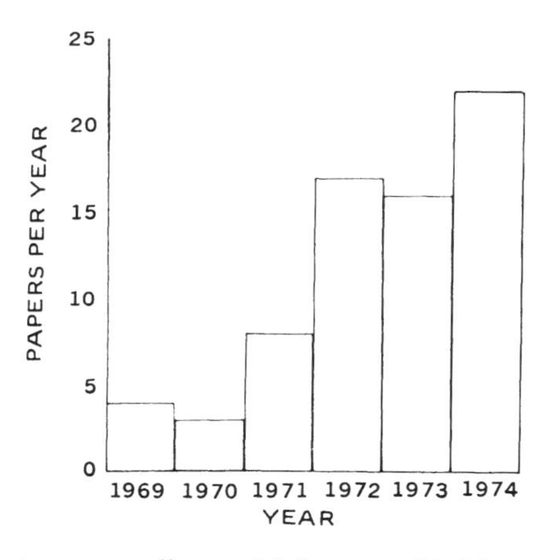

Figure 7. Early increase in articles utilizing the NSD concept published in the journals *Radiology, American Journal of Roentgenology,* and *Cancer.*[68] The total for 1974 was extrapolated from the number of articles published during the first six months of the year.

tumor radiosensitivity was highly variable and in most cases unpredictable, the aim of a course of therapy should be to take the normal tissues surrounding the tumor to their tolerance level, thus delivering maximum possible dose to the tumor and, presumably, obtaining maximum tumor destruction. Ellis[64] further postulated that, although the NSD formula had been derived from analysis of skin tolerance data, it could also be used for connective tissue stroma and indeed for "local tissue tolerance" in general. Although providing no evidence that the equation could be used for other tissues, Ellis[64] published NSD values for a wide variety of tissues and organs.[64] Thus began the extensive application of the NSD equation, without alteration of the exponents of N and T, for all tissues and organs, a practice which has led to many criticisms of the NSD concept and to several proposals for changing the exponents for certain tissues. Some of these will be reviewed later.

Calculation of the NSD for a course of radiotherapy is simple, provided that the treatments take place in a uniform manner (constant dose per fraction, constant number of fractions per week, no rest periods, etc.). The NSD is determined simply by substitution of D, N, and T values in the equation

$$\mathrm{NSD} = DN^{-0.24}T^{-0.11}$$

If the calculated NSD value is less than that corresponding to tolerance for the tissue or organ in question, then more treatments can be given until the tolerance value is reached. The equation is also useful for comparison of alternative treatment regimes, either inter- or intra-institutional.

Unfortunately, the problem of handling a situation in which the course of radiotherapy is not delivered uniformly requires a more complex calculation. This is because NSD is not linearly proportional to N and hence NSDs are not linearly additive. This was solved by Ellis by the linear addition of fractions of the tolerance NSD,[64] a method which was later formalized as the partial tolerance (PT) model,[69,70] where

$$\mathrm{PT} = (N/N_{\mathrm{tol}})\mathrm{NSD}$$

where N is the number of fractions actually delivered, N_{tol} is the number of fractions required to reach full tolerance, using the same fractionation schedule, and PT is a number which represents the biological effect of a nontolerance radiotherapy regime and, like NSD, has units of rets.

Then, since PT *is* linearly proportional to N, partial tolerances *are* linearly additive, and

$$\mathrm{PT}_{\mathrm{tot}} = \mathrm{PT}_1 + \mathrm{PT}_2 + \cdots$$

where $\mathrm{PT}_{\mathrm{tot}}$ is the single number which represents the cumulative biological effect of a course of radiotherapy made up of several parts, $1, 2, \ldots,$ each with a partial tolerance, $\mathrm{PT}_1, \mathrm{PT}_2, \ldots.$

The partial tolerance model is not only useful for the determination of NSDs for nonuniform treatment schedules, it can also be used to calculate numbers, $\mathrm{PT}_1, \mathrm{PT}_2, \ldots,$ which represent the biological effectiveness of nontolerance regimes.

Furthermore, it can be used for split-course radiotherapy by the use of a "decay factor", which is applied to PT.[64,70] If a partial tolerance PT_1 is reached in time T and then followed by a rest period R, the PT after the rest period is given by

$$\mathrm{PT}_{\mathrm{after\ rest}} = \mathrm{PT}_1\left(\frac{T}{T + R}\right)^{0.11}$$

Ellis theorized that the recovery represented by this "decay" equation does not continue indefinitely, but reaches a limiting value after 100 days.[64]

The practical application of the partial tolerance model will be demonstrated by the following example, which is the same as that used to illustrate the application of the linear–quadratic model in Section 2.2.

EXAMPLE. *A patient has been treated with 10 fractions of 3 Gy each in 2 weeks. The treatment is to be completed in a further 10 fractions in 4 weeks. What dose per fraction should be used if the overall effect is to be equivalent to that of 30 fractions of 2 Gy each in 6 weeks?*

In order to solve this problem, the PT_{tot} of the given course of treatment needs to be equated with the PT of the regime to which it is to be equivalent. But to calculate partial tolerances, the NSD needs to be known. For simplicity, assume that the NSD is that corresponding to 30 fractions of 2 Gy in 6 weeks. Then

$$\mathrm{NSD} = 6000 \times 30^{-0.24} \times 42^{-0.11}$$

[Note: NSDs are usually calculated with doses expressed in rads (or cGy). This is not essential, but this will be done here in order to avoid confusion when reviewing the literature on NSDs.] Hence

$$\mathrm{NSD} = 1758 \text{ ret}$$

N_{tol} now needs to be calculated for the first part of the course of treatment at 300 cGy per fraction using the NSD equation:

$$1758 = [300 \times (N_{tol})_1](N_{tol})_1^{-0.24}(T_{tol})_1^{-0.11} \quad (2)$$

where $(T_{tol})_1$ is the time to deliver $(N_{tol})_1$ fractions at 5 fractions per week. This clearly is a complicated function of $(N_{tol})_1$ and also depends on the day of the week the course of treatment began.

Winston *et al.*[70] presented a detailed analysis of this problem. They proposed an approximate solution which is a least-squares fit for schedules starting on various days of the week. The relationship between T and N which they proposed was

$$T = KN^{1.13}$$

where K depends on the number of fractions per week. Values of K are shown in Table IV.

TABLE IV
Values of K in the Equation $T = KN^{1.13}$ for Different Numbers of Fractions per Week[70]

Fractions per week	K
0.5	9.15
1	4.61
2	2.29
3	1.51
5	0.89

Substitution in Eq. (2) gives

$$(N_{tol})_1 = (1758 \times 300^{-1} \times 0.89^{0.11})^{1.572} \tag{3}$$

or

$$(N_{tol})_1 = 15.93$$

The value of N_{tol} for the second part of the course, at the unknown d cGy per fraction in 10 fractions over 4 weeks is given by

$$1758 = [d \times (N_{tol})_2] \times (N_{tol})_2^{-0.24} \times (1.90)^{-0.11}(N_{tol})_2^{-0.124}$$

Hence

$$(N_{tol})_2 = 1.411 \times 10^5 d^{-1.572}$$

Then, equating the partial tolerances:

$$PT_{tot} = PT_1 + PT_2$$

or

$$1758 = 1758\left(\frac{10}{15.93}\right) + 1758\left(\frac{10}{1.411 \times 10^5 d^{-1.572}}\right)$$

Solving this for d gives

$$d = 232\,\text{cGy}$$

This is significantly different from the 262 cGy per fraction determined by the linear quadratic model for early reacting tissues, where α/β was taken as 10 Gy. On the other hand, if the typical α/β ratio for *late* reacting tissues had been used (=2.5 Gy), then the linear–quadratic solution would have been 222 cGy. This is in better agreement with the NSD solution. This is surprising since the NSD equation was originally derived from data on *early* skin reactions.

One of the many criticisms of the NSD equation is that it does not work well for all levels of skin reaction, let alone reactions in tissues other than skin.[24] Some of these and other criticisms of the NSD model will now be reviewed.

2.5.2. Criticisms of the NSD Model

Following are some of the specific criticisms that have been aimed at the NSD model.

2.5.2.1. Separation of N and T Exponents from Original Data. When Ellis first derived the NSD equation, he surmised that the difference in the slopes of the skin tolerance and skin cancer cure curves represented the "time" factor for normal tissue recovery.[64] This has been criticized on two fronts. Firstly, it was pure hypothesis. There are many who do not agree that the difference between the skin tolerance and skin cancer cure curves is simply a matter of time-dependent recovery of skin. Secondly, the exponent of T was artificially created and not supported by any direct experimental evidence.

It is difficult to refute either of these two arguments, since they are both true. Of course, this does not mean that the NSD model is wrong; it means that the way in which Ellis first justified his concept was, to say the least, not entirely supported by hard evidence.

2.5.2.2. Strandqvist Curves Are Not Straight Lines. The NSD model relies on Strandqvist curves being straight lines. If they are not, then there is not a simple power law relationship between isoeffect dose, the number of fractions, and overall time. The evidence, one way or the other, is not conclusive as far as human tissues are concerned, owing to the difficulty of obtaining sufficient data from human studies. The ways in which much of the early Strandqvist-line data were analyzed and plotted, however, have been subjected to frequent criticism.[71–74] In some instances, statistical analysis of the data was not attempted, which was not good science since the data were so sparse that they required statistical "help." On the other hand, when statistical analysis *was* performed on the data, it was often of a somewhat basic nature and did not take advantage of recent advances in statistical theory, which, if used, would have given quite different answers.[73]

One of the more cogent criticisms, however, is that experimental animal data indicate that Strandqvist curves are not straight lines. The advantage of using animal studies is clear: large numbers of animals can be irradiated under carefully controlled conditions over a wide range of N and/or T. The disadvantage is equally clear: the way in which human tissues react is not necessarily mimicked by animal tissues.

In summary, the evidence is unclear. It must be admitted, however, that it would be strange if nature were to work to such simple rules when so many variables are involved. It would seem prudent to assume that Strandqvist curves are, at best, only approximately straight lines, and that this approximation is good only within a limited range of N or T values. In his defense, Ellis did stipulate that the NSD model might be valid only for $N \geqslant 4$ and $3 \leqslant T < 100$ days.[64,70]

2.5.2.3. The Exponents for N and T Should Not Be Constant. Ellis argued that tolerance at all sites, with the exception of certain organs, depends on connective tissue integrity, and hence the NSD equation for connective tissue stroma is applicable to most local tolerance situations. He specifically stated that brain tissue was almost certainly different, and that bone and other tissues might also respond differently.[64] Nevertheless, he proceeded to utilize the NSD model for brain and numerous other tissues. This gave the impression that he supported the use of the original NSD equation for all tissues and organs. The result was widespread use of the NSD equation with the original, constant, exponents of N and T for all applications. As shown in Fig. 8, it has now become clear that Strandqvist lines for different tissues are not all parallel[75] and hence the NSD exponents should be tissue specific. It was only after several years that alternative exponents began to be proposed for certain tissues. A summary of some of these alternatives is given in Table V. It should be noted that most of these exponents are of unknown accuracy since they are derived from sparse data sets often using less-than-optimal statistical analysis. Sometimes they are even downright

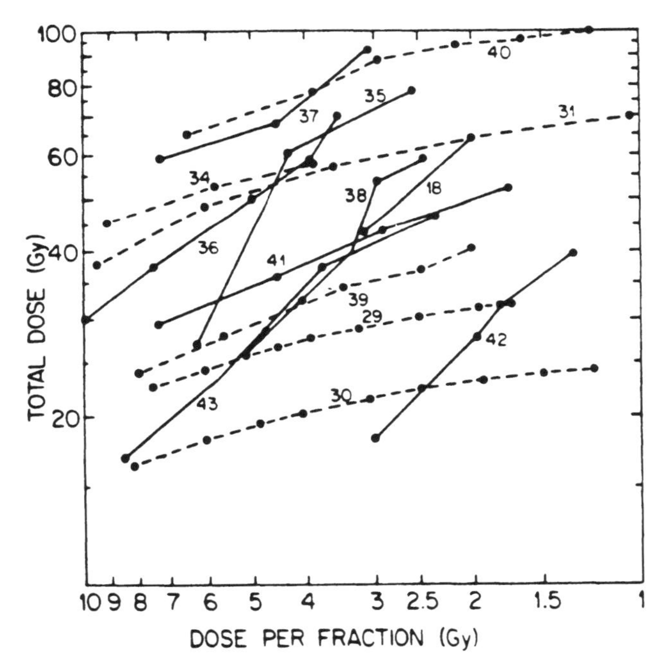

Figure 8. Strandqvist curves showing total isoeffect dose as a function of dose per fraction for a variety of acutely responding (broken lines) and late responding (solid lines) tissues.[75]

TABLE V
Alternative Exponents of N and T Proposed for the NSD Equation

Tissue	Reaction	Exponent of N	Exponent of T	Reference
Spinal cord	Myelopathy (human)	0.67	−0.52	Reinhold (1976)[76]
Spinal cord	Myelopathy (human)	0.377	0.058	Wara *et al.* (1975)[77]
Spinal cord	Myelopathy (human)	0.38 ± 0.04	0.07 ± 0.01	Cohen and Creditor (1981)[78]
Spinal cord	Myelopathy (human)	0.412	0.066	Schultheiss *et al.* (1984)[79]
Brain	Necrosis (human)	0.44	0.06	Scheline *et al.* (1980)[80]
Brain	Brain death (rats)	0.38	0.02	Hornsey *et al.* (1981)[81]
Brain	Necrosis (human)	0.45	0.03	Pezner and Archambeau (1981)[82]
Brain	Encephalopathy (human)	0.56	0.03	Cohen and Creditor (1983)[83]
Lung	Pneumonitis (human)	0.25	0.04	Cohen and Creditor (1983)[83]
Lung	Pneumonitis (mice)	0.377	0.058	Wara *et al.* (1973)[84]
Gut	Enteropathy (human)	0.29	0.08	Cohen and Creditor (1983)[83]
Kidney	Nephropathy (human)	0.25	0.19	Cohen and Creditor (1983)[83]

guesses (educated, of course!). Nevertheless, it is highly likely that the best-fitting exponents for several, if not most, tissues and organs in the body are different from the original NSD values, and that the appropriate exponents for early reactions differ from those for late reactions.[24,85–87] The problem is not whether the exponents need to be tissue specific, but how to obtain reliable values for each tissue. Furthermore, since the power-law format is only approximate, it is important to define the range of N and T values over which the new exponents are "valid."

This is a serious criticism of the NSD concept, since it will be very difficult indeed to make these determinations. However, it should be realized that this same criticism can be aimed at the best alternative concept presently available, namely, the linear–quadratic model: tissue-specific α/β values need to be determined before the model can be used reliably.[24]

2.5.2.4. Inappropriate Format of the "Time" Parameter. Based on acute skin-reaction experiments in animals, it appears that a simple power law equation is not representative of the way in which overall treatment time affects tolerance doses.[24,88] The experiments most frequently quoted are those of Denekamp[60] on mouse skin and Moulder and Fischer[62] on rat skin, in which it was shown that there is little, if any, increase in total dose necessary to obtain isoeffect during the first 2 or 3 weeks of treatment. On the other hand, a power law representation of this "time" effect predicts exactly the opposite, viz., the increase in dose per unit time is greater for T small than for T large.

A recent study of acute reactions in *human* skin has supported these findings.[87] For human skin the onset time, or latent period, for the appearance of an effect of T on isoeffect dose is about 4 weeks.

These observations have led several investigators to propose "time" corrections for time–dose models which include a delay interval or "latent period," which is subtracted from the overall treatment time.[51,63]

One problem with all these observations is that they refer to "peak" acute reactions which usually reach their "peak" value at times other than the last day of treatment. Typically, for short courses of treatment, the peak reaction occurs many days after the end of treatment, whereas for long courses the peak often occurs several days *before* the end of treatment. If, as is supposed in these experiments, acute tolerance is well represented by the "peak" skin reaction, then overall treatment time is clearly not the appropriate parameter to use in a relationship between isoeffect dose and time. Maybe a (predicted) time to maximum reaction should be used. Incidentally, many of the "peak" reactions for long treatment times observed in the experiments above occurred before all the treatments had been delivered. This raises some doubts as to precisely what were the isoeffect doses that gave rise to the observed "peak" reactions. The whole concept of what is meant by "isoeffect dose" and "time" for acute effects needs further study before improved time-correction models can be proposed.

As far as late effects are concerned, there are far fewer data, either animal or human. However, the general consensus is that, if a power-law format for T is appropriate, the exponent of T for late effects should be small compared with that which has been used for acute effects.

2.5.2.5. The NSD Model Is Difficult to Use and Is Confusing. The practical application of the NSD concept to the solution of clinical problems is complicated when treatment parameters are not constant throughout the course of therapy (see Example earlier). Unfortunately, the analysis of such complex schedules is one of the most important applications of any time–dose model. Despite several articles describing the correct application of the NSD concept,[70,89] many users were tempted to avoid complex calculations by simply adding NSDs linearly instead of trying to calculate linearly additive PTs.[68] Furthermore, since Ellis allocated the same units (rets) to both NSD and PT, these two numbers were frequently used interchangeably by those who did not realize that they were different. In the resulting confusion, the NSD concept was misunderstood and misused. One survey showed that about half of all those using the NSD model were applying it incorrectly.[68]

The NSD concept was a significant advance, yet owing to its complexity, it was often not used to its full potential. Subsequent introduction of two "simplifications" of the model helped the NSD concept to reach this potential. These were the cumulative radiation effect (CRE) and time–dose factor (TDF) models, which will be reviewed in the following sections.

2.6. The Cumulative Radiation Effect (CRE) Model

Kirk, Gray, and Watson[90] realized that the original NSD equation could be used to represent nontolerance levels of biological damage, without having to resort to the introduction of a second parameter, the partial tolerance, as had been proposed by Ellis. They based this assertion upon the fact that Strandqvist lines for nontolerance levels of skin damage were parallel to the tolerance Strandqvist line (Fig. 3), and hence the same equation applied to all levels of skin injury. This equation was

$$\text{CRE} = DN^{-0.24}T^{-0.11}$$

Kirk and his colleagues realized that, in order to simplify the application of this equation, some rearrangement was necessary. They proposed that the following substitutions be made:

$$d = D/N$$

where d is the dose per fraction, and

$$x = T/N$$

where x is the average time between fractions. Then the CRE equation reduces to

$$\text{CRE} = x^{-0.11}dN^{0.65} \tag{4}$$

where the unit of CRE is the reu (radiation equivalent unit).

Unfortunately, Kirk *et al.*[90] did not at first realize how simply this equation could be applied to the solution of practical problems. In their early papers on the CRE method, they resorted to a somewhat complex mathematical manipulation of the CRE formula in order to apply it.[90–94] It was not until six years after its introduction that they introduced the relatively simple equations that will be used here.[95] It was in this publication that they realized that, in order to be able to "add" CREs from different parts of a course of treatment, they needed to create a quantity that was linearly proportional to the number of fractions. Since

CRE was proportional to $N^{0.65}$, this quantity was therefore $\mathrm{CRE}^{1/0.65}$ or $\mathrm{CRE}^{1.538}$. Then, if a course of radiotherapy is divided into parts, $1, 2, \ldots$, the total CRE is given by the equation

$$(\mathrm{CRE_{tot}})^{1.538} = \mathrm{CRE}_1^{1.538} + \mathrm{CRE}_2^{1.538} + \cdots$$

or

$$\mathrm{CRE_{tot}} = (\mathrm{CRE}_1^{1.538} + \mathrm{CRE}_2^{1.538} + \cdots)^{0.65} \tag{5}$$

In order to illustrate the application of this equation, the problem used to illustrate the linear–quadratic and NSD methods will now be solved.

EXAMPLE. *A patient has been treated with 10 fractions of 3 Gy each in 2 weeks. The treatment is to be completed in a further 10 fractions in 4 weeks. What dose per fraction should be used if the overall effect is to be equivalent to that of 30 fractions of 2 Gy each in 6 weeks?*

In this problem, we can calculate what we want $\mathrm{CRE_{tot}}$ to be, and we can calculate CRE_1. Hence we can determine CRE_2 and use this to find the dose-per-fraction d. The following calculations use doses in cGy as in the original publications. From Eq. (4)

$$\begin{aligned}\mathrm{CRE_{tot}} &= x^{-0.11} d N^{0.65} \\ &= (42/30)^{-0.11} \times 200 \times 30^{0.65}\end{aligned}$$

which gives

$$\mathrm{CRE_{tot}} = 1758 \text{ reu}$$

Similarly, CRE_1 is calculated:

$$\begin{aligned}\mathrm{CRE}_1 &= (14/10)^{-0.11} \times 300 \times 10^{0.65} \\ &= 1291 \text{ reu}\end{aligned}$$

Then, using Eq. (5),

$$1758 = (1291^{1.538} + \mathrm{CRE}_2^{1.538})^{0.65}$$

which gives

$$\mathrm{CRE}_2 = 932 \text{ reu}$$

Then, from Eq. (4)

$$932 = (28/10)^{0.11} d 10^{0.65}$$

which gives

$$d = 234 \text{ cGy}$$

This, except for a small difference caused by the difficulty encountered in defining the relationship between T and N in the calculation of partial tolerances, is the same as the solution obtained by the NSD method. Of course, this is not surprising since both methods use the same basic equation. The CRE does, however, have the advantage that it is not necessary to go through the complicated procedure of calculating partial tolerances.

After the introduction of the CRE method in 1971, Kirk and his colleagues proceeded to thoroughly develop the concept for application to a wide variety of situations. Apart from developing CRE equations for brachytherapy, which will be considered later in Section 3.3.5, they provided methods of correction for the RBE of the radiation used, the area or volume of tissue irradiated, and time gaps in treatment regimes. Following are descriptions of each of these developments.

2.6.1. Correction for RBE

Kirk *et al.*[91] proposed that the effect of RBE could be accounted for by simply multiplying the total dose or dose per fraction by the RBE, q,

TABLE VI
RBE Values, q, for Commonly Used Teletherapy Beams for the CRE Model[95]

Radiation	q
50 kVp, 1 mm Al HVL	1.22
100 kVp, 4 mm Al HVL	1.19
100 kVp, 8 mm Al HVL	1.17
300 kVp, 2.5 mm Cu HVL	1.12
Cobalt-60	1.00
4 MeV bremsstrahlung	0.98
6 MeV bremsstrahlung	0.96
8 MeV bremsstrahlung	0.95
20 MeV bremsstrahlung	0.91
30 MeV bremsstrahlung	0.89

where q is determined from the equation

$$q = (E/E_{Co})^{-0.05}$$

where E is the equivalent monoenergetic energy of the radiation and E_{Co} is the average energy of Co-60 photons = 1.25 MeV. They further proposed that, for a bremsstrahlung spectrum of X radiation, E could be calculated from the peak energy, E_p, by the relationship

$$E = 0.46E_P$$

Values of q so calculated are shown in Table VI.[95]

2.6.2. Area and Volume Corrections

Volume effects will be reviewed in detail later in Section 4. In the present section, the CRE area and volume correction factors will be presented without justification or discussion.

Kirk *et al.*[93] correct the total dose, or dose per fraction, by area or volume multiplying factors, ϕ_a and ϕ_v, where

$$\phi_a = (A/100)^{0.24}$$

and

$$\phi_v = (V/1000)^{0.16}$$

2.6.3. Time Gaps in Therapy

The correction for time gaps, or rest periods, is handled in the CRE model by the introduction of an exponential CRE "decay factor," $\gamma(G)$, where[94]

$$\gamma(G) = e^{-0.008G}$$

where G is the rest period in days. Then, if a cumulative radiation effect, CRE_1, is reached before a rest period G days, the CRE after the rest period is given by

$$\text{CRE}_{\text{after gap}} = \text{CRE}_1 e^{-0.008G}$$

Since this decay factor continues to approach zero as G increases, Kirk *et al.*[94] proposed a limiting maximum value of G of 40 days.

2.7. The Time–Dose Factor (TDF) Model

The CRE method provided a partial simplification in the use of the NSD concept, but it still remained difficult to apply to practical problems, since it required nonlinear addition of each part of the course of therapy. This problem was solved by the introduction of time–dose factors (TDFs) by Orton[96] and Orton and Ellis.[97]

The TDF equation was derived from the basic NSD relationship:

$$\text{NSD} = DN_{\text{tol}}^{-0.24}T_{\text{tol}}^{-0.11}$$

which, if $T_{\text{tol}}/N_{\text{tol}}$ is represented by x, the average time between fractions for the course of therapy, becomes

$$\text{NSD} = N_{\text{tol}}dN_{\text{tol}}^{-0.35}x^{-0.11}$$

or

$$N_{\text{tol}} = (\text{NSD}d^{-1}x^{0.11})^{1.538}$$

where N_{tol} is the number of fractions of dose d required to reach tolerance. Then, if N is the number of fractions required to reach partial tolerance, PT,

$$\text{PT} = (N/N_{\text{tol}})\text{NSD}$$

Substitution for N_{tol} gives

$$\text{PT} = N(\text{NSD})^{-0.538}d^{1.538}x^{-0.169}$$

Clearly, the part of this equation which relates to time, dose, and fractionation is $Nd^{1.538}x^{-0.169}$, since NSD is a constant for each tissue. This part of the equation is thus all that is needed for a time–dose relationship, which leads to the definition of a time–dose factor (TDF), where

$$\text{PT} = (\text{NSD})^{-0.538} \times (\text{TDF}) \times 10^{3}$$

or

$$\text{TDF} = Nd^{1.538}x^{-0.169} \times 10^{-3} \tag{6}$$

The 10^{-3} term is simply a scaling factor, which makes Ellis' definition of skin tolerance (NSD = 1800) correspond to a TDF of approximately 100.

The beauty of the TDF concept is that, since TDFs are linearly proportional to PTs, they are linearly additive, i.e.,

$$\mathrm{TDF}_{\mathrm{tot}} = \mathrm{TDF}_1 + \mathrm{TDF}_2 + \cdots$$

Since the TDF model is just a simplification of the NSD concept, and hence also of the CRE method, it is not surprising that TDFs, NSDs, and CREs all give the same answer to the solution of straightforward fractionated radiotherapy problems. The example used previously will demonstrate this.

EXAMPLE. *A patient has been treated with 10 fractions of 3 Gy each in 2 weeks. The treatment is to be completed in a further 10 fractions in 4 weeks. What dose per fraction should be used if the overall effect is to be equivalent to that of 30 fractions of 2 Gy each in 6 weeks?*

Using the equation

$$\mathrm{TDF}_{\mathrm{tot}} = \mathrm{TDF}_1 + \mathrm{TDF}_2$$

and the TDF equation [Eq. (6)] gives

$$30(200)^{1.538} \times (42/30)^{-0.169} \times 10^{-3} = 10(300)^{1.538} \times (14/10)^{-0.169} \times 10^{-3} + 10d^{1.538}(28/10)^{-0.169} \times 10^{-3}$$

which simplifies to give

$$d = 234\ \mathrm{cGy}$$

This is exactly the same answer as obtained by the CRE method, but clearly the solution is much simpler.

As with the CREs, the TDF model has been developed to meet a number of practical situations. Apart from its application to brachytherapy, which is considered in Section 3.3.6, TDFs have been tabulated for simplicity of use,[97] have been applied to split-course treatments by use of a decay factor (Section 2.7.1), and have been modified for the effect of volume (Section 2.7.2), type of tissue (Section 2.7.3), and RBE (Section 2.7.4). Following are details of some of these developments.

2.7.1. Decay Factors

The TDF model makes use of the same decay factor as used with PTs, hence

$$\text{TDF}_{\text{after rest period}} = \text{TDF}_1 \left(\frac{T}{T + R}\right)^{0.11}$$

where T is the total time period from the day of the first treatment until the beginning of the rest period and R is the duration of the rest period.

2.7.2. Volume Correction Factors

Again, a detailed analysis of volume corrections appears later in Section 4. In the present section, only a basic description of the TDF volume correction method will be presented.

In the TDF volume effect model,[99] volume is represented by v, the "partial volume" of tissue irradiated. This is the volume irradiated represented as a fraction of some reference volume of tissue which might, for example, be the volume of an entire organ, in which case v would be the fraction of the organ irradiated. Then if TDF(1) is the TDF reached if the entire reference volume is treated with a given fractionation regime, and TDF(v) is the TDF for a partial volume, v, irradiated with the same regime, then

$$\text{TDF}(v) = \text{TDF}(1)v^{\phi}$$

where ϕ is a tissue-specific exponent. The TDF volume correction factor is thus similar to that employed in the CRE model, with the exception that different exponents are used for different tissues in the TDF model.

At first sight, this might appear to be a potential improvement on the CRE volume-effect model. Unfortunately, since there is little reliable data on which to base values of ϕ for different tissues, only rough

TABLE VII
Values of the TDF Volume Factor Exponent ϕ Determined Using Data Derived from the Cell Population Kinetic (CPK) Model[98][a]

Stroma	Brain	Spinal cord	Lung	Gut	Kidney
0.18	0.20	0.21	0.14	0.14	0.11

[a] These are initial estimates only and should be used with considerable caution.

estimates for a few select tissues can be given at this time, so the improvement is in concept only. These estimated values of ϕ, determined by analysis of data collected by Cohen and Creditor,[78,83] are given in Table VII.[98]

2.7.3. TDF Exponents for Different Tissues

One of the criticisms of the NSD model is that the exponents of N and T need to be allowed to vary with the tissues or organs irradiated. Values of these exponents were presented for a few tissues in Table V. The problem with simply changing the exponents in the NSD equation, or the corresponding exponents in the TDF formula is that a wide variety of NSD and TDF numbers results, with tolerance corresponding to widely different numbers for different tissues. This can be very confusing. What is needed is a model in which "tolerance" is represented by the same number for all tissues. This is the basis of the variable exponent TDF model.[98] In this model, the equation for TDF is

$$\mathrm{TDF} = K_1 N d^{\delta} (T/N)^{-\tau} v^{\phi}$$

where δ and τ are tissue-specific exponents and K_1 is a scaling factor for a reference volume ($v = 1$) of tissue irradiated, which is defined for each tissue such that TDF = 100 represents tolerance (e.g., 5% complication probability) for that reference volume.

As with the volume exponent ϕ in the previous section, each of the parameters K_1, δ, and τ needs to be determined, but there are few data to use for their determination. Again, Cohen's data have been used to make initial estimates of these values. These are shown in Table VIII.[98] Note that the scaling factor K_1 can serve a second purpose: it can be used to convert TDFs to SI units.[99]

TABLE VIII
Initial Estimates of TDF Parameters for Normal Tissue Tolerance (5% Risk)[98a]

Tissue	K_1 (doses in rad)	K_1 (doses in Gy)	δ	τ
Stroma	10^{-3}	1.19	1.538	0.169
Brain	8.48×10^{-6}	0.64	2.439	0.073
Spinal cord	1.70×10^{-4}	1.19	1.923	0.125
Lung	5.22×10^{-3}	3.42	1.408	0.056
Gut	8.24×10^{-4}	1.23	1.587	0.127
Kidney	1.50×10^{-3}	5.64	1.786	0.339

[a] The values of K_1 are for reference volumes defined by 10×10 cm fields.

As was stipulated earlier, owing to the sparsity of clinical data available for the derivation of these parameters, it is necessary to treat the values given in Table VIII as initial estimates only. They should be used with caution. It is recommended that they be used only as rough guides when used for patient calculations or prospective studies, although they are, of course, safe to use for retrospective analysis.

2.7.4. Correction for RBE

In the CRE model, RBE corrections were made simply by multiplying the dose (or dose per fraction) by the RBE of the radiation.[91] This is probably a reasonable approach for small changes in the RBE, i.e., for RBEs close to unity. However, since the shape of a cell survival curve is LET dependent, and hence RBE dependent, large changes in RBE will necessarily change the "shape" of an isoeffect relationship. This has been taken into account in the TDF model by allowing the exponents and scaling factor to vary with RBE.[100,101] For example, for neutrons the TDF equation is[101]

$$\mathrm{TDF} = K_1 N d^{1.176} (T/N)^{-0.129}$$

where $K_1 = 0.024$ for the high-energy p(66)Be(49) Fermilab beam, and $K_1 = 0.030$ for a low-energy d(15)Be unit.[102]

This relationship is useful for the estimation of the overall biological effect of combined high-LET and low-LET radiations, such as in combined neutron and X-ray therapy schedules, or to account for the γ-contamination component of a neutron therapy beam.[101]

2.8. Cell Kinetics Models

At about the same time that Ellis was developing the NSD model, several investigators were attempting to utilize cell-survival equations for radiotherapy following up on the earlier pioneering work of Fowler and his colleagues and Burns, mentioned previously,[48,54,57] and many others. Probably the most notable of these are the dual radiation theory of Wideröe,[103] and the cell population kinetic (CPK) model of Cohen.[58] In the original formulations, both are quite similar in concept, in that they use the product of a "high-LET" or single hit/single target cell survival equation, and a "low-LET" or multitarget equation.

2.8.1 The Wideröe Model

In the Wideröe model, the total surviving fraction is given by

$$S = \{e^{-\alpha D/D_{0\alpha}}[1 - (1 - e^{-(1-\alpha)D/kD_{0\beta}})^n]\}^N$$

where α is the fraction of the total dose due to the α (or high-LET) component; D cGy is the dose per fraction; $D_{0\alpha}$ cGy is the α-component mean lethal dose; $D_{0\beta}$ cGy is the β-component mean lethal dose; k is the Elkind recovery factor for the β component; n is the extrapolation number for the β component; and N is the number of fractions.

All of the above five parameters (α, $D_{0\alpha}$, $D_{0\beta}$, k, and n) are tissue specific. Before the model can be used to solve practical radiotherapy problems, the values of the three parameters $\alpha/D_{0\alpha}$, $(1 - \alpha)/kD_{0\beta}$, and n need to be determined for the tissue irradiated.

The Wideröe model does not take into account either volume or repopulation effects on cell surviving fraction. The CPK does.

2.8.2. The Cell Population Kinetic (CPK) Model

The CPK cell survival equation is[58]

$$S = \{e^{-Jd_1}[1 - (1 - e^{-Kd_1})^n]\}^N e^{LT}$$

where J cGy^{-1} is the single-target radiosensitivity constant; d_1 cGy is the dose per fraction modified by a field-size function; K cGy^{-1} is the multitarget radiosensitivity constant; L day^{-1} is the cellular regeneration growth constant; and T days is the overall treatment time.

The other parameters are the same as for the Wideröe equation. In the CPK model, the parameters that need to be determined for each tissue before clinical use are J, K, L, n, G, and Y, where G is the regenerative cycle limit and Y is the field size correction factor exponent.

Neither the Wideröe nor the CPK model has been developed to the point that it is safe to use clinically, although both have been applied to the retrospective analysis of clinical data. However, the CPK model is of special interest due to its extensive development.[104] Also a linear–quadratic form of the CPK cell survival equation has been developed, following upon the earlier proposals of Gray[105] and Gray and Scholes.[106]

Using the same nomenclature as for the target theory CPK equation, this linear quadratic form of the CPK model is[107]

$$S = e^{-(\alpha d_1 + \beta d_1^2)N} e^{LT}$$

In both forms of the CPK equation, the field-size modified dose per fraction, d_1, is related to the actual dose per fraction, d_z, by the equation[108]

$$d_1 = d_z Z^Y$$

where d_1 and d_z are the doses per fraction that will produce the same biological effect in the same number of fractions to volumes defined by field sizes with equivalent square sides 1 and z dm, respectively.

Clearly, the CPK model is significantly more complex than any of the empirical models described earlier. It requires the use of a computer to solve practical problems. It also needs considerably more parameters to be determined. Consequently, Cohen has developed a number of computer programs to aid in the application of the CPK method as well as to determine the "unknown" CPK parameters by the analysis of sets of clinical data. Some of these programs are shown in Table IX.[104]

The CPK model has several attractive features. First, it attempts to provide a time–dose model with a true biological basis, unlike most empirical models. Second, it allows the choice of three different radiation damage parameters[109]: either the $\log_{10}$ (critical surviving fraction), the equivalent single dose, or a probit value representative of the probability of a selected reaction. Finally, being a cell survival model, it can be used to determine cell survival for both normal tissues and tumors, and hence can potentially be applied to the specification of optimal treatment schedules.[110]

Notwithstanding all these advantages, the CPK model has not been found useful in clinical practice for the simple reason that it is too complex. It has too many parameters that need to be determined. Even with relatively sophisticated statistical analysis of large sets of clinical data, Cohen has been unable to determine any of the CPK parameters to

TABLE IX
Computer Programs for Use with the CPK Model[104]

RAD 1	Generation of isoeffect and therapeutic ratio tables. Optional determination of best-fit empirical time–dose equations.
RAD 2	Determination of CPK model parameters by the solution of 4–6 isoeffect equations defined by clinical or experimental data.
RAD 3	Determination of CPK model parameters from heterogeneous clinical data by the use of a "search" or iterative technique.
RAD 4	Plotting of cell survival and repopulation curves for normal tissues and tumors, and for the hypoxic component of a tumor, with or without reoxygenation.
RAD 5	Day-to-day interactive problem solving.
RAD 6	Searching for the optimal treatment schedule which maximizes the probability of uncomplicated local control.
RAD 7	Generation of tables of effective doses for brachytherapy with a variety of radionuclides.

an accuracy even close to that required to make the method clinically acceptable. Nevertheless, the model is of considerable theoretical interest, and, provided it is not used outside the limits of the data on which it is based, could conceivably be made clinically viable with an influx of new data. This, of course, is a criticism that can be aimed at all the time–dose relationships reviewed in this chapter.

3. DOSE-RATE EFFECTS

3.1. Introduction

Interest in the effect of dose rate in radiation therapy has been twofold. First was the concern that the dose rate at which each treatment was delivered in fractionated radiotherapy might influence the effectiveness of the radiation. This was of little concern in the early years of radiotherapy, despite the wide range of dose rates being used for external beam treatment. This lack of concern was due primarily to the fact that these dose rates were often difficult to control with the primitive X-ray equipment available, but also due in part to the lack of good dosimetry. Second was the realization that the effectiveness of radium therapy, and subsequently therapy with other isotopes, increased with increased dose rate.

These two aspects of the dose-rate effect will be considered separately.

3.2. Dose Rate during Fractionated Radiotherapy

The perfection of ionization methods of measuring radiation and the formalization of the definition by the ICRU of an international, consistent, unit of radiation (the roentgen) in 1928,[8] led to a flurry of activity in the study of dose-rate effects in fractionated radiotherapy. The most common studies were, of course, of effects on skin, either skin reactions or epilation. Although individual results varied somewhat, the general consensus was that, within the normal range of exposure rates used in external beam radiotherapy, there was no significant dose-rate effect. For example, by the observation of skin reactions, McKee and Mutscheller[111] showed that there was no significant effect of exposure rate between 12 and 435 R/min. Similar erythema observations by Pape,[112] Brunschwig and Perry,[113] and Fulton,[114] and epilation studies by Holthusen,[115] showed that only if the exposure rate was reduced to below about 5 R/min did the effectiveness of the radiation begin to decrease with decrease in dose rate. This lower limit was extended even lower by

McWhirter,[116] who reported that for fractionated radiotherapy there was no dose-rate effect on either skin reactions or cure of squamous cell skin cancer for exposure rates above 0.2 R/min. Recent analysis of the dose-rate effect by Hall[117] would, however, cast considerable doubt upon McWhirter's threshold for dose-rate effects, and possibly also the 5-R/min limit of the other observers. These early observations were probably not of sufficient precision to determine the small dose-rate effects which might occur in these studies, and it is believed that dose-rate effects probably start to become significant below about 50 cGy/min.[117] Since almost all modern external beam radiotherapy is performed at dose rates in excess of this limit, there is little need to be concerned about such effects.

It is worth noting here that there has been some interest expressed in the investigation of a possible *upper* threshold for dose-rate effects. This interest arose primarily in the 1960s as a result of radiobiological experiments at ultrahigh dose rates in the MGy/min range and above. It was hoped that treatments delivered at these high dose rates might eliminate the relative resistance of potential hypoxic cells in tumors. Unfortunately, the early promising demonstrations of such an effect by the work of Dewey and Boag,[118] Town,[119] Epp *et al.*,[120] Phillips and Worsnop,[121] Berry *et al.*,[122] and Griem *et al.*[123] was counteracted by conflicting experiments of Todd *et al.*[124] and Nias *et al.*[125] Although work is still continuing in the study of such ultrahigh dose rates and their effects, at this time there are no immediate plans to apply this technology to the treatment of patients. Hence, specification of an upper threshold to the dose-rate effect is of little interest at this time.

3.3. Dose Rate in Brachytherapy

As was the case with external beam radiotherapy, detailed study of the dose-rate effect in brachytherapy did not start in earnest until doses could be measured accurately and consistently. One of the earliest investigations of the effect of dose rate for continuous exposures was that of Pack and Quimby,[126] who observed that there was a certain critical overall treatment time below which there was no apparent dose-rate effect. They equated this threshold time with some type of "latent period" for recovery. Beyond this latent period, recovery of biological damage begins to occur during the treatment, and the effectiveness of the radiation per unit dose decreases as treatment time increases. They found that, for skin, this latent period was about 4 h. Hence, since treatment times exceed this latent period for just about all continuous radiotherapy, the effectiveness per unit dose decreases with increase in treatment time. Similar observations were made by many others, one of the most noteworthy studies being that of Gray.[105]

The impact of these observations upon practical radiotherapy was to encourage the establishment of fixed overall treatment times in order to avoid correction for the dose-rate effect. Standard treatment times formed the basis of several well-known treatment systems. Unfortunately, there were several schools of thought about what the standard treatment time should be, and this led to considerable variation in the activities of the radioactive sources used with the different systems. To some extent, these variations continue today. In very general terms, there is the relatively low dose-rate technique initially referred to as the Paris method but later known more universally as the Manchester System,[10] and then, at the other extreme, is the high dose-rate Stockholm technique.[27] The treatment times are about 7 days for the Manchester System and 1–2 days for the Stockholm method.

Since with practical implants the dose rates did not always work out as planned, and owing to interest in comparing the results of treatment at the various dose rates being used at different centers, several methods to quantify the dose-rate effect have been developed. One of the most used and better known of these dose-rate relationships was the isoeffect dose graph of Paterson[127] shown in Fig. 9. This was used to correct the total dose for the Manchester system if the dose rate varied from the nominal 35.7 cGy/h (or 6000 cGy in 7 days) and hence the treatment time needed

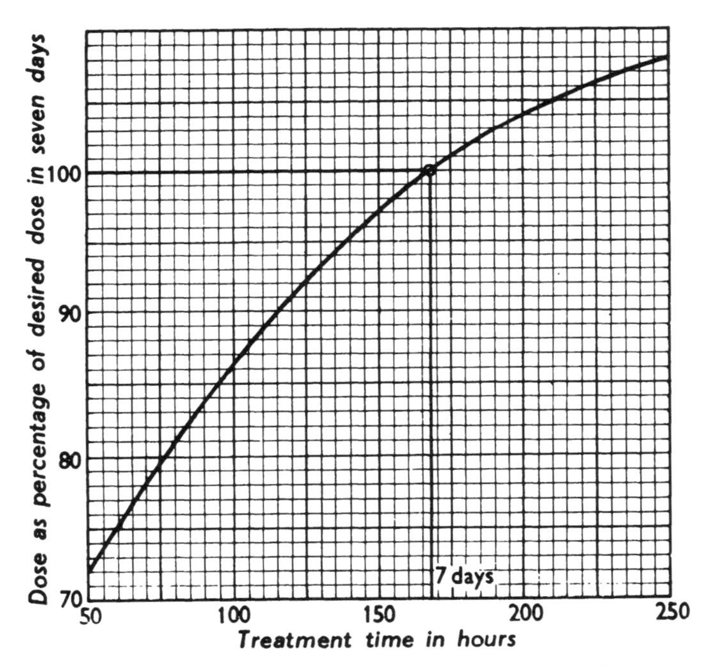

Figure 9. Paterson's dose-rate effect graph.[127]

to be changed from the nominal 7 days. This was not an easy curve to use in practice, so Paterson simplified the dose-rate effect by the following approximate rule of thumb:

1. Calculate a pseudo-treatment time, T' hours, from the dose rate r cGy/h by the equation

$$T' = 6000/r$$

2. Subtract 8 h for each 24 h that T' is less than 168 h (7 days).
3. Add 8 h for each 24 h that T' is greater than 168 h.

It is easily shown that this rule of thumb can be represented by the following equation for the new treatment time T hours[128]:

$$T = (8000/r) - 56$$

Hall[117] has demonstrated that this relationship appears to be quite good for treatment times in excess of about 3 days, but breaks down for shorter treatment times. It cannot therefore be used for the relatively high dose-rate techniques which have developed following the Stockholm method.

3.3.1. Power Law Equations

Clearly, what is needed is a dose-rate effect equation which is useful over a wider range of dose rates than the Paterson rule of thumb. One such equation, which, for reasons unknown, was little used in practice even though it appeared to provide a good fit to clinical observations, was that of Ellis[129]:

$$\text{isoeffect dose } D \propto T^{0.25}$$

Perhaps it was not used much due to its inconvenient form. What was really needed was an equation relating isoeffect dose to dose rate. It is unfortunate that Ellis did not rearrange his equation by substitution of D/r for T, which would have led to the relationship

$$D \propto r^{-0.33}$$

i.e., another cube root rule. Even more useful might have been a relationship between T and r for isoeffect. Substitution of (Tr) for D in the above equation leads to such a relationship:

$$Tr^{1.33} = \text{const}$$

TABLE X
Exponents, *a*, in the Power Law Dose-Rate Equation $D \propto r^{-a}$ Relating Isoeffect Dose, *D*, and Dose Rate, *r*

Reference	a
Ellis[129]	0.33
Kirk *et al.*[91]	0.41
Orton[130]	0.35
Wilkinson[128]	0.45

Other power law dose-rate relationships which have been proposed differ only slightly from the original Ellis equation. Some of these are shown in Table X. The Kirk *et al.*[91] and Orton[130] equations are used in the development of the cumulative radiation effect (CRE) and time–dose factor (TDF) brachytherapy models, respectively. These will be reviewed in detail later in Sections 3.3.5 and 3.3.6.

3.3.2. The Liversage Formula

An alternative formulation of the dose-rate effect has been proposed by Liversage.[131] This model is based upon the theoretical cell-survival curve analysis of Lajtha and Oliver,[132] who realized that continuous low dose-rate radiation could be simulated by an infinite number of small dose fractions. Using a theoretical equation for the cell surviving fraction combined with an equation for exponential "repair" of sublethal damage with a repair coefficient of μ h^{-1}, they derived the relationship

$$\text{equivalent acute dose } D_{\text{EQ}} = (r/\mu)(1 - e^{-\mu T})$$

Liversage stated that the average value of D_{EQ} over the treatment time T is given by

$$\begin{aligned}\bar{D}_{\text{EQ}} &= (1/T)\int_0^T (r/\mu)(1 - e^{-\mu T})\, dT \\ &= (r/\mu)[1 - (1/\mu T)(1 - e^{-\mu T})]\end{aligned}$$

He then surmised that, to a reasonable approximation, if the acute irradiation is given in N_{EQ} fractions each of magnitude $2\bar{D}_{\text{EQ}}$ the same reduction in cell surviving fraction per unit dose should be achieved as for a continuous irradiation in T h at r cGy/h., i.e.,

$$2\bar{D}_{\text{EQ}}N_{\text{EQ}} = Tr$$

or

$$N_{EQ} = \mu T/2[1 - (1/\mu T)(1 - e^{-\mu T})]$$

Liversage proposed that for any continuous course of brachytherapy of duration T h, the equivalent number of acute fractions could be calculated using this equation. The total dose could then be corrected for this "fractionation" effect using either a Strandqvist-type plot, the nomogram of Shuttleworth and Fowler[56] (Fig. 5), or the tabulated data of Fowler[54] (Table III). Analysis of data relating to the recovery from sublethal damage led Liversage[131] to propose a value for μ of $0.462\,h^{-1}$, although this has never been substantiated by human tissue reaction data so it needs to be considered only as a rough estimate if used for practical radiotherapy problems. It is not surprising, therefore, that this model does not work particularly well when used to compare brachytherapy regimes with equivalent courses of fractionated radiotherapy, as was demonstrated by Liversage himself in a study of high dose-rate remote afterloading.[32]

3.3.3. The CPK Model for Brachytherapy

One further method for the determination of isoeffect doses in brachytherapy, has been proposed by Cohen for use in the cell population kinetic (CPK) model.[104] Cohen simply uses his regular fractionated radiotherapy CPK equation for continuous treatments by making the assumption that each 24 h of treatment is equivalent to 5 fractions. This is very similar to the prediction of the Liversage formula which, for large values of T (e.g., greater than 24 h), reduces to

$$N_{EQ} \approx 0.23T$$

i.e., about (0.23×24) or 5.5 equivalent fractions for every 24 h. The Cohen equation was also shown to be not particularly accurate when used to compare brachytherapy and fractionated treatments for high dose-rate remote afterloading in the study by Liversage.[32]

3.3.4. The NSD Brachytherapy Model

Ellis and Sorensen[133] devised a method for the calculation of partial tolerances for low dose-rate irradiation. They based their correction for dose rate on an isoeffect curve of the dose equivalent to the Manchester standard of 6000 cGy in 7 days as a function of the overall treatment time. This graph, published several years earlier,[129] is reproduced in Fig.

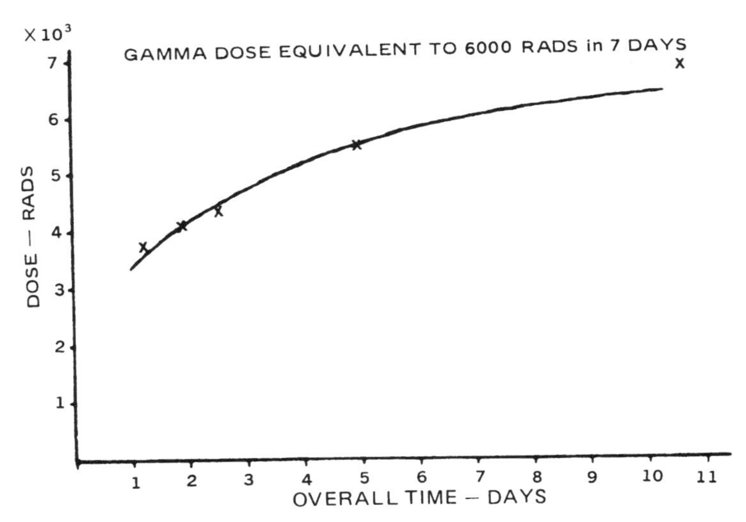

Figure 10. Isoeffect curve derived from clinical data of Green and used for the NSD brachytherapy model. The crosses represent points based on a survival curve with $D_0 = 160\,\text{cGy}$, $n = 2$, and a 30-min half-life for sublethal damage repair.[129] Points were calculated.

10. It is based on clinical experience of Green. For ease of use, Ellis and Sorensen presented these data in tabular form, giving a correction factor, F, by which total dose D should be divided in order to obtain an isoeffect at different dose rates (Table XI).

They then postulated that the equation for partial tolerance in brachytherapy, which was analogous to that for fractionated therapy, was

$$\text{PT} = \frac{\text{effective dose}}{\text{tolerance dose}}\,\text{NSD}$$

or

$$\text{PT} = \frac{D/F}{6000} \times 1800$$

where an NSD of 1800 ret has been assumed, i.e., they surmised that a tolerance NSD of 1800 ret for fractionated radiotherapy (about 6000 rad in 30 daily fractions) was isoeffective with 6000 rads in 7 days with radium ($F = 1$).

The following example with demonstrate the application of this method.

TABLE XI
The NSD Dose-Rate Correction Factor, F^a

Dose rate (cGy/h)	Dose factor F	Dose rate (cGy/h)	Dose factor F
25.0	1.06	70.0	0.780
27.0	1.051	75.0	0.758
30.0	1.033	80.0	0.737
32.0	1.019	85.0	0.717
35.0	1.001	90.0	0.702
35.7	1.000	95.0	0.687
40.0	0.967	100.0	0.672
45.0	0.931	110.0	0.642
50.0	0.897	120.0	0.622
55.0	0.864	130.0	0.602
60.0	0.832	140.0	0.584
65.0	0.805	150.0	0.567

[a] The dose equivalent to 6000 cGy in 7 days = $F \times 6000$ cGy.[133]

EXAMPLE. *What total dose of 6-MeV electrons delivered at 250 cGy per fraction, 5 treatments per week, is equivalent to a radium implant to the same area of superficial tissue delivering 6440 cGy in 7 days?*

Firstly, we should calculate the planned partial tolerance for the radium treatment. The dose rate is 38.33 cGy/h, which, from Table XI, gives a correction factor F of 0.978.

Then the partial tolerance is

$$\mathrm{PT} = \frac{6440/0.978}{6000} \times 1800$$
$$= 1975 \text{ ret}$$

For the fractionated electron treatment, it is first necessary to calculate the number of fractions, N_{tol}, required to reach tolerance at 250 cGy/fraction. This can be calculated using Eq. (3) in Section 2.5.1:

$$N_{\text{tol}} = (1800 \times 250^{-1}(0.89)^{0.11})^{1.572}$$
$$= 21.83$$

Hence, a PT of 1975 ret is reached in N fractions, where

$$1975 = (N/21.83)1800$$

Therefore

$$N = 24.0$$

and the total dose is $24.0 \times 250 = 6000$ cGy.

Rest periods are handled in precisely the same way as for fractionated radiotherapy partial tolerances, i.e.,

$$\text{partial tolerance after rest period} = \text{PT}_1\left(\frac{T}{T + R}\right)^{0.11}$$

where T is the time from onset of treatment to start of the rest period and R is the duration of the rest period.

The NSD brachytherapy model was never widely used, for two reasons. Firstly, it was not particularly convenient to use, since it required mixing equations and tables. Secondly, it could not be used if the dose rate varied appreciably during the treatment, i.e., for short-lived sources such as Rn-222, Au-198, and I-125. The CRE and TDF brachytherapy models to be described next did not have either of these deficiencies.

3.3.5. The CRE Brachytherapy Model

The CRE brachytherapy model is similar in concept to the NSD method, except that the tabulated dose-rate effect factors are represented in equation form. The equation used is derived from a log–log graph of isoeffect dose plotted against overall treatment time. Using data taken from the graphs of Ellis[129] and Paterson,[127] Kirk *et al.*[91] showed that the points fell on a straight line (Fig. 11) of slope 0.29, and hence

$$D \propto T^{0.29}$$

They then proposed that the CRE equation for brachytherapy, which was analogous to that for fractionated therapy, was

$$\text{CRE} = kqDT^{-0.29}$$

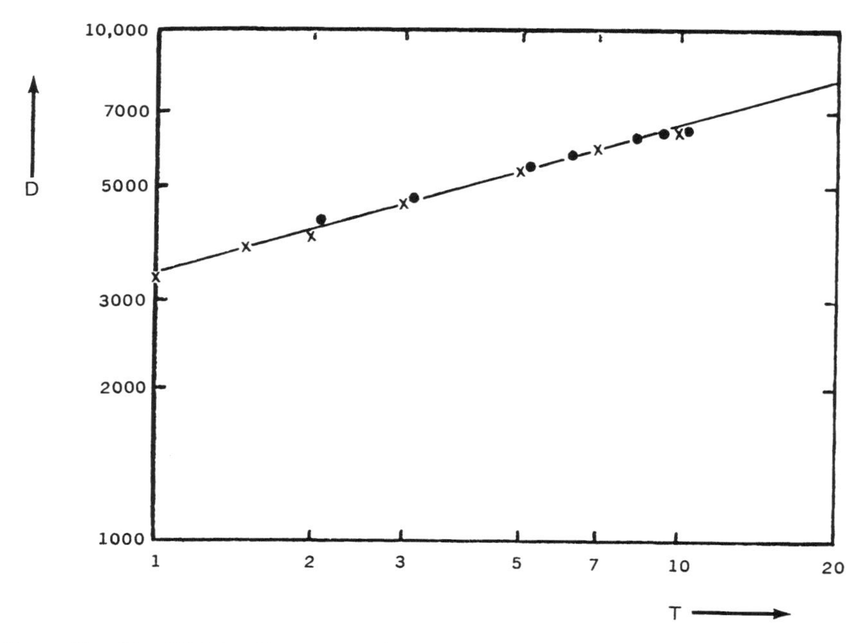

Figure 11. The CRE isoeffect curve of total dose vs. treatment time for radium gamma radiation.[91] Data points are taken from isoeffect doses of Ellis[129] and Paterson.[127]

or

$$\text{CRE} = kqrT^{0.71}$$

for long-lived sources, and

$$\text{CRE} = kqr_0\left[\frac{0.71}{\lambda}(1 - e^{-\lambda t/0.71})\right]^{0.71}$$

for short-lived sources, where k is a normalizing constant, q is the RBE of the radiation, r is the dose rate in cGy/day for long-lived sources, r_0 is the initial dose rate for short-lived sources, T is the treatment time in days, and λ is the decay constant of the isotope in days^{-1}. Values of q were calculated by the same method as used for fractionated CREs (Section 2.6.1). Some typical values for brachytherapy sources are given in Table XII.

The normalizing constant, k, was determined by equating the biological effectiveness of 6000 cGy in 7 weeks with radium ($q = 1.02$) to

TABLE XII
RBE Values, q, for Commonly Used Brachytherapy Sources for the CRE Model[95]

Source	q
Na^{24}	0.98
Co^{60}	1.00
Cs^{137}	1.03
Ta^{182}	1.02
Ir^{192}	1.06
Au^{198}	1.06
Rn^{222}	1.02
Ra^{226}	1.02

a CRE of 1830 reu for fractionated radiotherapy, i.e.,

$$1830 = k \times 1.02 \times 6000 \times 7^{-0.29}$$

which gives

$$k = 0.53$$

Rest periods and volume effects are handled in precisely the same way as for the fractionated radiotherapy CRE model (Section 2.6.3). The introduction of volume correction factors necessitated replacing the original normalization constant, k, by two factors, μ and ϕ, where

$$k = \mu\phi$$

ϕ being the area or volume correction factor, ϕ_a or ϕ_v, respectively. The value Kirk *et al.*[93] determined for μ was 0.77. Hence the final CRE equation was

$$\text{CRE} = \mu q \phi r T^{0.71}$$

Unfortunately, this equation sometimes gives ridiculous answers when applied to practical problems, as illustrated by the following example, which is the same one as used to demonstrate the use of the NSD brachytherapy model.

EXAMPLE. *What total dose of 6-MeV electrons delivered at 250 cGy per fraction, 5 treatments per week, is equivalent to a radium implant to the same area of superficial tissue delivering 6440 cGy in 7 days?*

Equating the CREs for the two alternative treatments gives

$$0.77 \times 1.02\phi(6440/7) \times 7^{0.71} = 250NN^{-0.24}(4/3)^{-0.11}N^{-0.11}\phi$$

Solving for N gives

$$N = 45.0$$

or

$$\text{Dose} = 11{,}250\ \text{cGy}$$

This is clearly far too high and significantly greater than the value of 6000 cGy determined by the NSD method. It is also more than double the dose of 5500 cGy advocated by Tapley[134] for electron beam treatments of this type.[135]

The problem with the CRE model lies with the field sizes used for the determination of the normalization constant.[93,135] Kirk and his colleagues assumed, incorrectly,[136] that the Manchester system tissue tolerance dose of 6000 cGy in 7 days applied to irradiated tissue volumes of 100 cm^3, whereas the appropriate volume should have been closer to 30 cm^3.

It has since been demonstrated[136] that a better value for the normalization constant is about 0.5, which is close to the original value of 0.53. Liversage[32] showed that use of the incorrect normalization constant leads to considerable errors when applied to the determination of appropriate doses to be used for high dose-rate remote afterloading for the treatment of carcinoma of the cervix. The CRE model predicts doses 75% too high, which, if used clinically, would result in disastrous overdoses and serious risks of complication.[32]

The "addition" of CREs for brachytherapy and fractionated radiotherapy is quite complicated, since CREs for both are nonlinear. For brachytherapy

$$\text{CRE}_{\text{tot}} = (\text{CRE}_1^{1.408} + \text{CRE}_2^{1.408} + \cdots)^{0.71}$$

and for fractionated radiotherapy

$$\text{CRE}_{\text{tot}} = (\text{CRE}_1^{1.538} + \text{CRE}_2^{1.538} + \cdots)^{0.65}$$

Then, for combined brachytherapy and fractionated radiotherapy, Kirk *et al.*[95] propose that there are two alternative ways of "adding" CREs: if brachytherapy is given last, then

$$\text{CRE}_{\text{tot}} = (\text{CRE}_{\text{brachy}}^{1.408} + \text{CRE}_{\text{fractionated}}^{1.408})^{0.71}$$

and if fractionated therapy is last

$$\text{CRE}_{\text{tot}} = (\text{CRE}_{\text{brachy}}^{1.538} + \text{CRE}_{\text{fractionated}}^{1.538})^{0.65}$$

Clearly, this is not satisfactory, and becomes really complicated when several sessions of the two types of therapy are interspersed, as is common with cervix cancer treatments. The TDF model, which uses linear addition, avoids this problem.

3.3.6 The TDF Brachytherapy Model

The TDF brachytherapy model is in many respects similar to the CRE model in that the tabulated dose-rate effect data of the NSD method is replaced by an equation. With TDFs, however, the equation is designed so as to make TDFs linearly additive.

Like the CRE model, the TDF dose-rate effect equation is based on the linearity of a log–log plot of isoeffect data (Fig. 12). In this case, the data of Mitchell[137] are used in addition to those of Green[129] and Paterson.[127] The isoeffect relationship derived from this graph is[130]

$$T = 2.1 \times 10^4 r^{-1.35}$$

which leads to the TDF equation

$$\text{TDF} = 4.76 \times 10^{-3} r^{1.35} T$$

for long-lived sources, or

$$\text{TDF} = 4.76 \times 10^{-3} r_0^{1.35} T_{\text{eq}}$$

for short-lived sources, where r is the dose rate in cGy/h for long-lived sources, r_0 is the initial dose rate for short-lived sources, T is the overall treatment time for long-lived sources (in hours), T_{eq} is the equivalent treatment time for short-lived sources, and 4.76×10^{-3} is a normalization constant which makes a TDF of 100 correspond to 6000 cGy in 7 days.

The equivalent treatment time, T_{eq}, is derived by integration of the

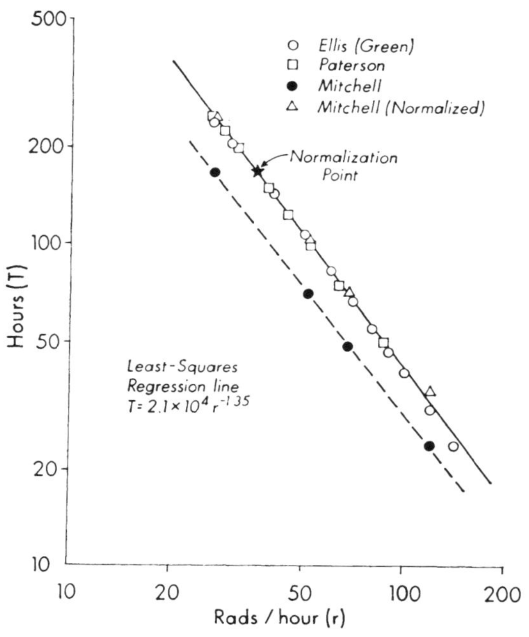

Figure 12. The TDF isoeffect curve showing treatment times at different dose rates that result in biological effects equivalent to those produced in 168 hours at 35.7 cGy/h.[130] The data are from Ellis,[129] Paterson,[127] and Mitchell.[137] The Mitchell data represented the equivalent of 113 h at 35.7 cGy/h (dashed line), and the time corresponding to each point has been normalized by a factor (168/113).

biological effectiveness over all dose rates as the isotope decays. Its value is given by

$$T_{eq} = (1 - e^{-1.35\lambda T})/1.35\lambda$$

where λ is the decay constant for the isotope in h^{-1}.

For permanent implants, $T = \infty$ and therefore

$$T_{eq} = 1/1.35\lambda$$

For split-course treatments, decay factors are identical to those used in the fractionated TDF model.

For simplicity of use, Orton[130] and Orton and Webber[138] provided tables of TDFs for long- and short-lived sources and permanent implants, and a table of T_{eq} values for a variety of short-lived sources.

The application of the TDF model is illustrated by the example used to demonstrate the NSD and CRE models.

EXAMPLE. *What total dose of 6 MeV electrons delivered at 250 cGy per fraction, 5 treatments per week, is equivalent to a radium implant to the same area of superficial tissue delivering 6440 cGy in 7 days?*

Equating the TDFs for the two alternative regimes gives

$$4.76 \times 10^{-3}(6440/168)^{1.35} \times 168 = N \times 250^{1.538}(4/3)^{-0.169} \times 10^{-3}$$

Solving for N gives

$$N = 23.6$$

or

$$\text{Dose} = 5900\,\text{cGy}$$

The addition of brachytherapy and fractionated radiotherapy TDFs is simple, since TDFs for both are linearly additive, i.e.,

$$\text{TDF}_{\text{tot}} = \text{TDF}_{\text{brachy}} + \text{TDF}_{\text{fractionated}}$$

Liversage[32] demonstrated that, unlike the CRE, CPK, and Liversage models, the TDF method worked well for the prediction of doses in high dose-rate remote afterloading.

4. VOLUME EFFECT MODELS

It was realized very early on in the history of radiotherapy that one of the reasons it was difficult to cure large tumors was that the dose required to sterilize a tumor increased with volume, whereas the tolerance dose of normal healthy tissues decreased with increase in volume.

4.1. Data on Volume Effects

Being by far the simplest tissue to study, most of the observations on the effect of volume (or area) on "isoeffect" doses have been made on skin. The results have been diverse to say the least. For example, some studies showed a very dramatic decrease in "isoeffect" dose with increase in field size,[139,140] whereas others have observed practically no effect.[141] Several reasons have been cited for these divergent conclusions. For example, several of the investigators used single-dose irradiations, which von Essen[142] argues should not be expected to show as large a volume effect as would be observed with fractionated radiation, although this is disputed by Hopewell and Young,[143] who show evidence that indicates no influence of fractionation on volume effect. Hopewell and Young and others[144] also point out a more important reason for the observed deviations between studies. They note that many of the investigations do not use a true isoeffect as the end point observed. For example, for skin some observers define a true isoeffect, such as the onset of moist desquamation or threshold erythema, whilst other use the term "skin tolerance" to mean different levels of damage acceptable for different field sizes. Two good examples of this are the following: "The dose for the small field will cause a moist desquamation . . . while that for the largest field will cause only a dry desquamation" (Ellis[145]); and "For the smaller areas these doses will ordinarily produce a moist desquamation . . . For larger areas the reaction will be rather less" (Paterson[127]). Unfortunately, many of the investigators do not make their definition of "tolerance" at all clear, which adds to the difficulty of using their data in order to define the effect of volume in radiotherapy.

The results of these volume effect investigations have been presented in a variety of formats, such as graphs (Figs. 13 and 14)[146,147] Tables (Tables XIII and XIV),[145,127] nomograms (Fig. 15),[127] and empirical equations (see following sections).

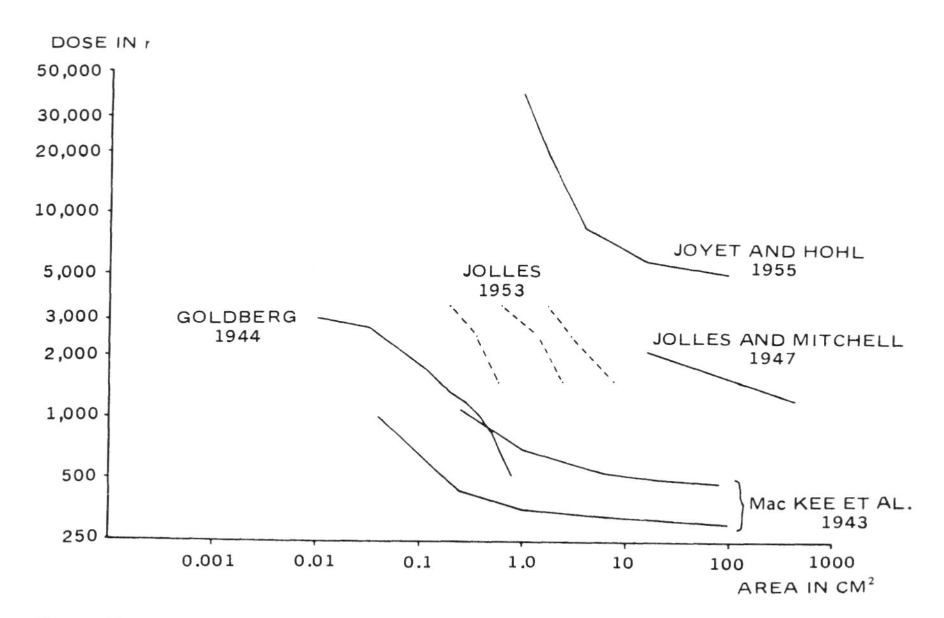

Figure 13. Berg and Lindgren's plot of the area factor for skin tolerance. The Joyet and Hohl[139] data refer to fractionated treatments, whereas all other graphs are for single dose irradiations.[146]

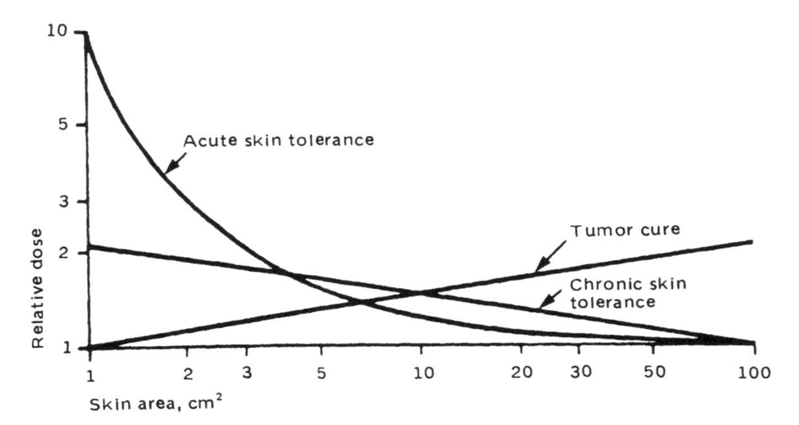

Figure 14. von Essen's area/dose effect curves for acute and chronic skin tolerance and skin cancer cure.[147]

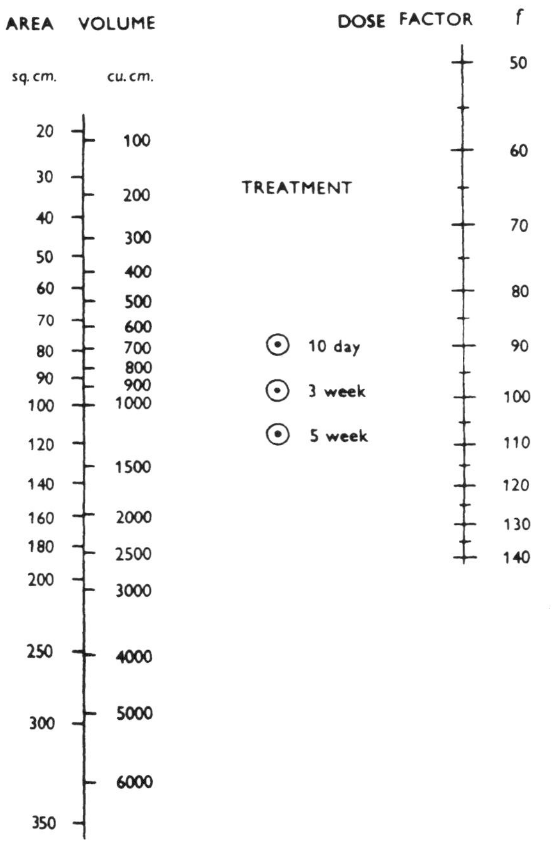

Figure 15. Paterson's equivalent dose nomogram for area, volume, and time.[127]

TABLE XIII
Variation of Tolerance Exposures for Skin According to Ellis[145a]

Time (nominal)	Field size (cm)				
	6 × 4	6 × 8	8 × 10	10 × 15	20 × 15
1 day	2000	1750	1450	1250	1100
1 week	3000	2600	2200	1900	1700
2 weeks	4400	4000	3300	2850	2650
3 weeks	5000	4500	3750	3300	2900
4 weeks	5500	5000	4100	3650	3200
5 weeks	5800	5250	4350	3850	3350
6 weeks	6000	5500	5400	4000	3500

[a] Exposures are in *R* for 200 kVp X rays.

4.2. Power Law Equations

One of the earliest recorded "volume" relationship was that of Meyer in 1939.[41] He observed that the skin tolerance dose varied with the area, *A*, of skin irradiated according to the equation

$$\text{skin tolerance dose} \propto A^{-0.25}$$

This was one of a long string of power law volume effect equations. Some of these are shown in Table XV.

As shown, most of these equations were derived from skin reaction observations and hence should strictly have only been used for the analysis of the "volume" effect on skin. However, since there was a lack of suitable data for tissues other than skin, these same equations and exponents were often used for other tissues and organs. Uses of power

TABLE XIV
Paterson's Skin Tolerance Dose Estimates[127a]

Time (nominal)	Field size (cm)				
	7 × 5	10 × 8	12 × 10	20 × 10	20 × 15
1 day	2000	1700	1500	—	—
4 days	3500	3000	2500	2000	—
10 days	4500	4000	3500	3000	2500
3 weeks	5250	4500	4000	3500	3000
5 weeks	6000	5000	4500	4000	3500

[a] Doses are in cGy for kilovoltage therapy.

TABLE XV
Various Published Linear, Area, and Volume Isoeffect Dose Equations

Reference data	Types of tissue	Linear correction factor	Area correction factor	Volume correction factor
Meyer[41]	Skin	$D \propto L^{-0.50}$	$D \propto A^{-0.25}$	
Jolles[148]	Normal tissues (implants)	$D \propto (V/S)^{-0.33}$	$D \propto A^{-0.17}$ (with 2-D symmetry)	$D \propto V^{-0.11}$ (with 3-D symmetry)
Jolles and Mitchell[149]	Skin	$D \propto (A/P)^{-0.33}$		
Cohen[47,107–110]	Skin	$D \propto L^{-(0.24 \rightarrow 0.36)}$	$D \propto A^{-(0.12 \rightarrow 0.18)}$	
von Essen[150]	Skin	$D \propto L^{-0.33}$	$D \propto A^{-0.17}$	
Ellis[129]	Normal tissues		$D \propto A^{-0.10}$	$D \propto V^{-0.10}$
Cohen[107,109,110]	Vascular stroma	$D \propto L^{-(0.10 \rightarrow 0.23)}$	$D \propto A^{-(0.05 \rightarrow 0.12)}$	
Paterson[127] and Jolles and Mitchell[149] as analyzed by Eads *et al.*[144]	Skin (but used for all tissues)	$D \propto L^{-0.367}$	$D \propto A^{-0.183}$	$D \propto V^{-0.157}$
Ellis[145] and Paterson[127] as analyzed by Kirk *et al.*[93]	Skin (but used for all tissues)		$D \propto A^{-0.24}$	$D \propto V^{-0.16}$
Busch and Rosenow[157]	Normal tissues			$D \propto V^{-(0.03 \rightarrow 0.17)}$
Ellis[145] and Paterson[127] as analyzed by Prasad[151]	Skin		$D \propto A^{-0.25}$	
Paterson[152] as analyzed by Brumm[153]	Skin		$D \propto A^{-0.19}$	
Walter and Miller[154] and von Essen[150] as analyzed by Gupta[155]	Skin		$D \propto A^{-0.18}$	

law volume effect relationships with either nonvarying or tissue-specific exponents in conjunction with the CRE, TDF, and CPK models were presented earlier.

At this juncture it is worth noting that the equivalent volume effect exponents shown in the final column of Table XV vary considerably from a low of −0.03 to a high of −0.17. Part of this imprecision is no doubt due to deviations caused by the imprecision of the data on which these values are based, especially with regard to the way in which isoeffect is defined for different volumes of tissue irradiated as explained in Section 4.1. Another reason why there is such a wide variation in the volume effect exponents shown in Table XV is dose uniformity. A single value is assumed for dose, whereas, in practice, the dose across the volume or area under study is inhomogeneous. A method for dealing with this dose inhomogeneity problem has been devised by Schultheiss *et al.*[156] using an "integral response" model. As shown below, this model also demonstrates another reason why the volume-effect exponents in Table XV are so variable. The model predicts that the exponential form of the volume effect correction factor is a first approximation to a more complex factor. The approximation is good only if probabilities of injury are kept low. Since probabilities of injury are usually not specified in the original data on which the exponents in Table XV are based, it is likely that a power law form for the volume effect correction is inappropriate for some of the studies reported, and hence the exponent reported is invalid. The following review of this integral response model should help clarify some of these points.

4.3. Integral Response Models

There are two important conceptual problems with the simple power law type of volume correction factors. Firstly, it is not obvious how such factors might be used in the "real world," where doses across the tissues or organs at risk are not normally homogeneous. Secondly, and probably more important, such correction factors imply that it is the volume of tissue irradiated that matters and *not* what organs or sensititive structures are contained within this volume. What is needed is a model that can account for inhomogeneous dose distributions and which "integrates" the probability of injury throughout the volume irradiated, for each structure contained therein.

Such "integral probability" models which attempt to address this problem have been proposed by Busch and Rosenow,[157] Dritschilo *et al.*,[158] and Schultheiss *et al.*[156] Each of these models is based upon the premise that the probability that a volume of tissue will remain injury free is the product of the probabilities that each subvolume of that tissue remains free of injury. It is implied, therefore, that in terms of injury each subvolume of tissue is independent of adjacent subvolumes. This is clearly not always going to be true and is a deficiency of each of these models, although it has been argued that such intratissue effects might be of only secondary importance for most tissues.[156]

Only the integral response model of Schultheiss *et al.*[156] will be reviewed in detail here, since all three models are based on the same basic premises, and since a first approximation of this model leads to the same volume effect corrections as obtained by the other two models.

In the integral response model, the volume V of tissue irradiated is divided into a number of subunits, each of which has a fractional (dimensionless) volume $v = V/V_0$, where V_0 is the volume of some specific reference volume of the tissue (or organ) irradiated for which the dose response function is known. It is then shown that the probability, $P(D, v)$, of inducing an injury in this fractional volume irradiated to dose D is related to the probability, $P(D, 1)$, of causing injury to the entire reference volume if it were to be irradiated uniformly to dose D, by the equation

$$P(D, v) = 1 - [1 - (P(D, 1)]^v \tag{7}$$

This equation is applicable only for uniform irradiation to volume v, since otherwise D could not be unambiguously defined. Inhomogeneous dose distributions will be considered later, but first it is interesting to see if the above equation can be used to find a convenient factor which will relate isoeffect doses for different volumes of tissue irradiated.

Firstly, it is necessary to define how probability of injury varies with dose. In the integral response model, a logistic form of this relationship is assumed:

$$P(D, 1) = \frac{1}{1 + (D_{50}/D)^k} \tag{8}$$

where D_{50} is the dose to the reference volume which would induce injuries in 50% of the treated population, and the slope of the dose–response curve at dose $D = D_{50}$ is k/D_{50}. The steepness of the dose–response curve thus defines k.

Then isoeffect is obtained if the probability of injury to the reference volume irradiated to dose D_1 is equal to that to volume v irradiated to dose D_v. Expressed mathematically that is

$$P(D_1, 1) = P(D_v, v)$$

But from Eq. (7)

$$P(D_v, v) = 1 - [1 - P(D_v, 1)]^v$$

Hence

$$P(D_1, 1) = 1 - [1 - P(D_v, 1)]^v$$

Then, using the logistic forms of the probability functions gives

$$\frac{1}{1 + (D_{50}/D_1)^k} = 1 - \left[1 - \frac{1}{1 + (D_{50}/D_v)^k}\right]^v$$

Solving this for D_v leads to the following relationship between the isoeffect doses to volume v and the reference volume:

$$D_v = D_{50}\{[1 + (D_1/D_{50})^k]^{1/v} - 1\}^{1/k} \quad (9)$$

This is clearly far more complicated than the simple power law equations described earlier. It also implies that the relationship between isoeffect doses for different volumes of tissue irradiated depends on the shape of the dose–response curve for that particular tissue.

Equation (9) can be simplified by making the assumption that in most practical radiotherapy situations, $P(D_v, 1) \ll 1$. Then, from the logistic dose–response equation (8), $(D_{50}/D_v)^k$ must be large, or conversely, $(D_v/D_{50})^k \ll 1$.

Rearrangement of Eq. (9) gives

$$[1 + (D_v/D_{50})^k]^v = 1 + (D_1/D_{50})^k$$

and the first binominal expansion of this equation for $(D_v/D_{50})^k \ll 1$ yields

$$D_v = D_1 v^{-1/k}$$

Clearly, this model predicts that the exponent for the simple power law volume effect relationship is tissue specific and is a function of the shape of the dose–response curve for each tissue. Furthermore, it is apparent that the power law volume effect correction is applicable only when $P(D_v, 1) \ll 1$. In other words, simple power law expressions for the volume effect correction can be applied only if the dose to the irradiated

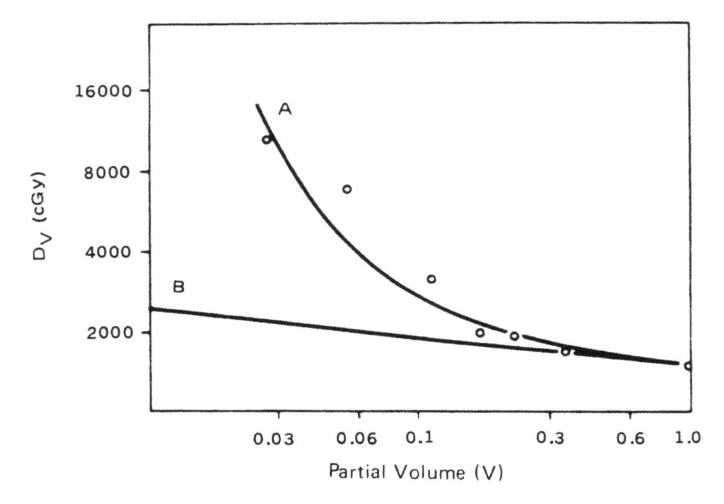

Figure 16. Variation of isoeffect dose with volume.[156] The data points are from Berg and Lindgren.[146] Curve A represents the integral response model Eq. (9) of Schultheiss *et al.*[156] with $1/k = 0.11$ and $D_{50} = 1570\,\text{cGy}$, while curve B represents the simple power-law model.

volume would result in a small probability of complication even if given to the entire reference volume of the organ or tissue.

Power law volume effect relationships would thus be expected to break down when very small volumes of tissue are irradiated to high doses, i.e., log–log plots of isoeffect dose vs. volume will diverge from linearity as the volume decreases. This is precisely what has been reported. For example, Fig. 13 illustrates that log–log plots of skin tolerance doses become nonlinear for field sizes less than about 10 cm^2.[146] Apparently, Eq. (9) needs to be used for such situations, and this is demonstrated in Fig. 16, which shows isoeffect doses for the induction of lesions in different volumes of the brain tissue of rabbits.[146] It seems from this analysis that the power law equation provides a reasonably good approximation to the volume effect relationship for partial volumes $v \gtrsim 0.15$. For smaller volumes, the more complex equation (9) seems to be appropriate. Of course, neither equation is useful if there is significant variation in dose across the tissue or organ at risk. The problem of inhomogeneous dose distributions will be addressed in Section 4.3.2, but before doing so it is interesting to examine values of the power law volume effect exponents which can be derived from dose–response data for various tissues, and to compare these values with those derived from direct volume effect observations.

4.3.1. Volume Effect Exponents for Different Tissues

There are numerous sources of dose–response data from which to determine volume effect exponents. Schultheiss *et al.*[156] chose to use the D_{50} and D_{05} doses quoted by Rubin,[159] and Rubin and Poulter[160] for these calculations, these being the doses that will produce 50% and 5% risks of complication, respectively, for patients treated at about 200 cGy per fraction, 5 fractions per week. Substitution of $P = 0.50$ and $P = 0.05$ in the logistic dose–response equation at doses D_{50} and D_{05}, respectively, yields

$$k = \log 19/\log (D_{50}/D_{05})$$

Table XVI shows $1/k$ values determined for several tissues.[98,156] It is interesting that some tissues (in this small sample, bone and thyroid) appear to exhibit much higher exponents than others.

Volume effect exponents for a wide variety of other tissues can be calculated from other D_{50} and D_{05} data in Rubin,[159] and Rubin and Poulter.[160]

4.3.2. Inhomogeneous Dose Distributions

It is clearly inappropriate to apply a volume effect correction factor to doses when dose distributions are not homogeneous, since "dose" is an ambiguous parameter. Taking an average dose will underestimate the probability of injury at low probabilities ($D < D_{50}$) and overestimate it at

TABLE XVI
Values of Volume Effect Exponents Determined from Dose–Response Data[156,159,160]

Tissue	Exponent
Stroma (muscle)	−0.10
Brain	−0.05
Cord	−0.07
Lung	−0.14
Gut	−0.10
Kidney	−0.07
Esophagus	−0.08
Skin	−0.08
Bladder	−0.10
Rectum	−(0.10–0.13)
Bone (adult)	−(0.17–0.31)
Thyroid	−0.41

high probabilities ($D > D_{50}$). Nor it is sufficient to take the average probability of injury over the irradiated volume, since probability is not a linear function of volume. As shown by Schultheiss *et al.*,[156] what needs to be done is to take the product of the probabilities of avoiding injury in each volume element Δv_i, each of which received a dose D_i, hence

$$P(\{D\}, v) = 1 - \prod_i [1 - P(D_i, 1)]^{\Delta v_i}$$

where $\{D\}$ represents a dose distribution and the product is taken over all i volume elements.

Then, if there is more than one organ or tissue irradiated, the total probability, P_T, of inducing an injury within the irradiated volume is given by

$$P_T = 1 - \prod_j \prod_i [1 - P_j(D_i, 1)]^{\Delta v_i}$$

where P_j is the probability of inducing an injury in organ j, and the products are taken over all organs j and all volume elements i.

Then, provided the dose–response parameters are known for all irradiated tissues, this overall complication probability can be determined using calculated dose distributions. This is relatively straightforward if three-dimensional CT data and treatment planning capabilities are available. Computer programs have been written to perform these calculations on treatment planning computers (Schultheiss *et al.*[156]).

One of the interesting features of this integral response approach is that it provides a means for intercomparing various treatment techniques, with the ultimate potential of optimization.[161–163]

5. SUMMARY AND CONCLUSIONS

In order to enhance the efficacy of radiation therapy, numerous attempts have been made, and are continuing to be made, to manipulate the characteristics of delivery of treatment, the ultimate aim being an improvement in the therapeutic ratio. Overall treatment time, total dose, dose per fraction, interfraction time, dose rate, and volume are all parameters commonly varied in this quest for the "optimal" schedule. In addition, a variety of nonuniform treatment regimes have been proposed, some of which are in common use today, the most obvious being the use of split courses, reducing fields, varying dose per fraction, and mixed external beam and brachytherapy.

All of these are characterized by a common difficulty, namely, the specification of *dose*. Yet it is important that we have a way of specifying dose in order to be able to compare treatment regimes, especially with regard to organ and tissue tolerance concerns and, ultimately, to be able to select the "optimal" schedule. This has become increasingly important in the last several years with the advent of three-dimensional treatment planning capabilities through the use of CT and MRI facilities. The physical doses to all tissues and organs can now be calculated rapidly and accurately. We now need to have a means of converting these physical doses into their biologically effective equivalents.

Clearly, such bioeffect dosimetry is important. Yet, despite significant efforts as reviewed in this chapter, no reliable time–dose model has yet been constructed. Even the most successful of the models so far developed have been seriously flawed and have been useful only under limited conditions. The reason is obvious. There have simply not been enough suitable human data on which to build time–dose models, and even the data that have been available have frequently not been analyzed by state-of-the-art statistical techniques. This was one of the conclusions at the Second International Conference on Time, Dose, and Fractionation in Radiation Oncology held in September 1984 in Madison, Wisconsin.[164] As an outcome of this conference, the Biological Effects Committee of the American Association of Physicists in Medicine formed a Task Group (Chairman, D. Herbert) to develop plans for an international cooperative Data Bank. It is planned that this Data Bank will be a repository for human data suitable for the development and testing of time–dose models. Furthermore, the Task Group will endeavor to employ and make available sophisticated statistical techniques for the analysis of this data.

It will be difficult, perhaps impossible, but there is optimistic hope and expectation that such efforts will bear fruit and that a major breakthrough in bioeffect dosimetry will occur in the next decade.

REFERENCES

1. J. T. Case, The early history of radium therapy and the American Radium Society, *Am. J. Roentgenol.* **82,** 574–585 (1959).
2. S. Krönig and W. Friedrich, Physikalische und biologisch Grundlagen der Strahlentherapie, *Strahlentherapie* (Sonderb.) (1918).
3. G. Schwarz, Heilung teifliegender Karzinome durch Roentgenbestrahlung von der Korperoberflache aus, *Münch. Med. Wochenschr.* **61,** 1733 (1914).
4. H. Wintz, Ergebnisse der Roentgentherapie des Mammakarzinoms, *Dtsch. Med. Wochenschr.* **57**(2), 1569–1573 (1931).

5. L. B. Kingery, Saturation in Roentgen therapy: Its estimation and maintenance: Preliminary report, *Arch. Derm. Syph.* **4,** 423–430 (1920).
6. G. E. Pfahler and B. P. Widman, Further observations on the use of the saturation method of radiation therapy in deep-seated malignant disease, with some statistics, *Radiology* **11,** 181–190 (1928).
7. H. Coutard, Roentgen therapy of epitheliomas of tonsillar regions, hypopharynx, and larynx from 1920 to 1926, *Am. J. Roentgenol.* **28,** 313–331 (1932).
8. ICRU, International X-ray unit of intensity, *Brit. J. Radiol.* **1,** 363–364 (1928).
9. E. H. Quimby, The grouping of radium tubes in packs or plaques to produce the desired distribution of radiation, *Am. J. Roentgenol.* **27,** 18–39 (1932).
10. R. Paterson and H. M. Parker, A dosage system for gamma ray therapy, *Brit. J. Radiol.* **7,** 592–632 (1934).
11. F. Baclesse, Clinical experience with ultrafractionated Roentgen therapy, in *Progress in Radiation Therapy,* Vol. 1 (F. Buschke, ed.), pp. 128–143, Grune and Stratton, New York (1958).
12. D. K. Sambrook, Theoretical aspects of dose-time factors in radiotherapy technique, *Clin. Radiol.* **14,** 433–441 (1963).
13. G. H. Fletcher, The scientific basis of the present and future practice of clinical radiotherapy, *Int. J. Radiat. Oncol. Biol. Phys.* **9,** 1073–1082 (1983).
14. A. Backstrom, P. A. Jacobsson, B. Littbrand, and J. Wersall, Fractionation scheme with low individual doses in irradiation of carcinoma of the mouth, *Acta Radiol. Ther. Phys. Biol.* **12,** 401–405 (1973).
15. J. C. Horiot, A. Nabid, G. Chaplain, S. Jampolis, W. van den Bogaert, E. van den Schueren, G. Arcangeli, D. Gonzales, V. Svoboda, and H. P. Hamers, Clinical experience with multiple daily fractionation (MDF) in the radiotherapy of head and neck carcinoma, *Cancer Bull.* **34**(6), 230–233 (1982).
16. H. D. Thames, L. J. Peters, H. R. Withers, and G. H. Fletcher, Accelerated fractionation vs. hyperfractionation: Rationales for several treatments per day, *Int. J. Radiat. Oncol. Biol. Phys.* **9,** 127–138 (1983).
17. V. H. J. Svoboda, Further experience with radiotherapy by multiple daily sessions, *Brit. J. Radiol.* **51,** 363–369 (1978).
18. C. C. Wang, Twice daily radiation therapy for carcinomas of the head and neck, *Int. J. Radiat. Oncol. Biol. Phys.* **7,** 1261–1262 (1975) (abs.).
19. W. Schumacher, Neue strahlenbiologische erkenntnisse zur verbesserung der strahlentherapie, *Strahlentherapie* **64,** 122–129 (1967).
20. H. J. Eichhorn, A. Lessel, and K. H. Rotte, Einfluss verschiedener Bestrahlungsrhythmen auf Tumor- und Normalgewebe in vivo, *Strahlentherapie* **143,** 614–629 (1972).
21. H. J. Habermalz and J. J. Fischer, Radiation therapy of malignant melanoma. Experience with high individual treatment doses, *Cancer* **38,** 2258–2262 (1976).
22. H. R. Withers, L. J. Peters, H. D. Thames, and G. H. Fletcher, Hyperfractionation, *Int. J. Radiat. Oncol. Biol. Phys.* **8,** 1807–1809 (1982).
23. H. D. Thames, H. R. Withers, L. J. Peters, and G. H. Fletcher, Changes in early and late radiation responses with altered dose fractionation: Implications for dose–survival relationships, *Int. J. Radiat. Oncol. Biol. Phys.* **8,** 219–226 (1982).
24. J. F. Fowler, What next in fractionated radiotherapy? *Brit. J. Cancer* **49** (Suppl. VI), 285–300 (1984).
25. B. Pierquin, F. Baillet, and C. H. Brown, Low dose irradiation in advanced tumors of head and neck, *Acta Radiol. Ther. Phys. Biol.* **14,** 497–504 (1975).
26. B. Pierquin, E. Calitchi, J. J. Mazeron, J. P. LeBourgeois, and S. Leung, A comparison between low dose rate radiotherapy and conventionally fractionated

irradiation in moderately extensive cancers of the oropharynx, *Int. J. Radiat. Oncol. Biol. Phys.* **11,** 431–439 (1985).
27. J. Heyman, The technique in the treatment of cancer uteri at Radiumhemmet, *Acta Radiol.* **10,** 49–64 (1929).
28. J. F. Utley, C. F. von Essen, R. A. Horn, and J. H. Moeller, High-dose-rate afterloading brachytherapy in carcinoma of the uterine cervix, *Int. J. Radiat. Oncol. Biol. Phys.* **10,** 2259–2263 (1984).
29. C. A. Joslin, The Cathetron as a part of the radical management of cervix cancer, in *High-Dose-Rate Afterloading in the Treatment of Cancer of the Uterus* (T. D. Bates and R. J. Berry, eds.), pp. 11–16, BIR Special Report No. 17, BIR, London (1980).
30. R. F. Mould, (ed.), *Brachytherapy* 1984, Nucletron, The Netherlands (1985).
31. V. Schulz, M. Busch, and V. Bormann, Interstitial high dose-rate brachytherapy: Principle, practice and first clinical experiences with a new remote-controlled afterloading system using Ir-192, *Int. J. Radiat. Oncol. Biol. Phys.* **10,** 915–920 (1984).
32. W. E. Liversage, A comparison of the predictions of the CRE, TDF and Liversage formulae with clinical experience, in *High-Dose-Rate Afterloading in the Treatment of Cancer of the Uterus* (T. D. Bates and R. J. Berry, eds.), BIR Special Report No. 17, BIR, London (1981).
33. C. C. Ling, L. L. Anderson, and W. V. Shipley, Dose inhomogeneity in interstitial implants using ^{125}I seeds, *Int. J. Radiat. Oncol. Biol. Phys.* **5,** 419–425 (1979).
34. R. Kienbrock, Über dosimeter und das quantimetrische verfahren., *Fortschr. Geb. Rontgenstr.* **9,** 276–295 (1905).
35. L. Freund, Ein wichtiger Fortschritt für die medizinische Lichtforschung, *Strahlentherapie* **10,** 1145–1161 (1920).
36. M. Strandqvist, Studien über die kumulative Wirkung der Röntgenstrahlen bei Fraktionierung, *Acta Radiol. Suppl.* **73,** 1–300 (1944).
37. A. Reisner, Hauterythem und Rontgenbestrahlung, *Ergebn. Med. Strahlenforsch.* **6,** 1 (1933).
38. E. Quimby and W. S. MacComb, Further studies on rate of recovery of human skin from effects of Roentgen or gamma-ray irradiation, *Radiology* **29,** 305–312 (1937).
39. E. Witte, Dosierung im biologischen Mass, *Strahlentherapie* **72,** 177–194 (1942).
40. R. K. Kepp, Ergebnisse von erythemversuchen mit fraktionierter rontgenbestrahlung verschiedener intensitat, *Strahlentherapie* **72,** 195–201 (1942).
41. W. H. Meyer, The co-relation of physical and clinical data in radiation therapy, *Radiology* **32,** 23–45 (1939).
42. L. A. DuSault, Time–dose relationships, *Am. J. Roentgen.* **75,** 597–606 (1956).
43. M. Friedman and A. W. Pearlman, Time–dose studies in irradiation of mycosis fungoides, iso-effect curve and tumor lethal dose, *Radiology* **66,** 374–379 (1956).
44. L. Cohen, Clinical radiation dosage, *Brit. J. Radiol.* **22,** 160–163 (1949).
45. L. Cohen, Clinical radiation dosage II. Inter-relation of time, area and therapeutic ratio, *Brit. J. Radiol.* **22,** 706–713 (1949).
46. L. Cohen, Radiation response and recovery, in *The Biological Basis of Radiation Therapy* (E. E. Schwartz, ed.) pp. 208–316 Lippincott, New York (1966).
47. L. Cohen and J. E. Kerrich, Estimation of biological dosage factors in clinical radiotherapy, *Brit. J. Cancer* **5,** 180–194 (1951).
48. J. F. Fowler and B. E. Stern, Dose–time relationships in radiotherapy and the validity of cell survival curve models, *Brit. J. Radiol.* **36,** 163–173 (1963).
49. J. F. Fowler, D. K. Bewley, R. L. Morgan, J. A. Silvester, T. Alper, and S. Hornsey, Dose effect relationships for radiation damage to organized tissues, *Nature* **199,** 253–255 (1963).
50. J. F. Fowler and B. E. Stern, Dose–rate effects: Some theoretical and practical considerations, *Brit. J. Radiol.* **31,** 389–395 (1960).

51. G. W. Barendsen, Dose fractionation, dose rate and iso-effect relationships for normal tissue responses, *Int. J. Radiat. Oncol. Biol. Phys.* **8,** 1981–1997 (1982).
52. B. G. Douglas, J. F. Fowler, J. Denekamp, S. R. Harris, S. E. Ayres, S. Fairman, S. A. Hill, P. W. Sheldon, and F. A. Stewart, The effect of multiple small fractions of x rays on skin reactions in the mouse, in *Cell Survival after Low Doses of Radiation,* Proc. 6th L. H. Gray Conf. (T. Alper, ed.), pp. 351–361 Institute of Physics, London (1975).
53. M. V. Williams, J. Denekamp, and J. F. Fowler, A review of α/β ratios for experimental tumors: Implications for clinical studies of altered fractionation, *Int. J. Radiat. Oncol. Biol. Phys.* **11,** 87–96 (1985).
54. J. F. Fowler, The estimation of total dose for different numbers of fractions in radiotherapy, *Brit. J. Radiol.* **38,** 365–368 (1965).
55. D. K. Sambrook, Theoretical aspects of dose–time factors in radiotherapy technique, *Clin. Radiol.* **14,** 290–297 (1963).
56. E. Shuttleworth and J. F. Fowler, Nomograms for radiobiologically-equivalent fractionated x-ray doses, *Brit. J. Radiol.* **39,** 154–157 (1966).
57. J. E. Burns, Nomogram for radiobiologically-equivalent fractionated doses, *Brit. J. Radiol.* **38,** 545–547 (1965).
58. L. Cohen, Theoretical "iso-survival" formulae for fractionated radiation therapy, *Brit. J. Radiol.* **41,** 522–528 (1968).
59. W. E. Liversage, A critical look at the ret, *Brit. J. Radiol.* **44,** 91–100 (1971).
60. J. Denekamp, Changes in the rate of repopulation during multifraction irradiation of mouse skin, *Brit. J. Radiol.* **46,** 381–387 (1973).
61. J. Denekamp, Changes in the rate of proliferation in normal tissues after irradiation, in *Radiation Research, Biomedical, Chemical and Physical Perspectives* (Nygaard, Adler, and Sinclair, eds.), pp. 810–825 Academic, New York (1975).
62. J. E. Moulder and J. J. Fischer, Radiation reaction of rat skin: The role of number of fractions and overall treatment time, *Cancer* **37,** 2762–2767 (1976).
63. S. Kozubek, A simple radiobiological model for fractionated radiation therapy, *Int. J. Radiat. Oncol. Biol. Phys.* **8,** 1975–1980 (1982).
64. F. Ellis, Fractionation in radiotherapy, in *Modern Trends in Radiotherapy* (Deeley and Wood, eds.), Vol. 1, pp. 34–51 Butterworth, London (1967).
65. L. Cohen, Radiation parameters. Ph.D. thesis, University of Witwatersrand, 1960.
66. F. Ellis, Relationship between log dose and log time in radiotherapy—the Strandqvist lines, *Brit. J. Radiol.* **49,** 651 (1976).
67. C. G. Orton and F. Ellis, Definition of T in the NSD equation, *Brit. J. Radiol.* **47,** 201–202 (1974).
68. C. G. Orton, Errors in applying the NSD concept, *Radiology* **115,** 233–235 (1975).
69. F. Ellis, Dose, time and fractionation: A clinical hypothesis, *Clin. Radiol.* **20,** 1–7 (1969).
70. B. M. Winston, F. Ellis, and E. J. Hall, The Oxford NSD calculator for clinical use, *Clin. Radiol.* **20,** 8–11 (1969).
71. R. E. Peschel and J. J. Fischer, Optimization of the time dose relationship, *Semin. Oncol.* **8,** 38–47 (1981).
72. J. J. Fischer and D. B. Fischer, The determination of time-dose relationships from clinical data, *Brit. J. Radiol.* **44,** 785–792 (1971).
73. D. Herbert, NSD forever? Night thoughts of a medical physicist, *Int. J. Radiat. Oncol. Biol. Phys.* **9,** 1099–1100 (1983).
74. J. C. Probert, Doubts about the nominal standard dose, *Brit. J. Radiol.* **44,** 648 (1971).
75. H. R. Withers, H. D. Thames, and L. J. Peters, Differences in the fractionation response of acute and late responding tissues, in *Progress in Radio-Oncology II,*

(Karcher, Kogelnik, and Reinartz, eds.), pp. 257–296, Raven Press, New York (1982).

76. H. S. Reinhold, J. G. Kaalen, and K. Unger-Gils, Radiation myelopathy of the thoracic spinal cord, *Int. J. Radiat. Oncol. Biol. Phys.* **1,** 651–657 (1976).
77. W. M. Wara, T. L. Phillips, G. E. Sheline, and J. G. Schwade, Radiation tolerance of the spinal cord, *Cancer* **35,** 1558–1562 (1975).
78. L. Cohen and M. Creditor, An iso-effect table for radiation tolerance of the human spinal cord, *Int. J. Radiat. Oncol. Biol. Phys.* **7,** 961–966 (1981).
79. T. E. Schultheiss, E. M. Higgins, and A. M. El-Mahdi, The latent period in clinical radiation myelopathy, *Int. J. Radiat. Oncol. Biol. Phys.* **10,** 1109–1115 (1984).
80. G. E. Sheline, W. M. Wara, and V. Smith, Therapeutic irradiation and brain injury, *Int. J. Radiat. Oncol. Biol. Phys.* **6,** 1215–1228 (1980).
81. S. Hornsey, C. C. Morris, and R. Myers, The relationship between fractionation and total dose for x ray induced brain damage, *Int. J. Radiat. Oncol. Biol. Phys.* **7,** 393–396 (1981).
82. R. D. Pezner and J. O. Archambeau, Brain tolerance unit: A method to estimate risk of radiation brain injury for various dose schedules, *Int. J. Radiat. Oncol. Biol. Phys.* **7,** 397–402 (1981).
83. L. Cohen and M. Creditor, Iso-effect tables for tolerance of irradiated normal human tissues, *Int. J. Radiat. Oncol. Biol. Phys.* **9,** 233–241 (1983).
84. W. M. Wara, T. L. Phillips, L. W. Margolis, and V. Smith, Radiation pneumonitis—A new approach to the derivation of time–dose factors, *Cancer* **32,** 547–552 (1973).
85. I. Turesson and G. Notter, The influence of the overall treatment time in radiotherapy on the acute reaction: Comparison of the effects of daily and twice-a-week fractionation on human skin, *Int. J. Radiat. Oncol. Biol. Phys.* **10,** 607–618 (1984).
86. I. Turesson and G. Notter, The influence of fraction size in radiotherapy on the late normal tissue reaction—I: Comparison of the effects of daily and once-a-week fractionation on human skin, *Int. J. Radiat. Oncol. Biol. Phys.* **10,** 593–598 (1984).
87. I. Turesson and G. Notter, The influence of fraction size in radiotherapy on the late normal tissue reaction—II: Comparison of the effects of daily and twice-a-week fractionation on human skin, *Int. J. Radiat. Oncol. Biol. Phys.* **10,** 599–606 (1984).
88. J. F. Fowler, Non-standard fractionation in radiotherapy, *Int. J. Radiat. Oncol. Biol. Phys.* **10,** 755–759 (1984).
89. R. L. Dixon, General equation for the calculation of Nominal Standard Dose, *Acta Radiol. Ther.* **11,** 305–311 (1972).
90. J. Kirk, W. M. Gray, and E. R. Watson, Cumulative radiation effect Part I: fractionated treatment regimes, *Clin. Radiol.* **22,** 145–155 (1971).
91. J. Kirk, W. M. Gray, and E. R. Watson, Cumulative radiation effect Part II: Continuous radiation therapy—long-lived sources, *Clin. Radiol.* **23,** 93–105 (1972).
92. J. Kirk, W. M. Gray, and E. R. Watson, Cumulative radiation effect Part III: Continuous radiation therapy—short-lived sources, *Clin. Radiol.* **24,** 1–11 (1973).
93. J. Kirk, W. M. Gray, and E. R. Watson, Cumulative radiation effect Part IV: Normalization of fractionated and continuous therapy—area and volume correction factors, *Clin. Radiol.* **26,** 77–88 (1975).
94. J. Kirk, W. M. Gray, and E. R. Watson, Cumulative radiation effect Part V: Time gaps in treatment regimes, *Clin. Radiol.* **26,** 159–176 (1975).
95. J. Kirk, W. M. Gray, and E. R. Watson, Cumulative radiation effect Part VI: Simple nomographic and tabular methods for the solution of practical problems, *Clin. Radiol.* **28,** 29–74 (1977).
96. C. G. Orton, Analysis and discussion of the time/dose/fractionation problem, *Q. Bull. Am. Assoc. Phys. Med.* **6,** 173–175 (1972).

97. C. G. Orton and F. Ellis, A simplification in the use of the NSD concept in practical radiotherapy, *Brit. J. Radiol.* **46,** 529–537 (1973).
98. C. G. Orton and L. Cohen, A variable exponent TDF model, Proceedings of the Second International Conference on Time, Dose, and Fractionation in Radiation Oncology, Madison, Wisconsin, September 1984 (Paliwal, Herbert, and Orton, eds.), AIP, New York (in press).
99. C. G. Orton, SI units of TDF, *Brit. J. Radiol.* **53,** 513–514 (1980).
100. L. Cohen, F. Hendricksen, J. Mansell, M. Awschalom, A. Hrejsa, R. Kaul, and I. Rosenberg, Late reactions and complications in patients treated with high energy neutrons p(66 MeV) Be(49 MeV), *Int. J. Radiat. Oncol. Biol. Phys.* **7,** 179–184 (1981).
101. L. Cohen and M. Awschalom, Fast neutron radiation therapy, *Ann. Rev. Biophys. Bioeng.* **11,** 359–390 (1982).
102. Y. Kutsutani-Nakamura, *Nippon Acta Radiol.* **38,** 950–960 (1978).
103. R. Wideröe, High-energy electron therapy and the two-component theory of radiation, *Acta Radiol. Ther.* **4,** 257–278 (1966).
104. L. Cohen, Biophysical Models in Radiation Oncology, CRC Press, Boca Raton, Florida (1983).
105. L. H. Gray, F. Ellis, G. C. Fairchild, and E. R. Paterson, Dosage-rate in radiotherapy, *Brit. J. Radiol.* **17,** 327–342 (1944).
106. L. H. Gray and M. E. Scholes, The effect of ionizing radiations on the broad bean root, *Brit. J. Radiol.* **24,** 285–291 (1951).
107. L. Cohen, Derivation of cell population kinetic parameters from clinical statistical data (program RAD 3), *Int. J. Radiat. Oncol. Biol. Phys.* **4,** 835–840 (1978).
108. L. Cohen, A cell population kinetic model for fractionated radiation therapy. I. Normal tissues, *Radiology* **101,** 419–427 (1971).
109. L. Cohen, An interactive program for standardization of prescriptions in radiation therapy, *Comput. Programs Med.* **3,** 27–35 (1973).
110. L. Cohen, Cell population kinetics in radiation therapy: optimization of tumor dosage, *Cancer* **32,** 236–244 (1973).
111. G. M. McKee and A. Mutscheller, *The Science of Radiology,* Thomas, Springfield, Illinois (1933).
112. R. Pape, Der Einfluss der Veränderung des Minuten-r-Zuflusses auf die Hautreaktion bei kontinuierlicher und geteilter Dosenapplikation, *Strahlentherapie* **45,** 475–486 (1932).
113. A. Brunschwig and S. P. Perry, High versus low intensity irradiation in the treatment of carcinoma, *Radiology* **26,** 706–716 (1936).
114. J. S. Fulton, *Report on Discussion of Society of Radiotherapists* (1937).
115. H. Holthusen, Vergleichende Untersuchungen über die Wirhung von Röntgen- und Radiumstrahlen, *Strahlentherapie* **46,** 273–288 (1933).
116. R. McWhirter, Radiosensitivity in relation to time intensity factor, *Brit. J. Radiol.* **9,** 287–299 (1936).
117. E. J. Hall, Radiation dose-rate: A factor of importance in radiobiology and radiotherapy, *Brit. J. Radiol.* **45,** 81–97 (1972).
118. D. L. Dewey and J. W. Boag, Modification of the oxygen effect when bacteria are given large doses of radiation, *Nature* **183,** 1450–1451 (1959).
119. C. D. Town, Effect of high dose-rates on survival of mammalian cells, *Nature* **215,** 847–848 (1967).
120. E. R. Epp, H. Weiss, and A. Santomasso, The oxygen effect in bacterial cells irradiated with high intensity pulsed electrons, *Radiat. Res.* **34,** 320–325 (1968).
121. T. L. Phillips and R. B. Worsnop, Oxygen depletion by ultra-high-dose-rate electrons in bacteria and mammalian cells, *Radiat. Res.* **35,** 545 (1968) (abs).

122. R. J. Berry, E. J. Hall, D. W. Forster, T. H. Storr, and M. J. Goodman, Survival of mammalian cells exposed to x rays at ultrahigh dose-rates, *Brit. J. Radiol.* **42,** 102–107 (1969).
123. M. L. Griem, L. S. Skaggs, L. H. Lanzl, and F. D. Malkinson, Experience in radiobiological dosimetry with high dose-rate electrons, *Ann. N.Y. Acad. Med.* **161,** 317–322 (1969).
124. P. W. Todd, H. S. Winchell, J. M. Feola, and G. E. Jones, Irradiation by pulsed high-intensity x rays of human cells cultured *in vitro, Radiat. Res.* **31,** 644 (1967) (abs.).
125. A. H. W. Nias, A. J. Swallow, J. P. Keene, and B. W. Hodgson, Survival of HeLa cells from 10 nanosecond pulses of electrons, *Int. J. Radiat. Biol.* **17,** 595–598 (1970).
126. G. T. Pack and E. H. Quimby, Time-intensity factor in irradiation, *Am. J. Roentgenol.* **28,** 650–667 (1932).
127. R. Paterson, *The Treatment of Malignant Disease by Radiotherapy,* 2nd edn, Williams and Wilkins, Baltimore (1963).
128. J. M. Wilkinson, Interstitial radiotherapy at low dose-rate, *Brit. J. Radiol.* **45,** 708 (1972).
129. F. Ellis, Dose–time relationships in clinical radiotherapy, in *Cancer, Progress Volume* (R. W. Raven, ed.), pp. 163–176, Butterworths, London (1963).
130. C. G. Orton, Time–dose factors (TDFs) in brachytherapy, *Brit. J. Radiol.* **47,** 603–607 (1974).
131. W. E. Liversage, A general formula for equating protracted and acute regimes of radiation, *Brit. J. Radiol.* **42,** 432–440 (1969).
132. L. G. Lajtha and R. Oliver, Some radiobiological considerations in radiotherapy, *Brit. J. Radiol.* **34,** 252–257 (1961).
133. F. Ellis and A. Sorensen, A method of estimating biological effect of combined intracavitary low dose rate radiation with external radiation in carcinoma of the cervix uteri, *Radiology* **110,** 681–686 (1974).
134. N. Tapley, *Clinical Applications of the Electron Beam,* Wiley, New York (1976).
135. C. G. Orton, Time, dose, fractionation, and volume relationships in radiotherapy, in *Handbook of Medical Physics,* Vol. 1. (Waggener, Kerieakes, and Shalek, eds.), pp. 265–293, CRC Press, Boca Raton (1982).
136. C. G. Orton, Re-assessment of normalization between fractionated and continuous radiotherapy for the CRE and TDF equations, *Brit. J. Radiol.* **53,** 374–375 (1980).
137. J. S. Mitchell, *Studies in Radiotherapeutics,* p. 234, Blackwell, Oxford (1960).
138. C. G. Orton and B. Webber, Time–dose factor (TDF) analysis of dose rate effects in permanent implant dosimetry, *Int. J. Radiat. Oncol. Biol. Phys.* **2,** 55–60 (1977).
139. G. Joyet and K. Hohl, Die biologische Hautreaktion in der Tiefentherapie als Funktion der Feldgrosse; ein Gesetz der Strahlentherapie, *Fortschr. Geb. Röntgen.* **82,** 387–400 (1955).
140. M. Garcia, Further observations on tissue dosage in cancer of cervix uteri, *Am. J. Roentgen.* **73,** 3560 (1955).
141. J. W. Hopewell and C. M. A. Young, The effect of field size on the reaction of pig skin to single doses of x rays, *Brit. J. Radiol.* **55,** 356–361 (1982).
142. C. F. von Essen, Effect of field size on the reaction of pig skin to single doses of x rays, *Brit. J. Radiol.* **55,** 936 (1982).
143. J. W. Hopewell and C. M. A. Young, Effect of field size on the reaction of pig skin to single doses of x rays, *Brit. J. Radiol.* **55,** 936–937 (1982).
144. D. L. Eads, J. M. Vaeth, and D. G. Baker, To rec or to ret, that is the question, *Radiol. Clin. Biol.* **43,** 21–39 (1974).
145. F. Ellis, Tolerance dosage in radiotherapy with 200 kV x rays, *Brit. J. Radiol.* **15,** 348–350 (1942).

146. N. Berg and M. Lindgren, Relationship between field size and tolerance of rabbit brain in roentgen irradiation (200 kV) via a slit-shaped field, *Acta Radiol.* **1,** 147–168 (1963).
147. C. F. von Essen, Clinical radiation tolerance of the skin and upper aerodigestive tract, in *Frontiers of Radiation Therapy Oncology* (Vaeth, ed.), Vol. 6, pp. 148–159, University Park Press, Baltimore (1972).
148. B. Jolles, Quantitative biological dose control in interstitial radium therapy, *Brit. J. Radiol.* **19,** 143–144 (1946).
149. B. Jolles and R. G. Mitchell, Optimal skin tolerance dose levels, *Brit. J. Radiol.* **20,** 405–409 (1947).
150. C. F. von Essen, A spatial model of time–dose–area relationship in radiation therapy, *Radiology* **81,** 881–884 (1963).
151. S. G. Prasad, Relation between tolerance dose and treatment field size in radiotherapy, *Med. Phys.* **5,** 430–433 (1978).
152. R. Paterson, *The Treatment of Malignant Disease by Radium and X-Rays,* Arnold, London (1948).
153. P. Brumm, On the validity of the NSD concept, *Brit. J. Radiol.* **56,** 957–962 (1983).
154. J. Walter and H. Miller, *A Short Textbook of Radiotherapy for Technicians and Students,* p. 240, J. and A. Churchill, London (1950).
155. M. K. Gupta, Reconsideration of area correction factor for CRE and TDF models, *Brit. J. Radiol.* **57,** 188–190 (1984).
156. T. E. Schultheiss, C. G. Orton, and R. A. Peck, Models in radiotherapy: Volume effects, *Med. Phys.* **10,** 410–415 (1983).
157. M. Busch and V. Rosenow, Dose–volume relationships, in *Computer Applications in Radiation Oncology* (E. S. Sternick, ed.), pp. 279–291, University Press of New England, Hanover, New Hampshire (1976).
158. A. Dritschilo, J. T. Chaffey, W. A. Bloomer, and A. Marck, The complication probability factor: A method for selection of radiation treatment plans, *Brit. J. Radiol.* **51,** 370–374 (1978).
159. P. Rubin, *Radiation Biology and Radiation Pathology Syllabus,* pp. 2–5, ACR, Chicago (1975).
160. P. Rubin and C. Poulter, in *Clinical Oncology for Medical Students and Physicians, a Multi-disciplinary Approach* (P. Rubin, ed.), pp. 35–39, ACS, New York (1978).
161. A. B. Wolbarst, E. S. Sternick, and A. Dritschilo, Optimized radiotherapy treatment planning using the complication probability factor (CPF), *Int. J. Radiat. Oncol. Biol. Phys.* **6,** 723–728 (1980).
162. A. B. Wolbarst, Optimization of radiation therapy II: The critical voxel model, *Int. J. Radiat. Oncol. Biol. Phys.* **10,** 741–745 (1984).
163. T. E. Schultheiss and C. G. Orton, Models in radiotherapy: Definition of decision criteria, *Med. Phys.* **12,** 183–187 (1985).
164. Proceedings of the Second International Conference on Time, Dose, and Fractionation in Radiation Oncology, Madison, Wisconsin, September 1984 (Paliwal, Herbert, and Orton, eds.), AIP, New York (in press).

Chapter 2

A Comparison of National and International Megavoltage Calibration Protocols

Peter R. Almond

1. INTRODUCTION

Although the American Association of Physicists in Medicine (AAPM) protocol for the determination of absorbed dose from high-energy photon and electron beams was published in 1983, it was only one of several national and international protocols that have appeared since 1980. These protocols are concerned with the absorbed dose calibration procedures for megavoltage X-ray and electron beams by ionization chamber measurements. The ionization chambers are calibrated at a single photon energy against a national standard, and since this would be the same for all users in a given country, intercomparison of the national standards determines the consistency of megavoltage dose calibrations after differences in the protocols are taken into account between countries.

Ionization chambers are universally recommended because of their availability, excellent reliability, precision (a few tenths of one percent), and over 50 years of extensive experience. If the assumption is made that the national standards agree to within a known value (generally $< \pm 1\%$ for Co-60 γ rays, the standard most often used) then any differences between the protocols will indicate the overall difference to be expected when using a specific protocol and national standard.

Peter R. Almond • Department of Therapeutic Radiology, James Graham Brown Cancer Center, School of Medicine, University of Louisville, Louisville, Kentucky 40292.

Although all the protocols appear quite similar, there are some fundamental differences in the approaches used, and because they have appeared over a number of years there are differences in the interaction coefficients used. It is therefore not surprising if differences in the final calibrations exist.

2. THE PROTOCOLS

Table I lists those protocols and standards which have been published since 1980 and which will form the basis for this comparison. In addition, several other protocols are in press or in the process of final review. At the Third Annual Meeting of the European Society for Therapeutic Radiology and Oncology, September 9–15 1984 in Jerusalem, Israel, details of these additional protocols were given. They are listed in Table II. They will not be discussed here, but in essence all of them were very similar in concept to the Nordic Association Protocols.

The overall description of the protocols is shown in Fig. 1. Each protocol can be divided into two main components: (i) the calibration of the ionization chambers with the standard radiation and (ii) the absorbed dose determination for the user's radiation. In order to go from one section to the other, each protocol makes some basic assumptions, and in

TABLE I
Protocols and Standards Published Since 1980

- DIN (1980). Deutsches Institut für Normung, Dosimessverfahren in der radiologischen Technik—Ionisationsdosimetrie, No. 6800, Teil 2; and Klinische Dosimetrie: Therapeutische Anwendung gebündelter Röntgen-, Gamma- and Elektronenstrahlung, No. 6809, Teil 1.
- NACP (1980). Procedures in external radiation therapy dosimetry with electron and photon beams with maximum photon energies between 1 and 50 MeV, Recommendation by the Nordic Association of Clinical Physics, *Acta Radiol. Oncol.* **19,** 55.
- NACP (1981). Electron beams with mean energies at the phantom surface below 15 MeV, Supplement to the recommendations by the Nordic Association of Clinical Physics (NACP) 1980, *Acta Radiol. Oncol.* **20,** 6.
- NCRP (1981). National Council on Radiation Protection and Measurements, Dosimetry of X-ray and gamma-ray beams for radiation therapy in the energy range 10 keV to 50 MeV, NCRP report No. 69.
- HPA (1983). The Hospital Physicists Association, Revised code of practice for the dosimetry of 2 to 25 MV X-ray, and of caesium-137 and cobalt-60 gamma-ray beams, *Phys. Med. Biol.* **28,** 1097.
- AAPM (1983). A protocol for the determination of absorbed dose from high-energy photon and electron beams, Task Group 21 Radiation Therapy Committee, American Association of Physics in Medicine, *Med. Phys.* **10,** 741.

TABLE II
Protocols under Development

- ICU. International Commission on Radiation Units and Measurements, Radiation dosimetry: Electron beams with energies between 1 and 50 MeV, ICRU report to be published in 1984.
- HPA. The Hospital Physicists Association, A practical guide to electron dosimetry up to 50 MeV for radiotherapy purposes.
- BNM. Bureau National de Metrologie, The French dosimetry protocol.
- SEFM. Sociedad Espanola de Fisica Medica, The Spanish dosimetry protocol.

each section further assumptions and conditions are also applied for the use of specific ionization chambers.

The NCRP report No. 69 is included in the list of protocols, not only because it is tied to the historical concepts of C_λ but because it represents what is widely being practiced for calibration procedures. It is similar in concept and scope to a large number of documents which preceded the

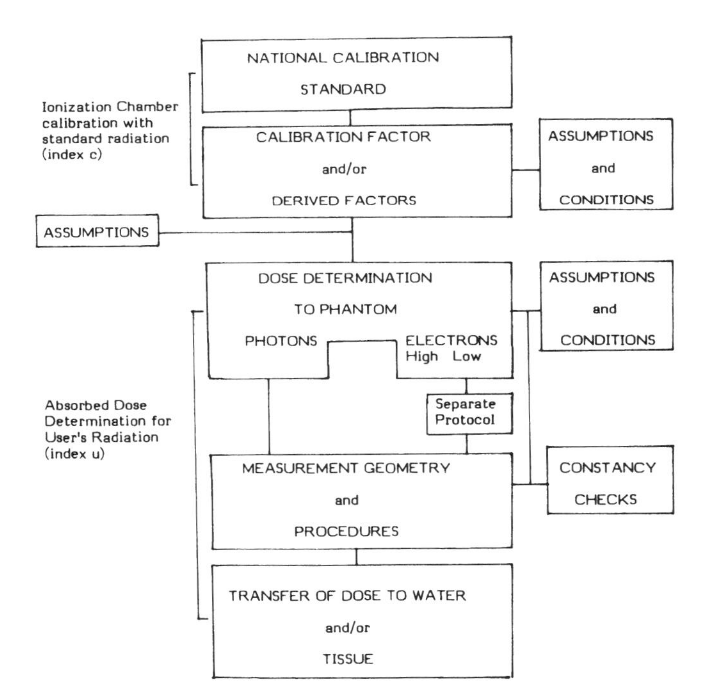

Figure 1. Block diagram of the general organization of calibration protocols.

TABLE III
Summary of the DIN Calibration Standard

Operation	Calibration parameters	Comments
Chamber calibration	Standard radiation: Co-60γ Calibration factor N_{Co} Exposure in roentgen Derived factor cavity ion dose $J_c = k_c N_{Co} M$	k_c correction factors for chamber walls of graphite and Plexiglas
Assumptions	—	—
Dose determination	$D_m = gJ_c$ (rad) $= gk_c N_{Co} M$ Water phantom Chamber at recommended depths	g values listed for several materials For electrons perturbation factors given Cylindrical chambers Standard precautions $W/e = 33.73$ J/C Buildup cap off

current recommendations (e.g., HPA,[1] ICRU,[2] NACP,[3] AAPM,[4] etc.) and therefore provides a basis for comparison to what was done in the past for photon beams.

Many of the previous protocols were empirical in nature and treated photons and electrons separately. All of the new documents, with the exception of the HPA Codes of Practice, treat both photon and electrons together. All of them are based upon theoretical analysis of dosimetry concepts and are therefore much more self consistent than the previous recommendations, and all the new documents are in terms of S.I. units.

Tables III–VII show the basic approach of each protocol following the general outline of Fig. 1. In Table VIII the symbols and units have been listed, and in Tables IX and X the basic equations for photon beam and electron beam calibrations, respectively, are given. The equations have been rewritten from the protocols to express the equations in terms of the chamber Co-60 exposure calibration factors. The ratio $D_w/N_x M$ is then a constant, at a given energy, for each protocol and can be determined by the parameters at the stated energy taken from each protocol.

3. COMPARISON OF PROTOCOLS

Table XI gives the wall material for cyclindrical chambers (and whether plane parallel chambers are also recommended) for each of the

TABLE IV
Summary of the NACP Protocols

Operation	Calibration parameters	Comments
Chamber calibration	Standard radiation: Co-60γ Chamber: Free in air Quantity calibrated: Air kerma, $K_{\mathrm{air},c}$ Calibration factor: $N_K = K_{\mathrm{air},c}/M_c$ Derived factor: $N_D = \bar{D}_{\mathrm{air},c}/M_c = N_K(1-g)k_m k_{\mathrm{att}}$	k_m and k_{att}, defined for air equivalent graphite or tissue equivalent (A150) chambers only Buildup cap same as wall material Cylindrical chamber 4–6 mm diam. and 25 mm long
Assumption	$N_D = \bar{D}_{\mathrm{air},c}/M_c = D_{\mathrm{air},u}/M_u$	
Dose determination	$D_{w,u} = D_{\mathrm{air},u}((S/\rho)_{w,\mathrm{air}})_u P_u$ $= M_u N_D((S/\rho)_{w,\mathrm{air}})_u P_u$ (gray) Dose to water only Water phantom chamber center at the reference points (specified) Chamber without buildup cap Illustrative examples given	$((S/\rho)_{w,\mathrm{air}})_u$ restricted values for electrons $\Delta = 15\,\mathrm{keV}$ Nonrestricted for photons P_u for electrons is a perturbation factor, P_u X rays includes wall material effect and point of measurement. Electrons below 10 MeV go to supplement (flat chamber). Standard precautions. Measured energy (mean). Constancy check in plastic phantoms

TABLE V
Summary of the NCRP Handbook 69 Procedures[a]

Operation	Calibration parameters	Comments
Chamber calibration	Standard radiation: Co-60γ Chamber: Free in air Quantity calibrated: Exposure, $X_{air,c}$(R) Calibration factor: $N_c = X_{air,c}/M_c$	Chamber response is material and dimension independent
	No derived factors	
Assumption	$M_c/Q_c = M_\lambda/Q_\lambda$ i.e., electrical sensitivity (scale units per unit charge) of instrument is independent of energy	
Dose determination	$D_{w,\lambda} = D_{air,\lambda}((S/\rho)_{w,air})_\lambda p_\lambda$ $= M_\lambda N_c C_\lambda$ (rad) Water phantom, dose to water only Chamber center at recommended depths $D_{muscle} = 0.99 D_w$ for up to 25 MV $D_{muscle} = 0.98 D_w$ for 25–50 MV	C_λ values listed Assume all chambers are water equivalent Use nominal machine energy $p_\lambda = 1$ Standard precautions Constancy check in plastic phantom

[a] All stopping power ratios are unrestricted.

TABLE VI
Summary of the HPA Photon Calibration Code of Practice

Operation	Calibration parameters	Comments
Chamber calibration	Standard radiation: 2-MV X rays Quantity calibrated: Air kerma or exposure (roentgen) Calibration factor $N_K = K_{air,c}/M_c$ $N_X = X_{air,c}/M_c$ No derived factors	(Co-60γ rays) One chamber only recommended: N.E. 2561
Assumptions	$M_c/Q_c = M_\lambda/Q_\lambda$, i.e., electrical sensitivity (scale units per unit charge) of instrument is independent of energy	
Dose determination	$D_{w,\lambda} = 1.139 M_\lambda N_K C_\lambda$ (gray) or $D_{w,\lambda} = 0.01 M_\lambda N_x C_\lambda$ (gray) Water phantom, waterproof buildup cap (lucite) Chamber center at recommended depths (SSD or SAD setup)	C_λ values listed calculated using unrestricted stopping power Shiragai formula Nominal machine energy (i.e., electron energy on target) Displacement factor = 0.98 $W/e = 33.85\ \mathrm{J\ C^{-1}}$

TABLE VII
Summary of the AAPM Protocol

Operation	Calibration parameters	Comments
Chamber calibration	Standard radiation: Co-60 Chamber: Free in air or in water for absorbed dose Quantity calibrated: Exposure, X or absorbed dose, D_w Calibration factor $N_x = X/M$ or $N_D = D_w/M$ Derived factor $N_{gas} = D_{gas}A_{ion}/M$	Formula for N_{gas} is chamber dependent for material and dimension Calculated from N_x or N_D
Assumptions	N_{gas} is constant with energy, i.e., W/e is a constant over energy range	
Dose determination	$D_{med} = D_{gas}(\bar{L}/\rho)^{med}_{gas}P_{ion}P_{repl}P_{wall}$ $= MN_{gas}(\bar{L}/\rho)^{med}_{gas}P_{ion}P_{repl}P_{wall}$ (gray) Water or plastic phantoms Dose to water or dose to plastic Conversion and fluence factors given	All factors are chamber dependent for material and dimension Measured energy for photons and electrons Electrons use mean energy Cylindrical chambers for photons and electrons Flat chamber low-energy electrons Standard precautions Buildup caps off $W/e = 33.73\ \mathrm{J\ C^{-1}}$ All stopping powers are restricted $\Delta = 10\ \mathrm{keV}$ Examples and worksheets provided

TABLE VIII
List of Symbols and Units

Symbol	Definition
A_{ion}	Ion-collection efficiency in the user's chamber obtained at the time of Co-60 exposure calibration at NBS or an ADCL[a]
A_{wall}	Correction for attenuation and scatter in the wall and buildup cap of the user's chamber when exposed in air to Co-60 gamma rays
C_E	A conversion factor for electrons, which is a function of electron energy
C_λ	A conversion factor for photons, which is a function of radiation quality
β	Quotient of absorbed dose and the collision part of kerma
Δ	Cutoff energy in Spencer–Attix formulation of the Bragg–Gray equation (keV)
g	$(W/e)(S/\rho)_{w,\mathrm{air}}$
k_1	Charge produced in air per unit mass per unit exposure (2.58×10^{-4} C kg^{-1} R^{-1})
k_{att}	Attenuation and scattering factor, correcting for attenuation and scattering in the ionization chamber material at the calibration in the Co-60 gamma-ray beam
k_m	Chamber material dependent factor correcting for the lack of air equivalence of the ionization chamber material at the calibration in the Co-60 gamma-ray beam
k_c	Overall factor accounting for attenuation and scattering and lack of air equivalence of the chamber material at the calibration in the Co-60 gamma-ray beam
$\bar{L}/\rho$	Mean restricted collision mass stopping power
M	Electrometer reading normalized to 22°C and one standard atmosphere (C or scale division) and corrected for ion recombination
N_x	Exposure calibration factor (R C^{-1} or R per scale division or C kg^{-1} or C kg^{-1} per scale division)
N_{gas}	Cavity-gas calibration factor (Gy/C or Gy/scale division)
p	Total perturbation factor including correction for lack of water equivalence in the ionization chamber material at the user's radiation quality, perturbation of the fluence due to the insertion of the air cavity, and location of the effective point of measurement of the cylindrical chamber due to the curved ionization chamber wall (only for photon beams)
P_{ion}	Ion-recombination correction factor applicable to the calibration of the user's beam
P_{repl}	A factor that corrects for replacement of phantom material by an ionization chamber
P_{wall}	Perturbation factor correcting for the lack of medium equivalence of the ionization chamber wall material
$\bar{\mu}en/p$	Mean mass energy-absorption coefficient
W/e	Mean energy expended per unit charge in room at usual humidity (33.73 J/C for humid air or 33.85 J/C for dry air)
$(S/\rho)_{w,\mathrm{air}}$	Mass stopping power ratio, water to air, at the reference point at the user's radiation quality

[a] Accredited dosimetry calibration laboratory.

TABLE IX
X ray Basic Equations[a]

DIN	$D_w = MN_x g k_c$
NACP	$D_w = MN_D((S/\rho)_{w,\mathrm{air}})p$
	$= MN_x k_{\mathrm{att}} k_m (W/e) k_1 ((S/\rho)_{w,\mathrm{air}})p$
NCRP	$D_w = MN_x C_\lambda$
HPA	$D_w = 0.01 MN_x C_\lambda$
AAPM	$D_w = MN_{\mathrm{gas}}(\bar{L}/\rho)^w_{\mathrm{gas}} P_{\mathrm{repl}} P_{\mathrm{wall}}$
	$= MN_x k_1 (W/e)\left[\frac{A_{\mathrm{ion}} A_{\mathrm{wall}} \beta_{\mathrm{wall}}}{(\bar{L}/\rho)^{\mathrm{wall}}_{\mathrm{gas}} (\bar{\mu}_{\mathrm{en}}/\rho)^{\mathrm{air}}_{\mathrm{wall}}}\right]_c (\bar{L}/\rho)^w_{\mathrm{gas}} P_{\mathrm{repl}} P_{\mathrm{wall}}$

[a] At a given energy $D_w/N_x M$ = const for each protocol. Account must be taken of absorbed dose and calibration units. NACP, HPA, and AAPM express absorbed dose in grays, DIN and NCRP in rad. DIN, NCRP, and HPS are for calibration factors related to exposure in terms of roentgens. NACP and AAPM allow for exposure in roentgens or C kg^{-1} with $k_1 = 2.58 \times 10^{-4}$ or 1, respectively.

protocols. It can be seen that only graphite-walled chambers are allowed by all the protocols. An intercomparison can only be made therefore for such chambers, and the results of such calculations for D_{water}/MN_x for 25 and 6 MV X-ray energies are shown in Table XII and Figs. 2 and 3. Although some protocols allow calibration at dose maximum, all of them also give specified depths for calibration, and these were used in the calculations.

The AAPM and the NACP protocols specify measuring beam quality by determining ionization chamber reading ratios. The AAPM is

TABLE X
Electron Beam Basic Equation[a]

DIN	$D_w = MN_x g k_c$
NACP	$D_w = MN_D((S/\rho)_{w,\mathrm{air}})p$
	$= MN_x k_{\mathrm{att}} k_m (W/e) k_1 (S_{w,\mathrm{air}})p$
AAPM	$D_w = MN_{\mathrm{gas}}(L/\rho)^w_{\mathrm{gas}} P_{\mathrm{repl}} P_{\mathrm{wall}}$
	$= MN_x k_1 (W/e)\left[\frac{A_{\mathrm{ion}} A_{\mathrm{wall}} \beta_{\mathrm{wall}}}{(L/\rho)^{\mathrm{wall}}_{\mathrm{gas}} (\bar{\mu}_{\mathrm{en}}/\rho)^{\mathrm{air}}_{\mathrm{wall}}}\right]^c (L/\rho)^w_{\mathrm{gas}} P_{\mathrm{repl}} P_{\mathrm{wall}}$

[a] At a given energy $D_w/N_x M$ = const for each protocol. Account must be taken of absorbed dose and calibration units. NACP and AAPM express absorbed dose in grays and the exposure calibration in terms of roentgens or C kg^{-1} with $k_1 = 2.58 \times 10^{-4}$ for exposure expressed in roentgens and $k_1 = 1$ for exposure in terms of C kg^{-1}. For the DIN protocol, absorbed dose is in rad and exposure in roentgens.

TABLE XI
Wall Material for Cylindrical Chambers Specified by Protocols

DIN	Graphite, acrylic (or plane parallel chamber)
NACP (1980)	Air equivalent, graphite, tissue equivalent
NACP (1981)	(Plane parallel chamber only specified)
NCRP	Not specified
HPA (1983)	Graphite (NE 2561)
AAPM	Graphite, acrylic tissue equivalent, air equivalent nylon (or plane parallel chamber)
ICRU (1984)	Air equivalent, graphite, acrylic, tissue equivalent (or plane parallel chamber)
HPA (1984)	Graphite (NE 2571) or plane parallel chamber

for a fixed source-to-detector distance and the NACP is for a fixed source-to-surface distance. A nominal 25-MV X-ray beam was used for the comparison, which proved to be closer to 19 MV when the effective energy was measured by either of the two protocols. These calculations are also shown in Table XII and as the additional white bars in Fig. 2. It can be seen from these data that the NCRP values are lower by 1% or 2% from the other protocols, and if this value is excluded from the

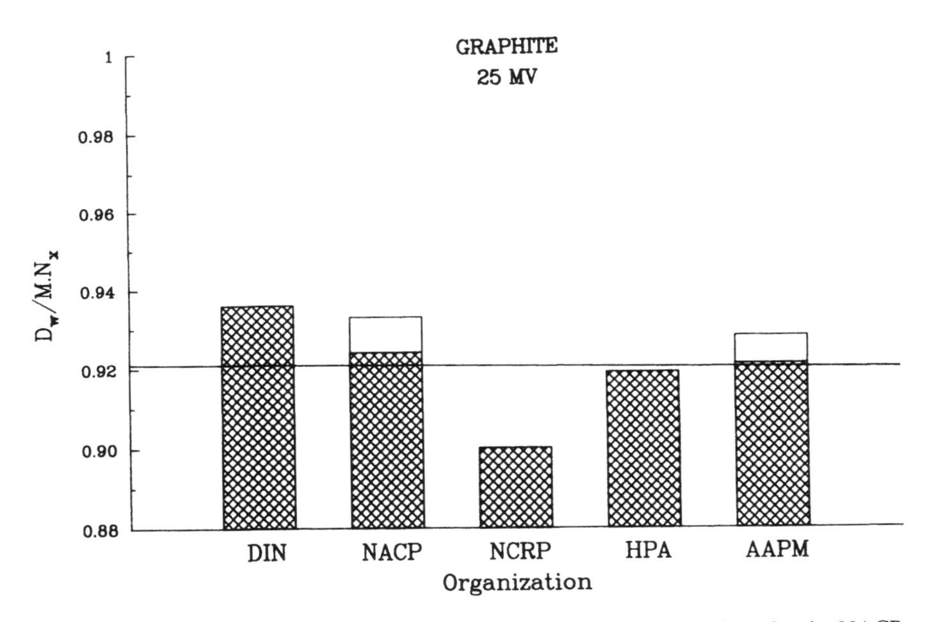

Figure 2. D_w/MN_x calculation for 25-MV X rays. The unshaded extensions for the NACP and AAPM calculations shown the values obtained when the beam quality is measured according to the protocol recommendation. The horizontal line is the average of the shaded areas.

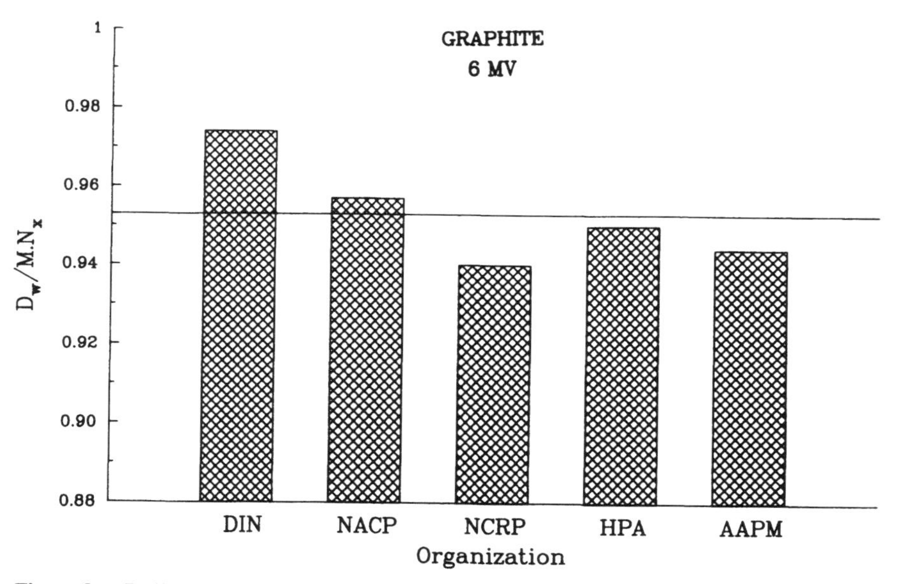

Figure 3. D_w/MN_x calculation for 6-MV X rays. The horizontal line is the average value.

25-MV data the standard deviations between the protocols reduces to less than 1%.

At the present time only three protocols have been investigated for electron beams. However, the other four listed in Table II follow in outline the recommendation for the NACP document, so that calculations using them are expected not to differ too much. Agreement between the protocols is again less than 1% standard deviation as shown in Table XIII and Fig. 4.

TABLE XII
Protocol Comparison[a]

Protocol	25 MV		6 MV
DIN	0.936		0.974
NACP	0.929	0.933[b]	0.957
NCRP	0.90		0.94
HPA	0.919		0.95
AAPM	0.921	0.927[b]	0.945
Average	0.921 ± 0.014		0.953 ± 0.013

[a] Nominal 25-MV X rays and 6-MV X rays, for graphite chamber values of D_w/MN_x calculated for each protocol.
[b] Corrected for beam quality. Only NACP and AAPM recommend determining beam quality for X rays. Sagittaire 25 MV X rays (nominal) are closer to 19 MV X rays in beam quality.

TABLE XIII
Protocol Comparison[a]

Protocol	D_w/MN_x
DIN	0.836
NACP	0.849
AAPM	0.834
Average	0.840 ± 0.008

[a] For 20-MeV electrons, measurement depth 30 mm. Values of D_w/MN_x for HPS recommended graphite chamber.

The reason for the differences can be seen in Table XIV. Although there are small differences in the W values since the protocols either assume dry air (W/e = 33.85) or humid air (W/e = 33.73), and parameters such as the perturbation factors, the main differences are in the stopping power values used.

In Chapter 3, Svensson and Brahme suggest using the product of $(W/e) \times (S/\rho)_{m,\text{air}} = \omega_{m,\text{air}}$ instead of determining (W/e) and $(S/\rho)_{m,\text{air}}$ separately. As pointed out by Svensson and Brahme, ω is the quantity which is determined experimentally by calorimetric and ionometric

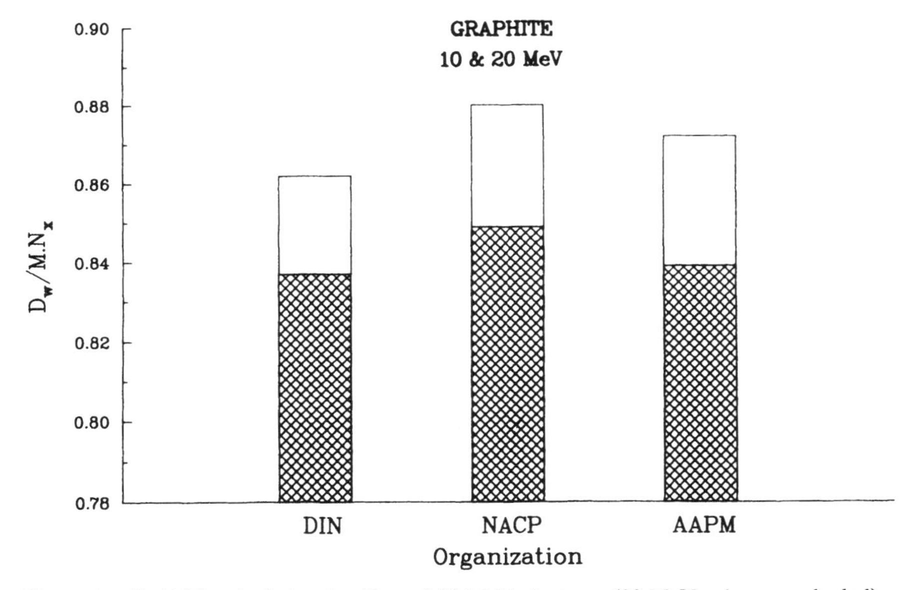

Figure 4. D_w/MN_x calculation for 20- and 10-MeV electrons (20-MeV values are shaded).

TABLE XIV
Protocol Parameters for 10-MeV Electrons[a]

Protocol	W/e ($J\,C^{-1}$)	Stopping power	Δ	$(S/\rho)_{w,\mathrm{air}}$ 10 MeV	p
DIN	33.73	Unrestricted	—	1.048	0.964
NACP (1980)	33.85	X rays unrestricted	—		
		Electrons restricted	15 keV	1.053	0.975
NACP (1981)	33.85	Electrons restricted	10 keV		
HPA (1983)	33.85	Unrestricted	—		
AAPM	33.73	Restricted	10 keV	1.038	0.984

[a] I values have changed (especially for air and water) between 1980 and 1984, resulting in changes in the stopping powers.

measurements and would help remove uncertainties in the calculated values of S/ρ and possible variations of W/e with energy. The quantity ω is the same as the quantity of g in the DIN standards.

REFERENCES

1. Hospital Physicist's Association, *Phys. Med. Biol.* **14,** 1 (1969).
2. ICRU report No. 14, *Radiation Dosimetry: X-Rays and Gamma Rays with Maximum Photon Energies Between 0.6 and 50 MeV,* ICRU, Washington, D.C. (1969); ICRU Report No. 21, *Radiation Dosimetry: Electrons with Initial Energies Between 1 and 50 MeV,* ICRU, Washington, D.C. (1972).
3. Nordic Association of Clinical Physics, Procedures in radiation therapy dosimetry with 5 to 50 MeV electrons and roentgen and gamma rays with maximum photon energies between 1 MeV and 50 MeV, *Acta Radiol. Ther. Phys. Biol.* **II**, 603 (1972).
4. Subcommittee on Radiation Dosimetry, AAPM, *Phys Med. Biol.* **16,** 379 (1971).

Chapter 3

Recent Advances in Electron and Photon Dosimetry

Hans Svensson and Anders Brahme

1. INTRODUCTION

The possibilities to improve radiation therapy have increased during recent years, not only because of the use of new or improved tools such as computed tomography and dose planning, and high-quality electron and photon beams from therapy accelerators, but also because of increased knowledge in fields like clinical radiation biology about dose fractionation and dose–response relations. These new developments increase the demand for accurate dosimetry as illustrated by the following examples.

1. One department may today use several different radiation qualities to treat a special type of cancer, and the choice of radiation quality is individually optimized to the anatomy of each patient. To give consistent treatments it is therefore necessary to be able to give the same absorbed dose independent of the beam quality;
2. The CT information together with accurate computed dose planning systems facilitates the correction of the dose distribution for the density and composition of the various tissues in the body. Accurate input dose distribution data are needed if the advantages of a precision computer system are to be exploited fully;

Hans Svensson • Radiation Physics Department, University of Umeå, S-901 85 Umeå, Sweden. **Anders Brahme** • Department of Radiation Physics, Karolinska Institute, and University of Stockholm, S-104 01 Stockholm, Sweden.

3. Finally, a more accurate analysis of dose–effect curves will be possible as corrections for various fractionation schedules are better known today (e.g., from calculations of the NSD, TDF, or CRE). The accuracy in the dose determination and dose delivery is in many cases the limiting factor in such an analysis.

The ICRU 24[75] has concluded that there is evidence, for certain types of tumors, that the accuracy in the determination of absorbed dose to the target volume should be within ±5% or even lower if an eradication of the tumor is sought. Recent intercomparisons in the Scandinavian countries[82] show that this accuracy is not even obtained at the reference depth in a water phantom. Even larger uncertainties have been shown in other surveys (e.g., Refs. 48, 124, 148).

There is thus a great need for improving the dosimetry. However, today this is a fairly complicated task as no single dominant reason for the large uncertainty can be found. Instead, Johansson and Svensson[82] showed that the total uncertainty is due to a combination of a number of different systematic errors and random uncertainties. In this analysis it was assumed that the ionization chamber method was used for the absorbed dose determination. An improvement of the dosimetry can therefore only be achieved if most of the uncertainties in the calibration procedure can be reduced or if a completely different method is applied. The aim in the present report is to analyze the complete dosimetrical procedure for the determination of absorbed dose in a phantom.

The report contains both original material and a review of recent publications. It starts by describing the basic dosimetric quantities and their physical relations (Section 2). The special problems of spatially extended detectors including such problems as fluence perturbation and effective point of measurement are dealt with in Section 3. The basic physical data and the implications of the most recent stopping power values[11] for the determination of absorbed dose are discussed in Section 4. Finally, more practical problems in the calibration of a medical accelerator are analyzed in Section 5.

2. BASIC DOSIMETRIC QUANTITIES AND THEIR PHYSICAL RELATIONS

2.1. Energy Deposition

A principal problem in the description of the interaction of electron and photon beams with matter is that all real beams consist of a finite number of discrete particles, whereas they are most conveniently described by continuously variable probability distributions. For this reason the most fundamental concepts in radiation dosimetry are stochastic quantities. This is the case with the energy imparted ε, defined as the difference in energy of particles incident on and emerging from the volume under consideration plus the decrease in rest mass energy that might occur inside that volume.[77] When the energy imparted is due to a very large number of energy depositions it is often sufficient to use the expectation value of the stochastic quantity ε, i.e., the mean energy imparted $\bar{\varepsilon}$.

The response of most radiation dosimeters is proportional to the energy imparted. Such a detector, when appropriately calibrated, can be used to determine the energy imparted in the radiation sensitive volume according to

$$\varepsilon = N_{\varepsilon} M \tag{1a}$$

where M is the value of the detector signal and N_{ε} is the energy imparted calibration factor. For the most common case in dosimetry, the gas-filled ionization chamber, $M = Q_g$, that is the recombination corrected charge collected in the gas volume, and $N_{\varepsilon} = W_g/e$ is the mean energy needed to generate an ion pair in the gas. For this simple case Eq. (1a) becomes

$$\varepsilon_g = \frac{W_g}{e} Q_g \tag{1b}$$

The stochastic quantity for the energy absorbed per unit mass is called the specific energy imparted, z, and is obtained by dividing the energy imparted by the mass of the volume under consideration $z = \varepsilon/m$. The expectation value of the specific energy imparted, $\bar{z}$, is by definition equal to the mean value of the absorbed dose in the volume considered:

$$\bar{z} = \frac{\bar{\varepsilon}}{m} = \bar{D} \tag{2}$$

The absorbed dose at a point is thus defined as the differential quotient:

$$D = \frac{d\bar{\varepsilon}}{dm} = \lim_{m \to 0} \bar{z} \tag{3}$$

which thus coincides with the limiting value of the specific energy when the volume of interest or the mass approaches zero (cf. Ref. 77).

By using Eq. (2), Eq. (1) may be rewritten in the following form:

$$\bar{D} = N_D \bar{M} \tag{4a}$$

where $\bar{M}$ is the expectation value of the detector signal, $\bar{D}$ the mean absorbed dose to the detector, and N_D the absorbed dose calibration factor, which is related to N_ε by $N_D = N_\varepsilon / m$.

When we apply these equations on a normal air ionization chamber Eq. (4a) takes the form

$$\bar{D}_{\text{air}} = \frac{W_{\text{air}}}{e m_{\text{air}}} Q_{\text{air}} \tag{4b}$$

which may be rewritten

$$\bar{D}_{\text{air}} = \frac{W_{\text{air}}}{e} J_{\text{air}} \tag{4c}$$

where J_{air} is the specific ionization or mass ionization in the air volume ($Q_{\text{air}}/m_{\text{air}}$) assuming the detector signal is measured in this unit.

For uncharged particles like photons it is useful to define an auxiliary field quantity K, the kerma, defined as the initial kinetic energy of all charged particles liberated per unit mass.[77] Because part of this kinetic energy will be converted back to photons through bremsstrahlung and annihilation in flight processes, it is of interest to distinguish that part of the kerma that remains as kinetic energy of charged particles, the collision kerma, K_{col}.[8] This latter quantity is of special interest as the collision kerma in air is directly related to the traditional unit exposure, X, through the relation

$$K_{\text{col,air}} = X \frac{W_{\text{air}}}{e} \tag{5}$$

where W_{air} is the mean energy expended in air per ion pair formed and e is the electron charge. Because the collision kerma is an expression of the energy in a photon beam that is permanently converted to kinetic energy of charged particles, the three quantities mean energy imparted, collision kerma, and absorbed dose are related through the equality

$$\bar{\varepsilon}_V = \iiint_V K_{\text{col}}(\mathbf{r})\rho(\mathbf{r})\, dV = \iiint_V D(\mathbf{r})\rho(\mathbf{r})\, dV \tag{6}$$

where $\rho(\mathbf{r})$ is the density distribution in the irradiated volume V. This relation can be used to determine the kerma distribution in a medium where the absorbed dose distribution is known.[111]

2.2. Fluence and Absorbed Dose

The physical quantity which describes the radiation field is the fluence, Φ, defined by the ICRU as the quotient of the number of particles dN incident on a sphere and its cross-sectional area da:

$$\Phi = \frac{dN}{da} \tag{7}$$

This definition can be generalized by multiplying both dN and da by the mean cord length $\bar{l} = 4dV/dA$ of the sphere; see Fig. 1. Because $\bar{l}\,dN$ is just equal to the total path length, $d\bar{s}$, and $\bar{l}\,da$ is the volume dV, dA being the surface area, the fluence can also be defined as

$$\Phi = \frac{d\bar{s}}{dV} \tag{8}$$

This definition is more general than Eq. (7) as it is independent of the shape of the volume considered and avoids the problem of whether the incident particles just enter the volume or really cross it. By the latter definition the fluence is easily visualized as the total path length per unit volume.

It is important that the definitions (7) and (8) contain all particles independent of energy and direction. However, it is easy to generalize the fluence concept and define a fluence differential in energy and angle

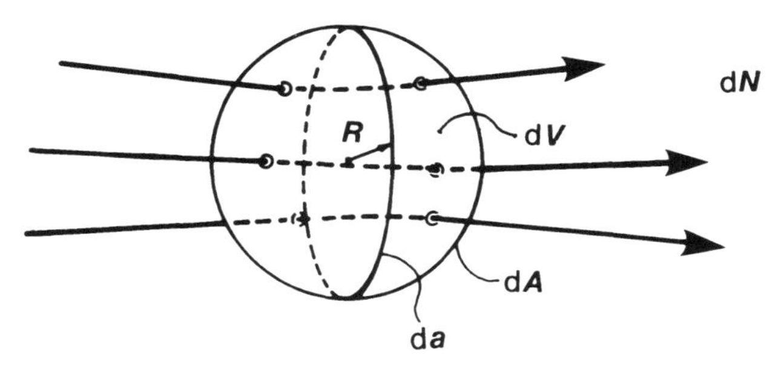

Figure 1. Definition of the fluence of particles as dN/da the number crossing a sphere of cross section da or $d\bar{s}/dV$, the path length per unit volume. When the radius of the sphere is R, then $da = \pi R^2$, $dV = \frac{4}{3}\pi R^3$, and $dA = 4\pi R^2$.

$\Phi_{E,\mathbf{\Omega}}$ such that $\Phi_{E,\mathbf{\Omega}}\, dE\, d\Omega$ contains all the particles of energy between E and $E + dE$ with direction in the interval $\mathbf{\Omega}$ to $\mathbf{\Omega} + d\Omega$ of solid angles. Based on this differential fluence all other fluence concepts can be defined, e.g., the fluence differential in energy:

$$\Phi_E = \int \Phi_{E,\mathbf{\Omega}}\, d\Omega \tag{9}$$

and the vectorial energy fluence:

$$\mathbf{\Psi} = \iint \mathbf{\Omega} E \Phi_{E,\mathbf{\Omega}}\, dE\, d\Omega \tag{10}$$

This latter expression is of special interest as it can be used to calculate the mean energy imparted and the absorbed dose from first principles.

When, for simplicity, the changes in rest mass energy can be disregarded, the mean energy imparted inside a closed surface A is given by the net energy flow across this surface:

$$\bar{\varepsilon}_V = \oiint_A \mathbf{\Psi n}\, dA = -\iiint_V \operatorname{div} \mathbf{\Psi}\, dV \tag{11}$$

where Gauss' law has been used in the last equality, $\mathbf{n}$ is a unit vector normal to A, and V is the volume enclosed by A. By allowing A and V to decrease to zero, Eq. (3) can be used to express the absorbed dose at a point $\mathbf{r}$:

$$D(\mathbf{r}) = -\frac{1}{\rho(\mathbf{r})} \operatorname{div} \mathbf{\Psi}(\mathbf{r}) \tag{12}$$

The absorbed dose distribution can now be calculated from the spatial distribution of the vectorial energy fluence of all contributing particles. By using Eq. (10), assuming $\mathbf{\Phi}$ includes the total fluence of electrons, and by changing the order of the integration and the divergence operations, the absorbed dose becomes

$$D = \frac{-1}{\rho} \iint (\mathbf{\Omega} \Phi_{E,\mathbf{\Omega}} \operatorname{grad}(E) + \mathbf{\Omega} E \operatorname{grad}(\Phi_{E,\mathbf{\Omega}}))\, dE\, d\Omega \tag{13}$$

because $\operatorname{div} \mathbf{\Omega} \equiv 0$ as $\mathbf{\Omega}$ by definition is an independent variable. When E is regarded as a variable depending on the space coordinates along the particle path l and after the $\mathbf{\Omega}$ integration is performed, Eq. (13) becomes

$$D = -\int\left(\frac{1}{\rho}\frac{dE}{dl}\Phi_E + \frac{E}{\rho}\frac{d\Phi_E}{dl}\right)dE \tag{14}$$

This relation shows that two types of energy depositions by the electrons can be distinguished: quasicontinuous energy losses described by the first term, or track ends described by the second term. In the Spencer–Attix formulation Eq. (14) is closely approximated by

$$D = \int_\Delta^\infty \frac{L_\Delta(E)}{\rho}\Phi_E\,dE + \frac{\Delta}{\rho}S_{\text{col}}(\Delta)\Phi_E(\Delta) \tag{15}$$

where the last term gives the track end contribution (cf. Refs. 31 and 103 and also the discussion by Svensson and Brahme[135]).

Under assumption of an equilibrium in the transport of secondary electrons, such that the outflow through each volume element is compensated by an equal inflow, Eqs. (14) and (15) can be simplified considerably

$$D_{\text{eq}} = \int \frac{S_{\text{col}}(E)}{\rho}\Phi_E^p\,dE \tag{16}$$

where Φ_E^p is the fluence of primary electrons. An alternative expression, when the electrons are set in motion by photons, is obtained if the energy fluence differential in energy of the photons, Ψ_E, is used in Eq. (13):

$$D_{\text{eq}} = \frac{-1}{\rho}\iint \mathbf{\Omega}\,\text{grad}\,\Psi_E\,dE\,d\Omega \tag{17}$$

which in analogy with Eq. (14) becomes

$$D_{\text{eq}} = -\int \frac{1}{\rho}\frac{d\Psi_E}{dl}dE = \int \frac{\mu_{\text{en}}(E)}{\rho}\Psi_E\,dE \tag{18}$$

where $\mu_{\text{en}}(E)$ is energy absorption coefficient of photons of energy E. Under equilibrium conditions both Eqs. (16) and (18) can be used, whereas in absence of complete equilibrium in the electron transport the basic expressions Eqs. (14) and (15) have to be applied.

3. GENERAL THEORY OF RADIATION DOSIMETERS

3.1. Background

Since the earliest theories about the response of radiation detectors (cf, e.g., Ref. 19) it has been well known that the material surrounding the radiation sensitive part of a small detector is of critical importance. This is so because, independent of the energy of the beam, about 30% of the absorbed dose is due to low-energy secondary electrons, the range of which is only about 0.5 mm or less in unit density material. Over such distances there will be a transition zone near the interface between two materials, where the slowing down spectrum of electrons changes from that characteristic of one material to that of the next (cf. Refs. 77 and 14). This implies that the response of a small detector will depend on the atomic number of the surrounding medium.

There are at least three different approaches to dosimetry depending on how this central problem is handled. The traditional Bragg–Gray type approach is to use such a small detector that the electron spectrum existing in the medium will penetrate the detector and not be affected by its presence. Strictly speaking, this is theoretically impossible, as no interactions can then be allowed in the detector and thus no signal will be obtained. In particular, it is a very poor assumption when the elemental composition of the detector differs from that of the medium, as interface phenomena will then have a strong influence and the important low-energy part of the spectrum will be affected (cf. Ref. 150).

The second approach is to make the dosimeter of finite size with only the radiation sensitive part different from the medium. This simplifies the theoretical treatment, and, for example, a Spencer–Attix type theory could be applied.[13,31,103,126] A typical example is the use of a water equivalent ionization chamber in water with only the air volume as the nonwater equivalent part (see Fig. 2a).

The third possibility is to use a detector which is surrounded by a wall of the same material as used in the radiation sensitive part of the detector. This approach will ensure that the interface effects will mainly disturb the dose distribution in the wall, and the low-energy spectrum in the radiation sensitive part of the detector will be fully built up. In the first approximation, the simple theory of Harder covers this situation (Ref. 62; cf. also Refs. 20 and 35). A typical example here is an air ionization chamber with air equivalent walls when used in water (see Fig. 2b).

In many cases the most accurate dosimetry is performed when one of the above approaches is applied to a detector which is as similar as possible in elemental composition to the medium. This is the case with

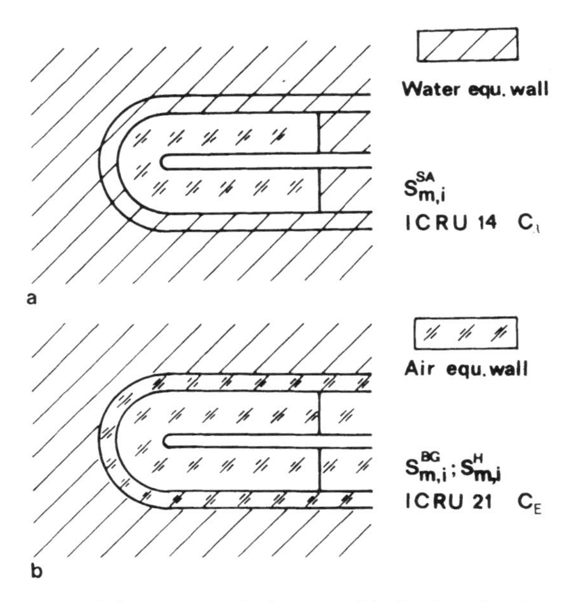

Figure 2. Illustration of the two principal types of ionization chambers used in electron and photon dosimetry. The two-media detector (a), with phantom equivalent walls (m) around the radiation sensitive material (i), has generally been used in photon beams (cf. ICRU 14) and should strictly be used with a Spencer–Attix type stopping power ratio [cf. Eq. (40)]. The matched detector (b), with walls of the same material as the radiation sensitive part of the detector, is often assumed in electron beams (cf. ICRU 21) and could be used with a Bragg–Gray stopping power ratio [Eq (38)], or often equally well with the approximation due to Harder [eq. (39)]. In photon beams a difference in mass ionization of about 3% may result between these two detector types (cf. Ref. 2).

detector systems like ferrous sulfate, liquid ionization chambers, and tissue equivalent or water calorimeters. These detector systems are of special importance when consistent dosimetry at quite different radiation qualities is required.

One of the most important requirements of a theory of the response of radiation dosimeters is that all manipulations of the detector signal are performed with internal consistency. For example, it must be assured that the stopping power ratio, normally calculated for the same energy spectrum in the detector material and in the medium, really is applicable. This requires that the electron spectrum at the effective point of measurement (see Sections 3.4.2 and 3.4.3) in the medium is the same as that averaged over the radiation sensitive part of the detector. Similarly, it is essential that necessary fluence perturbation factors (see Sections 3.4.2 and 3.4.4) really are based on the fluence in the medium at the effective point of measurement where the determined absorbed dose later is described.

Most published cavity theories have shortcomings in these respects. One of the aims with the present treatment is to give a consistent set of equations defining the most important quantities already identified by ICRU 14[73] and 35[74] in a stringent and consistent way, namely, the stopping power ratio, the perturbation correction factor, and the effective point of measurement.

Because the user of a calibrated dosimeter will make measurements in many different beam qualities, the definitions will sometimes differ from those used by the standard laboratories where mainly ^{60}Co γ photons are used. For example, in the present treatment a more general reference volume concept (see Section 3.4.2) will be used instead of the chamber center which in clinical use necessitates more or less complex gradient corrections to be performed on the detector signal.

3.2. Fano's Theorem

The fundamental properties of the secondary electron spectrum in low- (or high-) density cavities are made use of in many radiation dosimeters as was described in detail already by Bragg.[19] However, a strict formal proof was first given by Fano.[52] A slightly generalized formulation of "Fano's" theorem, which is approximately applicable to radiation dosimeters, is as follows: In a medium of uniform composition, but of varying density, and with a local source density of generated particles which is a given function of the density, the fluence of these particles is uniform and independent of the density at least one particle range inside the volume where this condition is fulfilled, provided the interaction cross sections for these particles with the medium have the same density dependence as the source density (cf. Refs. 51 and 65).

This implies that the fluence of secondary electrons in photon or electron beams (photoelectric, Compton, and pair electrons and positrons in photon beams, and δ rays in electron beams) is uniform and independent of density variations in a volume V, provided the fluence of primary photons or electrons is uniform in V and at least one maximum secondary electron range outside V. This very general statement relies on the fact that the cross section for secondary electron production is proportional to the electron density N_e and so is the interaction cross section for the secondary electrons. Intuitively this can be understood as follows. When the density is high the number of secondary electrons produced per unit volume is high but their range is short so their fluence or path length per unit volume (cf. Section 2.2) is the same as if the density were low, with a long range of the secondary electrons but a small number produced per unit volume.

A more formal proof is almost as straightforward if the time-independent Boltzman equation describing the spatial distribution of the differential fluence $\Phi_{E,\mathbf{\Omega}}(\mathbf{r})$ is considered (cf. Ref. 44):

$$\mathbf{\Omega}\,\mathrm{grad}\,\Phi_{E,\mathbf{\Omega}}(\mathbf{r}) = \iint_{E}^{\infty} N_e(\mathbf{r})\Phi_{E',\mathbf{\Omega}'}(\mathbf{r})\sigma(E', \mathbf{\Omega}', E, \mathbf{\Omega})\,dE'\,d\Omega' - \Phi_{E,\mathbf{\Omega}}(\mathbf{r}) \iint_{0}^{E} N_e(\mathbf{r})\sigma(E, \mathbf{\Omega}, E', \mathbf{\Omega}')\,dE'\,d\Omega' + S_{E,\mathbf{\Omega}}(\mathbf{r}) \qquad (19)$$

Here $N_e(\mathbf{r})$ is the electron density, the term $S_{E,\mathbf{\Omega}}(\mathbf{r})$ describes the source density, the term $\mathbf{\Omega}\,\mathrm{grad}\,\mathbf{\Phi}$ describes the collision free motion of the particles, the positive collision integral describes the inflow of particles of higher energies to the energy E and direction $\mathbf{\Omega}$, and the corresponding negative integral the collisional loss of particles from this energy and angle to lower energies. In the case of a uniform collision and source density, $N_e(\mathbf{r})\sigma = N_e^u\sigma$ and $S_{E,\mathbf{\Omega}}(\mathbf{r}) = S_{E,\mathbf{\Omega}}^u$ there is no spatial dependence and the fluence will also be uniform in space, $\Phi_{E,\mathbf{\Omega}}(\mathbf{r}) \equiv \Phi_{E,\mathbf{\Omega}}^u$, and consequently $\mathrm{grad}\,\mathbf{\Phi}^u \equiv 0$.

However, when the collision density is proportional to the density or some function of the density, $N_e(\mathbf{r})\boldsymbol{\sigma} = f(\boldsymbol{\rho}(\mathbf{r}))N_e^u\sigma$, and the source density is the same function of the density, $S_{E,\mathbf{\Omega}}(\mathbf{r}) = f(\rho(\mathbf{r}))S_{E,\mathbf{\Omega}}^u$, the term $f(\rho(\mathbf{r}))$ is common to all terms on the right-hand side in the Boltzman equation:

$$\mathbf{\Omega}\,\mathrm{grad}\,\Phi = f(\rho(\mathbf{r}))\left\{\iint_{E}^{\infty} N_e^u\Phi_{E',\mathbf{\Omega}'}(\mathbf{r})\sigma(E', \mathbf{\Omega}', E, \mathbf{\Omega})\,dE'\,d\Omega' - \Phi_{E,\mathbf{\Omega}}(\mathbf{r}) \iint_{0}^{E} N_e^u\sigma(E, \mathbf{\Omega}, E', \mathbf{\Omega}')\,dE'\,d\Omega' + S_{E,\mathbf{\Omega}}^u\right\} \qquad (20)$$

When this is observed it is immediately clear that $\Phi_{E,\mathbf{\Omega}}^u$ will also make the right-hand side of the nonuniform equation zero, as the expression inside the brackets is just the right-hand side of the uniform equation. Thus $\Phi_{E,\mathbf{\Omega}}^u$ is also a solution to the whole nonuniform equation since $\mathrm{grad}\,\Phi^u$ is already known to be zero. According to the uniqueness theorem this is also the only solution of the nonuniform case.

This very general result relies on at least three conditions that are rarely strictly met in practical dosimetry:

1. The fluence of primary electrons or photons is seldom constant

over a sufficiently large volume to make $N_e(\mathbf{r})\sigma$ and $S_{E,\Omega}(\mathbf{r})$ proportional to each other.
2. The Möller cross section σ is modified by Fermi's density effect at relativistic energies and this makes the density dependence of the collision integrals and the source term somewhat different.
3. The wall around the radiation sensitive material of the detector is often not strictly detector equivalent, which again violates the proportionality between the collision integrals and the source term.

In these cases the fluence of secondary electrons in the dosimeter will differ from that in the surrounding uniform medium and a fluence perturbation correction factor should in principal be introduced to compensate for this. By designing the detector appropriately the perturbation of the secondary electron fluence can often be sufficiently reduced to bring the dosimetric uncertainty below the 1% level without using a perturbation correction factor.

3.3. Bragg–Gray Detectors

For a detector which is small relative to the spatial variation of the electron fluence in the uniform medium and also small enough not to disturb the electron fluence in the medium as it penetrates the detector, the Bragg–Gray relation

$$D_m = D_i s_{m,i} \tag{21a}$$

can be used for calculating the absorbed dose D_m to the medium at the point of interest from the absorbed dose D_i to the detector, where $s_{m,i}$ is the collision mass stopping power ratio which accounts for the differences in energy deposition per unit mass in materials m (medium) and i (detector). For such a small detector it is sufficient to evaluate $s_{m,i}$ for the electron spectrum at the point of interest in the uniform medium, as this, by definition, is identical to the spectrum inside the detector.

For medium equivalent detectors it is in the first approximation only necessary that the detector be small compared to fluence variations in the medium, as the fluence in the detector will be independent of density according to Fano's theorem.

If the slowing down properties of the detector material differ from those of the medium, it has recently been shown by Zheng-Ming[150] that, no matter how small the cavity is, the change in the low-energy part of the electron spectrum can never be neglected because the spectrum is proportional to E^{-2}.

For a small medium equivalent gas ionization chamber the Fano condition can often be fulfilled with fairly high accuracy. The Bragg–Gray relation for a small gas cavity can thus be written using Eq. (4c)

$$D_m = \frac{W_i}{e} s_{m,i} J_i \tag{21b}$$

When the influence of different basic quantities on the dosimetric accuracy is investigated it is convenient to combine the factors in front of the specific ionization according to

$$D_m = \omega_{m,i} J_i \tag{21c}$$

The ratio of the absorbed dose to the specific ionization is thus given by the quantity $\omega_{m,i}$.

A further advantage of this simplified version of the Bragg–Gray relation is that ω is exactly the quantity determined in experiments where D_m and J_i are known, e.g., by calorimetric and ionometric measurements (cf. Refs. 43 and 108).

3.4. Extended Detectors

3.4.1. Definitions

In practical dosimetry it is generally not possible to construct a sufficiently small detector with adequate sensitivity. Most practical detectors are thus large relative to fluence variations in the medium and they perturb the fluence in the medium. The signal from an extended detector of different slowing-down and/or scattering properties therefore differs from that of a Bragg–Gray detector of negligible spatial extension, and the basic assumptions in the Bragg–Gray relation are violated in two principal respects:

1. The electron fluence will be sampled over an extended volume, which means that the response of the detector is proportional to the mean value of the absorbed dose distribution inside its radiation sensitive volume ($\bar{D}_i$) rather than being a well-defined point value (D_i).
2. The electron fluence averaged over the radiation sensitive part of the detector volume will in general differ from that at the point of interest in the uniform medium in absence of the detector. Three principally different types of fluence perturbations can be identified as discussed in detail in Section 3.4.4 and illustrated schematically in Fig. 5.

These two effects of spatially extended detectors have to be taken into account in the generalization of the Bragg–Gray relation.

3.4.1.1. Reference Volume. The averaging property of an extended detector can be taken into account by relating the mean absorbed dose inside the detector volume ($\bar{D}_i$) to that inside a similarly shaped reference volume in the unperturbed medium ($\bar{D}_m$) (see Fig. 3). This is the natural generalization of an extended medium-equivalent detector, which will necessarily measure a mean value over its radiation sensitive volume.

It is desirable that the basic form of the Bragg–Gray relation is retained for extended detectors. Therefore the dimensions of the reference volume should be chosen such that the shape of the electron spectrum (i.e., the relative electron energy distribution, irrespective of absolute normalization) averaged over the detector volume is as similar as possible to the spectrum averaged over the reference volume in the unperturbed medium. This requirement is due to the fact that most formulations of stopping power ratios are based on the same electron

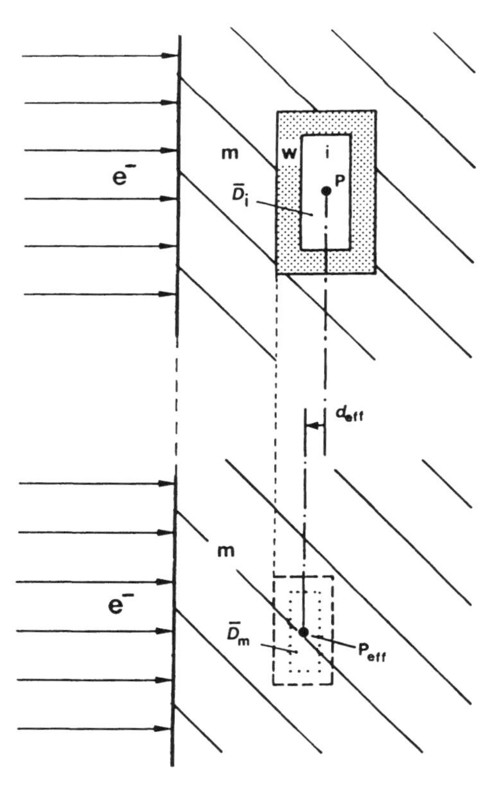

Figure 3. Definition of the reference volume in the uniform medium, the dash-dotted line in the lower half of the figure, and the effective point of measurement, P_{eff}, for solid and liquid detectors. In this figure it is assumed that the radiation sensitive material of the detector (i) is surrounded by a wall (w) of the same material, and that the linear stopping power is lower than for the medium (m). The mean absorbed dose inside the detector is $\bar{D}_i$ and the one inside the dash–dotted reference volume is $\bar{D}_m$. Under the assumption of a linear dose variation over the detector, the effective point of measurement will be the center of the reference volume.

energy spectrum, both for the detector and the medium. The present choice of reference volume will therefore allow the use of existing tabulations of stopping power ratios.

The reference volume is particularly simple to identify when the energy straggling in detector and medium is similar and a negligible disturbance of the energy distribution in the detector is produced by electrons that are scattered laterally into the detector from the medium (e.g., a flat detector, cf. Fig. 5c). The reference volume in the uniform medium should simply be chosen such that the mean energy loss by an electron in crossing the detector is the same as in crossing the reference volume in the medium. This implies that all linear dimensions of the reference volume in the uniform medium, t_m, are obtained by multiplying the corresponding dimension of the detector volume, t_i, by the ratio of the total linear stopping power of detector to medium evaluated at the mean energy of the primary electrons in the detector volume:

$$t_m = \frac{S_{\mathrm{tot},i}}{S_{\mathrm{tot},m}} t_i \tag{22}$$

This scaling is illustrated in Fig. 3 and was used already by Gray[58,59] and more recently by Brahme and Lax.[25] This choice will thus make the first two moments of the energy distribution, the mean energy and the energy spread, equal in the detector and the reference volume. The special importance of the mean energy of the primary electrons for electron and photon dosimetry due to the energy dependence of the stopping power ratio, through the influence of the density effect in gas-filled detectors, is well known.[6,20,62,135]

When parameters other than the absorbed dose in a broad uniform beam are of prime interest the reference volume must be chosen in a different way. For example, when the relative dose profile across the beam in a certain direction is determined the reference volume should be chosen equal to the true dimension of the detector in the direction of measurement.[22]

3.4.1.2. Effective Point of Measurement. It is in practice more convenient to use an effective point of measurement, to which the absorbed dose in the medium may be allocated, rather than a reference volume. When the distribution of absorbed dose in the uniform medium is a continuously varying, monotonic function of the space coordinates there will exist one and only one point inside the reference volume (according to the mean value theorem) where the absorbed dose is exactly equal to the mean dose inside that volume:

$$D_m(P_{\mathrm{eff}}) = \bar{D}_m \tag{23}$$

This point, P_{eff}, is therefore defined as the effective point of measurement and the "upstream" distance between P and P_{eff} is defined as the shift of the effective point of measurement, d_{eff}:

$$d_{\text{eff}} = z_p - z_{p_{\text{eff}}} \tag{24}$$

3.4.1.3. Fluence Perturbation. There will always be a perturbation of the electron fluence of the medium in the presence of an extended detector when the electron slowing down and/or scattering properties of the detector differ from those of the medium. The different types of fluence perturbation are discussed in detail in Section 3.4.4.

The total effect of the fluence perturbations is that even if the shape of the electron energy spectrum in the detector volume is not significantly different from that in the reference volume, the absolute values of the mean electron fluences in these two volumes may differ. The mean fluence inside the reference volume, $\bar{\Phi}_m$, can be related to that inside the detector volume, Φ_i, by defining a fluence perturbation factor $p_{m,i}$ according to

$$\bar{\Phi}_m = p_{m,i}\bar{\Phi}_i \tag{25}$$

The perturbation factor is unity for a small "Bragg–Gray" detector or a detector of the same composition and density as the medium but departs from unity with increasing size and difference in linear electron stopping and scattering properties between detector and medium.

3.4.1.4. Generalized Absorbed Dose Relation. By the above choice of the reference volume the basic Bragg–Gray relation, Eq. (21), can be generalized to apply to extended detectors by introducing the mean dose values $\bar{D}_m$ and $\bar{D}_i$, averaged over the reference and detector volumes, respectively, and by multiplication by the fluence perturbation correction factor $p_{m,i}$, which accounts for the difference in mean fluence in the two volumes:

$$\bar{D}_m = \bar{D}_i s_{m,i} p_{m,i} \tag{26}$$

Using the concept of an effective point of measurement the Bragg–Gray relation for extended detectors can also be written

$$D_m(P_{\text{eff}}) = \bar{D}_i s_{m,i} p_{m,i} \tag{27}$$

It should be pointed out here that the above procedure of defining a perturbation correction factor and an effective point of measurement in connection with a reference volume is not the only way of generalizing the Bragg–Gray relation to extended detectors. An important advantage of this method, however, is that it allows a strict separation of the definition of the effective point of measurement from that of the perturbation correction factor in a straightforward way and further that it is a general approach which is applicable both for low-density gaseous detectors and solid or liquid detectors. It is also of fundamental importance that the shape of the energy distribution spectrum of the electrons in the detector volume and that at the effective point of measurement in the uniform medium should be as similar as possible to validate the use of existing theoretical expressions and tabulations of stopping power ratios.

3.4.2. Effective Point of Measurement for Gaseous Detectors

3.4.2.1. Theory. The effective point of measurement, as defined in Section 3.4.1, may be located at an appreciable distance from the detector center when the linear stopping power of the detector differs substantially from that of the medium. This is the case particularly for gaseous detectors.

As an illustrative example the distance from the effective point of measurement to the central axis of a cylindrical gas-filled detector of radius r, with the axis perpendicular to the direction of the electron beam (Fig. 4), will be evaluated under two different sets of simplifying assumptions.

In the first set it is assumed that the electron fluence varies linearly in the direction of the beam over the diameter of the gas volume but that the mean electron energy does not vary appreciably over the diameter of the gas volume. These two conditions are generally met in photon beams and approximately also over the steep dose fall-off section of the electron depth–dose curve.

In the second set it is assumed that the variation of the electron fluence or absorbed dose over the diameter of the gas volume is fairly small and that the mean electron energy in the medium decreases linearly over the diameter of the gas volume. These latter conditions are complementary to the first set of conditions and are generally met over the therapeutically useful plateau part of the electron depth dose curve.

The following four assumptions are common to both sets:

1. all electrons entering the gas volume move parallel to the beam direction;
2. the number of ions generated by an electron crossing the gas volume is proportional to the chord length;

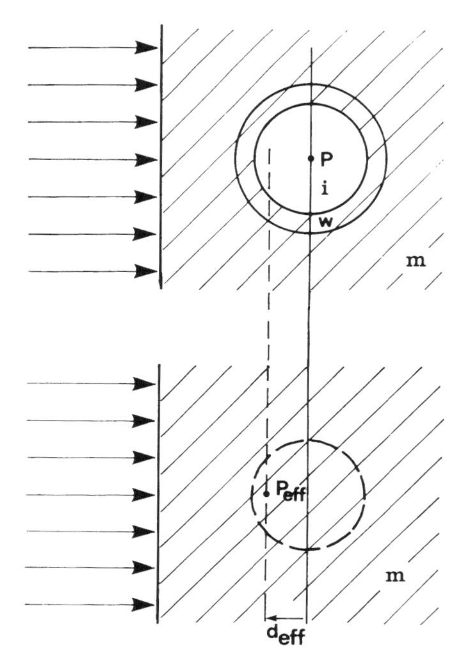

Figure 4. The shift of the effective point of measurement P_{eff}, from the center P, of a cylindrical gas-filled detector. The shift d_{eff} is defined as the upstream distance from the detector center P to P_{eff}. Owing to the low density of the gas relative to the medium, the size of the reference volume (cf. Section 3.4.2) will be negligible for gaseous detectors and it will be located at P_{eff}.

3. the wall of the detector is medium equivalent;
4. the influence of the central electrode can be neglected.

For the first set of conditions it can be calculated that the average mass ionization in the cylindrical probe corresponds to an effective point of measurement located upstream at a distance $d_{\text{eff}} = 8r/3\pi \sim 0.85r$ from the cylinder axis.[46,66,125] This result holds for a decreasing as well as an increasing fluence.

For the second set of conditions the effective point of measurement will be located at that point in the unperturbed medium where the mean electron energy is equal to that in the gas volume. This is so since the density of the gas is only 10^{-3} of that in the medium and thus the reference volume will be very small and in practice coincide with the effective point of measurement. Based on the second set of conditions, it can be shown that the mean electron energy in the gas volume will be equal to that at a point in the uniform medium, P_{eff}, located a distance $d_{\text{eff}} = 8r/3\pi$ upstream of the cylinder axis, P (Fig. 4).

For both sets of assumptions the same effective point of measurement is thus obtained. The reason for this is that in both cases a linear variation of the variable under consideration is assumed. This indicates that, in the transition region between the first and second set of assumptions, the same effective point of measurement could also be used.

Indeed, a simultaneous assumption of linear variation in energy and fluence results in a small additional shift of P_{eff} depending on the steepness and sign of the fluence variation relative to the energy variation.

However, several other factors will in practice influence the location of the effective point of measurement. The most important factor is that the electrons generally have a wide angular distribution. To make the above analysis applicable this has to be taken into account. For example, if a fraction f of the electrons enters the gas volume parallel to the beam direction, whereas a backscattered fraction g enters the volume in an antiparallel direction, and assuming the fluences at both the front and the back surfaces have constant negative or positive gradients in the beam direction, the location of the effective point of measurement is $(8r/3\pi)(f - g)$[66] upstream from the chamber center of a cylindrical chamber.

An even more realistic description is obtained if the true angular distribution of the electrons is taken into account. The shift of the effective point of measurement then becomes $(8r/3\pi)\overline{\cos\Theta} \approx (8r/3\pi)[1 - \overline{\Theta^2}(z)/2]$, where $\overline{\Theta^2}(z)$ is the mean square angular spread of the electrons in the forward hemisphere at the depth z of the chamber. When both the angular spread and the backscattered electrons are included simultaneously the distance may be estimated from

$$d_{\text{eff}} = \frac{8r}{3\pi}\left[1 - \frac{\overline{\Theta^2}(z)}{2} - \eta_B\right] \tag{28}$$

where η_B is the backscattering coefficient. The backscattering coefficient is a lower limit for the fluence backscatter which strictly should be used. However, η_B is only a few percent for electrons with energies above 2 MeV in water.[47,64] In electron beams $\overline{\Theta^2}(z)$ increases quasilinearly with depth and reaches a saturation value of about 0.6 rad^2 near the dose maximum.

In low-energy photon beams (^{60}Co, 5 MV) the value increases from about 0.45 rad^2 near the surface to 0.6 rad^2 at the depth of transient equilibrium. At higher photon energies both values are lower. At 20 MV, for example, the value increases from 0.15 rad^2 close to the surface and saturates at 0.45 rad^2 at the depth of transient equilibrium.[24]

At the dose maximum and beyond, the effective point of measurement will under the above assumptions therefore be located about $0.55r$ upstream from the center of a cylindrical chamber in electron beams and low-energy photon beams.

Other factors that have an influence on the location of the effective point of measurement are as follows:

(i) The actual shape of the gas volume. In the case of a thimble chamber, the contribution of the hemispherical part of its gas volume to the total ionization will move the effective point of measurement closer to the central axis compared to its value for a cylindrical volume.

(ii) The presence of the central electrode. Since the number of ions produced in the gas in the cavity is reduced by the insertion of a central electrode, the relative weight of the electrons passing through the peripheral parts of the gas volume is increased. Again this results in a smaller shift of the effective point of measurement from the chamber axis.

(iii) The actual weighting factor. The number of ions generated by an electron entering the gas-filled volume need not be exactly proportional to the chord length along the entrance direction. Ionization due to secondary electrons generated in the chamber wall, gas, or central electrode will influence the location of the effective point of measurement. It has been calculated that the shift becomes $(\pi/4)r = 0.79r$, if the weighting factor for electrons entering a cylindrical volume is assumed to be independent of the point of entry instead of being proportional to the chord length.[66]

3.4.2.2. Experimental Results for Electrons. Experimental determinations of the effective point of measurement can be made with cylindrical ionization chambers of varying radii, using an extrapolation to $r = 0$. Alternatively, the effective point of measurement can be found by use of flat detectors like film emulsions, Fricke solution in flat irradiation cells, or plane-parallel ionization chambers.

Dutreix and Dutreix[46] found a depth-independent upstream shift of $0.65r$ for 20-MeV electrons, whereas the upstream shift for 10-MeV electrons appeared to increase with depth. Harder[63] showed that after the application of the perturbation correction, the 10-MeV values showed an almost constant shift of $0.50r$. Hettinger *et al.*[67] found a shift of $0.75r$ for 13- and 34-MeV electrons when no perturbation correction was performed. The displacement measured by Johansson *et al.*[81] for 19- and 27-MeV electrons was about $0.55r$ independent of depth, whereas the upstream shift appearing in their 4.5- and 9.5-MeV beams increased with depth in a similar manner as the uncorrected 10-MeV values of Dutreix and Dutreix.[46] After administration of the perturbation correction, the value of Johansson *et al.*[81] is consistent with a constant shift of $0.40r$.

Weatherburn and Stedeford[141] recommended a constant value of $0.45r$ to be used from 3 to 10 MeV after the perturbation correction according to Harder[63] is made. Similarly they recommended $0.6r$ between 10 and 30 MeV. Up to 10 MeV their recommendations are based on their own data, while at higher energies they are based on their data at 30 MeV together with recalculations of data from Dutreix and Dutreix[46] and Hettinger *et al.*[67] This recommendation is also in approximate agreement with the data of Johansson *et al.*[81] and also with the theoretical value after correction for the angular spread.

Based on the experimental data available, a constant value of $(0.5 \pm 0.1)r$ can therefore be used in the whole energy range from 3 MeV up to 30 MeV. When the uncertainty stated is unacceptable a chamber of smaller radius or a flat chamber should be used.

For flat, thin-window chambers with the window perpendicular to the beam and with no in-scattering perturbation, the effective point of measurement lies just inside the plane of the entrance wall of the air cavity [Ref. 125 and Eq. (31)].

3.4.2.3. Experimental Results for Photons. A very extensive set of measurements of the effective point of measurement in photon beams was made by Johansson *et al.*[81] They found, as a general trend, that the surface value was quite high and close to the theoretical value in a monodirectional beam, but the value decreased rapidly with depth over the first few centimeters. This result is in agreement with the present theory as the angular spread of the secondary electrons increases with depth up to the depth of transient equilibrium.[24] The same trend should be expected from the increase in backscatter with depth as the complete slowing down spectrum is being built up. Johansson *et al.* also found a slightly larger value at depth in high-energy photons (42 MV) than at ^{60}Co and 5 MV, again in reasonable agreement with the lower saturation value in angular spread in high-energy photon beams.[24] An additional factor responsible for this effect could be the larger fraction of scattered photons in low-energy photon beams. Experimental results essentially in agreement with the above statements were reported by Hettinger *et al.*,[67] Casanovas,[35] and Dutreix.[45]

3.4.3. Effective Point of Measurement for Nongaseous Detectors

For solid and liquid detectors, the variation of the absorbed dose within the detector, rather than the variation of the electron fluence over the entrance surface as in a gas-filled detector, will determine the effective point of measurement. The dimensions of nongaseous detectors such as ferrous sulfate dosimeters, film dosimeters, TLD rods, or disks should generally be chosen in such a way that any expected displacement of the effective point of measurement from the detector center can be neglected. However, in a practical case a compromise has to be made between the two conflicting aims of a well-defined effective point of measurement and a high detector sensitivity.

3.4.3.1. Medium-Equivalent Detectors. For extended, medium-equivalent detectors such as the ferrous sulfate dosimeter used in a water phantom, the effective point of measurement may be significantly shifted from the center of the detector volume if the dose distribution is varying in a strongly nonlinear fashion over the dosimeter volume. When the thickness, t, of the detector in the direction of measurement is constant, the measured dose profile may be corrected in a simple way for the finite detector size by using the measured data. The corrected dose distribution is given by[22]

$$D_{\text{corr}}(x) = \bar{D}_{\text{meas}}(x) - \frac{t^2}{24}\bar{D}''_{\text{meas}}(x) + \cdots \tag{29}$$

where $\bar{D}_{\text{meas}}(x)$ is the measured mean absorbed dose with the detector center at a point x in the medium and x is the variable along the direction of measurement e.g., parallel or perpendicular to the direction of the beam. The resultant upstream shift of the effective point of measurement from the detector center at P is therefore given in the first approximation by

$$d_{\text{eff}} = -\frac{t^2}{24}\frac{\bar{D}''_{\text{meas}}(P)}{\bar{D}'_{\text{meas}}(P)} \tag{30}$$

The first and second derivatives of the measured dose distribution, $\bar{D}'_{\text{meas}}(x)$ and $\bar{D}''_{\text{meas}}(x)$, may be obtained by fitting any three consecutive data points by a second-order polynominal.

3.4.3.2. Non-Medium-Equivalent Detectors. When the detector is non-medium-equivalent, the situation is more complicated as fluence perturbation phenomena also have to be considered. For flat detectors (cf. Figs. 5c and 5d), where the influence of the inscattering effect on the energy distribution in the detector is small, the shift of the effective point of measurement is given by

$$d_{\text{eff}} = \left(1 - \frac{S_{\text{tot},i}}{S_{\text{tot},m}}\right)\frac{t}{2} \tag{31}$$

where t is the thickness of the detector. This expression is based on the definition of the reference volume and the effective point of measurement [Eqs. (23) and (24); cf. Fig. 3] and is valid when the dose variation is linear [$\bar{D}''_{\text{meas}} \approx 0$; cf. Eq. (29)].

An analytic expression for the effective point of measurement in

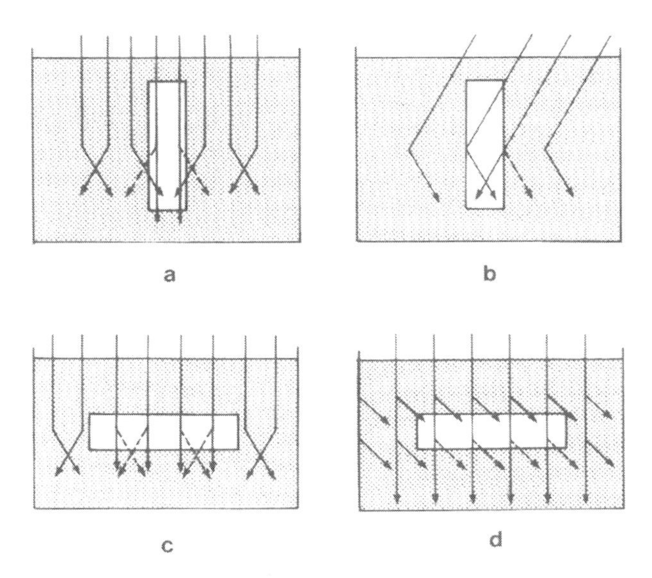

Figure 5. Disturbances of the electron fluence by scattering effects for the case of a low-density detector, such as an air ionization chamber. The multiple scattering of the electrons is schematically illustrated by one large-angle single scattering interaction. Dashed electron paths would occur in the absence of the detector. Panel (a) illustrates the in-scattering, (b) the quasireflection and integrating effects, and (c) the build-up of primary electrons, whereas (d) illustrates the lack of secondary electron equilibrium. (Panels a and c are adapted from Ref. 65.)

electron beams which also takes the fluence perturbation into account was derived by Svensson and Brahme[135] for a cylindrical detector of radius r and thickness t when irradiated in a region of linearly increasing fluence with the beam axis parallel to the axis of the cylinder. The distance to the effective point of measurement as measured from the cylinder center can be determined from

$$d_{\text{eff}} = \frac{t}{2}\left(1 - 0.43\frac{t}{r}\right)\left(1 - \frac{T_i}{T_m}\right) \tag{32}$$

where T_i and T_m are linear scattering powers of the detector and the medium, respectively. In the first approximation the shift of P_{eff} from the detector center is negligible if either the linear scattering powers of detector and medium are equal or the detector shape is selected such that $t \approx 2.3r$. Furthermore, for a low-density dosimeter the effective point of measurement is very close to the front surface, as is expected for the limit of a flat gas ionization chamber ($T_i/T_m \approx 0$) provided $t/r \ll 1$.

The effective point of measurement used in deriving Eq. (32) differs from that of Section 3.4.2 and Eq. (31) in that it defines that point in the uniform medium where the electron fluence is equal to the mean fluence inside the detector volume. By this choice, no additional perturbation correction is needed, but instead different mean electron energies should be used for the detector and for the effective point of measurement when calculating mean stopping power values[135]:

$$\bar{E}_i = \bar{E}_m + (S_{\text{tot},m} - S_{\text{tot},i})\frac{t}{2} - S_{\text{tot},m}d_{\text{eff}} \tag{33}$$

where $\bar{E}_m$ is the mean electron energy at the effective point of measurement in the uniform medium, and $\bar{E}_i$ is the mean energy in the detector.

3.4.4. Fluence Perturbation Correction Factor

3.4.4.1. Perturbation Mechanisms. There will necessarily be a perturbation of the electron fluence of the medium as it penetrates an extended detector of different electron slowing-down and/or scattering properties compared to those of the surrounding medium. This fluence perturbation can be of three principal types as specified below and illustrated schematically in Fig. 5 for the case of a low-density cavity.

(*i*) *Lack of lateral scattering equilibrium* across the boundaries of the detector volume (cf. Figs. 5a and 5b). This effect is due to different linear scattering powers in the detector and the medium. At least two separate contributions can be recognized. Firstly, owing to the low density there is a lack of scattering out of the side walls of the detector, and hence a net scattering into the cavity from the medium with a resultant increased fluence in the detector volume. This "inscattering effect" is illustrated by Fig. 5a. Secondly, there is an apparent reflection of cavity electrons by the walls of the detector (cf. Ref. 1), which is not compensated by a corresponding apparent reflection of medium electrons by the radiation sensitive material in the detector. This "quasireflection" effect, illustrated by Fig. 5b, will increase the fluence in the detector when the fraction of cavity electrons meeting the walls of the detector at small angles is large. This is the case, for instance, near the surface in electron beams with a small angular spread. A third effect in a low-density cavity is due to the integrating or accumulation of inscattered electrons from different depths in the medium as their absorption in the cavity is negligible. This integrating effect of a low-density cavity will also increase the mean electron fluence in the detector volume.

(*ii*) *Difference in buildup or attenuation* of the electron fluence in detector and medium. A difference in fluence buildup is obtained when the angular distribution of primary electrons in the detector volume and in the reference volume of the medium differs (cf. Fig. 5c). The "buildup effect" is thus due to the lack of scattering inside the detector volume compared to that in the medium. This results in a reduced buildup of the primary electron fluence near the surface and a reduced fluence decrease near the end of the electron range in an electron beam. This effect arises when the values of the linear stopping and scattering power ratios for detector to medium differ. The fluence buildup or attenuation in the detector and in the reference volume will generally be very similar when the reference volume is chosen appropriately (see Sections 3.4.2 and 3.4.3.2) and the detector material is not too different from the medium.

(*iii*) *Lack of secondary electron equilibrium* is due to differences in the slowing down properties of the detector and the medium. In an extended detector a significant part of the low-energy secondary electron spectrum is characteristic of the detector material rather than of the medium. Often, the scattering properties of the two materials will also differ, and a considerable disturbance of the dose distribution near the interface between detector and medium may result (cf. Section 3.2 and Refs. 14, 15, 112, and 120). This perturbation of the fluence of secondary electrons is not discussed further for electron beams, but in principle it could be taken into account either by adjusting the stopping-power ratio or by applying a perturbation factor. For photon beams this effect is very important (cf. text of Fig. 2), and it is generally treated by a generalized type of cavity theory as discussed in Section 3.5.

3.4.4.2. Gaseous Detectors. (*i*) *Theory.* The inscattering effect is of great importance for gas-filled detectors in a solid or liquid medium. It increases with the surface area of the detector that is parallel to the initial beam direction and is therefore largest in slim detectors elongated in the direction of the beam (Figs. 5a and 5b).

A theoretical treatment of the perturbation correction for plane, *monodirectional electron beams,* based on the small-angle multiple scattering approximation, has been given by Harder.[63] For a coin-shaped, gas-filled detector with its axis parallel to the incident beam, the inside plane of the circular entrance window defines the effective point of measurement because the detector essentially samples the electron fluence from the window. Accordingly, inscattering through the side walls of the gas volume (Fig. 5a) disturbs the measured values. The resulting formula[63] can be written as

$$p_{m,i} = 1-0.13\frac{t}{r}(tT_m)^{1/2} \qquad (34)$$

where t is the thickness and r the radius of the coin-shaped volume, and T_m is the linear scattering power for the medium m. The perturbation correction factor is material and energy dependent through its dependence on T_m. Owing to the change of the electron spectrum with depth the energy dependence can also be described as a depth dependence.

It is important to note that the above expression, Eq. (34), is strictly applicable only at very small depths in beams with negligible angular divergence at the surface. At slightly greater depths or in clinical beams with a significant angular spread the treatment of inscattering from the medium (Fig. 5a) must take account of the angular divergence of the primary electrons, and allowance should also be made for the quasi-reflection effect (Fig. 5b).

The inscattering effect can be taken into account more accurately by generalizing Eq. (34) to take the angular spread of the electrons incident on the cavity into account. Interestingly enough, this changes the size dependence of the perturbation correction factor and it becomes instead

$$p_{m,i} = 1-0.11 T_m t \frac{t}{r}\left(1 - \frac{T_i}{T_m}\right)\left\{\frac{[3\overline{\Theta^2}(z)]^{1/2} + [3\overline{\Theta^2}(z)]^{-1/2}}{2}\right\} \tag{35}$$

where $\overline{\Theta^2}(z)$ is the mean square angular spread of the electrons incident on the front surface of the cavity and T_i is the linear scattering power of the detector gas. This expression should be used instead of Eq. (34) when the depth of the detector is larger than one third of the detector thickness, or rather when $\overline{\Theta^2}(z) > T_m t/3$. In practice this means that the condition is nearly always fulfilled since there is generally a substantial angular spread already in the incident beam.[21,22] The new term containing the angular spread $[\overline{\Theta^2}(z)]$ differs by less than 10% from unity provided $\overline{\Theta^2}(z)$ is larger than 0.15 rad^2.

The more general expression taking the angular spread into account shows the direct proportionality between cavity size t and the perturbation factor assuming the shape parameter t/r remains constant. This general result is consistent with the experimental results of Johansson *et al.*[81] for cylindrical thimble ionization chambers when irradiated perpendicular to the chamber axis.

However, for practical dosimetry the perturbation by scattered electrons entering a coin-shaped gas volume from the sides can in practice be reduced to a negligible magnitude when the sensitive volume of the collecting electrode is defined by a surrounding guard electrode of sufficient width (see Fig. 6). An estimate of this width is $2t[\overline{\Theta^2}(z) + T_m t/3]^{1/2}$.

(*ii*) *Experimental.* In carrying out experimental determinations of the perturbation factor it is desirable to avoid any complications caused

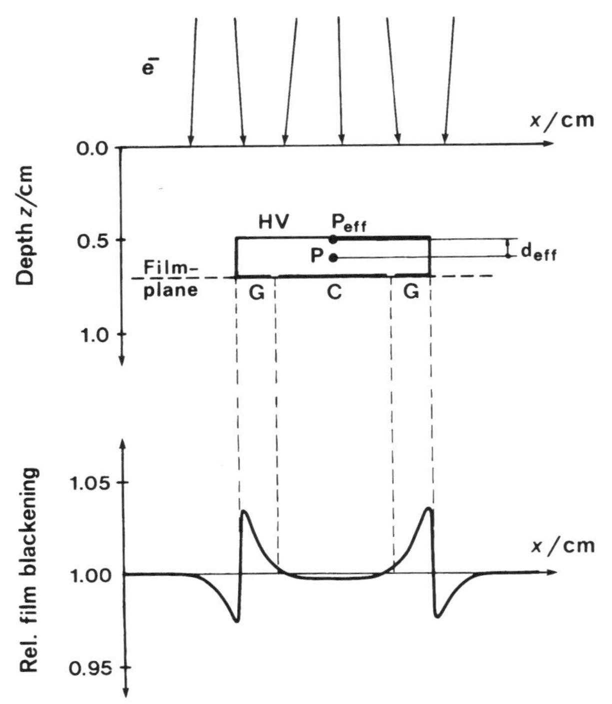

Figure 6. Thin-window, coin-shaped ionization chamber in a weakly scattered electron beam; HV, high-voltage electrode; C, collecting electrode; and G, guard ring. With a well-designed guard ring the scattering effect (cf. Figs. 5a and 5b) mainly influences the electron fluence over the guard. Therefore the chamber can be considered to measure the electron fluence at the entrance window as both the energy loss and scattering in the gas are negligible. This is illustrated by the relative film blackening profile in the lower half of the figure. The measurements were carried out at $\bar{E}_0 = 6\,\text{MeV}$ in PMMA (cf. Refs. 101 and 56).

by the uncertainty in the position of the effective point of measurement. The perturbation effect due to inscattering is therefore determined from dose measurements made at or near a wide dose maximum, where the exact position of the effective point of measurement is immaterial.

Johansson *et al.*[81] made measurements of the perturbation corrections for cylindrical chambers with their axes perpendicular to the electron beam for mean electron energies between 2.5 and 22 MeV at the depth dose maxima in PMMA. These results (Fig. 7) show the increase of the perturbation with increasing chamber diameter and decreasing electron energy, and demonstrate the fact that the theoretical values given by ICRU[74] underestimate the effect. The strong energy variation of the perturbation effect implies that relative dose measurements are also affected when made at different depths in a phantom (cf. Ref. 6).

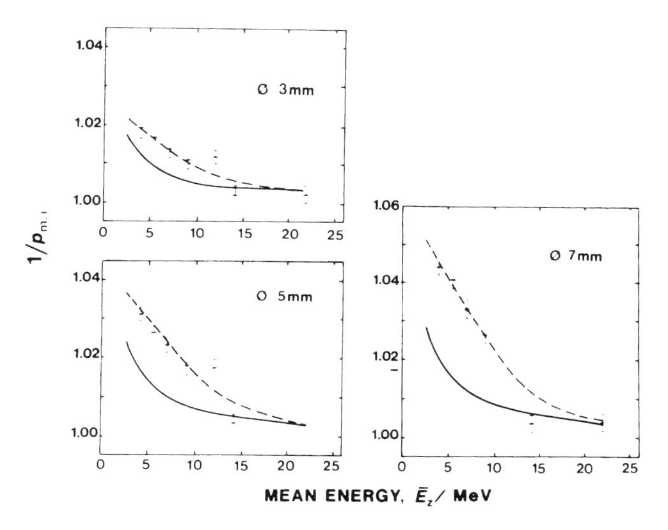

Figure 7. The reciprocal of the perturbation correction factor ($1/p_{m,i}$) for cylindrical ionization chambers with the cylinder axis perpendicular to the beam axis. The chamber diameters were d = 0.3, 0.5, and 0.7 mm, respectively. Experimental data: from Ref. 81 measured at the depth of dose maximum (in PMMA); calculated data for water: from Ref. 63 and Ref. 74, solid line). The deviations from the experimental results are mainly caused by the quasireflection and integrating effects (see Fig. 5b). The fairly large spread in the measured points can be explained by the build-up of large electrostatic fields in insulating plastic after irradiations with electron beams. This field disturbs the electron fluence in the phantom (see Section 5.4.1).

For plane-parallel chambers the perturbation factor has been determined by an extrapolation method.[89,96] The first reference found that the effect was negligible for electron beams with mean energies of 2.2 and 14.3 MeV at the point of measurement; the second reference reported that the perturbation correction was less than 0.5% at electron energies above 0.5 MeV. No fluence perturbation correction is thus needed for plane-parallel ionization chambers that have a thin air cavity and a sufficiently wide guard ring when the entrance window is taken as the effective point of measurement.

However, flat chambers could have an angular dependent response as they do not possess the same symmetry as a spherical chamber or a cylindrical thimble chamber (cf. Ref. 24).

3.4.4.3. Nongaseous Detectors. For solid and liquid detectors the density of the radiation sensitive material can no longer be disregarded and the reference volume will be of finite extension. This implies, for example, that the effective point of measurement will no longer be at the front surface of a flat detector [cf. Eq. (31)].

Very few theoretical and experimental investigations of the perturbation of the primary electron fluence by solid and liquid detectors have been published. Svensson and Brahme[135] treated the general case of an arbitrary detector material (i) in a uniform medium (m), for the case of flat circular detectors with the axis parallel to the beam axis. The buildup effect (cf. Fig. 5c) was treated in detail as it dominated in their application whereas the inscattering effect was only treated in an approximate manner. The total perturbation correction factor derived by Svensson and Brahme has therefore been modified here to take the inscattering effect into account in a more accurate fashion [cf. Eq. (35)] and also to make it applicable to a reference volume reduced in size by the detector to medium stopping power ratio (see Section 3.4.2):

$$p_{m,i} = 1 + \frac{T_m t}{4}\left\{\frac{S_{\text{tot},i}}{S_{\text{tot},m}} - \frac{T_i}{T_m} - 0.43\frac{t}{r}\left(1 - \frac{T_i}{T_m}\right)\right.$$
$$\left.\times\left[\frac{[3\overline{\Theta^2}(z)]^{1/2} + 1/[3\overline{\Theta^2}(z)]^{-1/2}}{2}\right]\right\} \quad (36)$$

where t is the thickness of the cylinder in the direction of the beam and r the cylinder radius, $\overline{\Theta^2}(z)$ the mean square angular spread of the electrons in a uniform medium at the entrance surface of the cylinder, $S_{\text{tot},m}$ and $S_{\text{tot},i}$ the total linear stopping powers, and T_m and T_i the linear scattering powers of the medium and the detector, respectively. The expression is valid for $\overline{\Theta^2}(z) > T_m t/3$, which is always fulfilled provided that $z > t/3$. The first two terms inside the curly brackets are due to the buildup effect (Fig. 5c), whereas the remainder is due to the inscattering effect (Fig. 5a). The term within the large square brackets differs by less than 10% from unity provided that $\overline{\Theta^2}(z) > 0.15$. It is seen that the perturbation of the primary electron fluence due to the buildup effect is negligible when the linear stopping power ratio equals the linear scattering power ratio, a theoretically interesting result as this is one of the most important conditions for material equivalence. If instead the linear scattering powers are identical ($T_i = T_m$), no perturbation due to in- or outscattering of the primary electrons should be expected. Furthermore, the inscattering effect decreases with increasing angular spread of the electrons [$\overline{\Theta^2}(z)$] whereas the buildup effect is not affected in the first approximation. These two principal types of fluence perturbation have different signs. Therefore, by appropriate choice of the shape parameter t/r, they may compensate each other. It should also be noted that the buildup effect is only valid for depths smaller than the therapeutic range because the angular distribution then becomes saturated at larger depths (cf. Refs. 119 and 23).

It can be noted that also in this general case the fluence perturbation [Eq. (36)] is directly proportional to the linear dimension of the detector provided the shape parameter t/r has a constant value. This general result agrees phenomenologically well with the experimental results of Johansson *et al.*[81] even though these pertain strictly to thimble chambers with the cylinder axis perpendicular to the beam. From the general expression for the perturbation correction for a flat dosimeter [Eq. (36)] it can also be concluded that thin, disk-shaped detectors such as TLD chips will introduce a negligible in-scattering effect (actually an out-scattering effect since the linear scattering power of the TLD detector is generally larger than for the medium). However, the buildup effect may become quite large since the difference between the linear stopping- and scattering-power ratios is large. For a 0.4-mm-thick LiF-Teflon TLD disk in a polystyrene phantom, the buildup effect introduces a perturbation of the primary electron fluence of about 0.5% in the surface region of a 5-MeV electron beam.

3.5. Review of Recent Dosimetric Theories

The theoretical work for gas detectors in photon beams was reviewed by Burlin.[32] Many new ideas have evolved since then.

A very interesting and new approach to dosimetry has been published by Zheng-Ming.[149,150] By studying the Boltzman transport equation [cf. Eq. (19)] for a uniform medium with a small enclosed gas volume he derived an explicit expression for the equivalent electron source which is responsible for the perturbation of the electron fluence in the region of the gas volume. This allows a more accurate calculation of the energy deposition than the Spencer–Attix theory including second-order effects and taking into account bremsstrahlung and the energy deposited in the cavity by electrons from the equivalent source with energies above Δ.

During the last decade the group in Ghent[78–80] have extensively treated the theory of small and large cavities. Their theory could also be regarded as a modified form of the Spencer–Attix theory with a more accurate energy deposition model than the simple two-step approximation of Spencer and Attix.

Interface effects in solid (TLD) domimeters were measured by Bertilsson.[15] She interpreted the results in the form of a generalized Burlin-type equation where the atomic numbers of the medium and the detector were taken into account more accurately by considering the electron scattering phenomena near the interfaces.

By defining restricted energy absorption coefficients the local energy deposition by photon interactions in a cavity was calculated by Brahme.[21] The contribution to the direct energy deposition by photons in a detector was obtained by multiplying the traditional mass energy absorption coefficient ratio by the ratio of the restricted to the unrestricted energy absorption coefficients of the detector. The restricted mass energy absorption coefficients were calculated in a manner similar to the energy deposition function also used by Janssens.[80] By studying the mean path length of insiders (cf. Ref. 36), Horowitz and Dubi[69] introduced a modification of the detector generated dose contribution.

Based on the method of treatment of electron scatter at an interface suggested by Bernard,[14] Kearsley[83] introduced further modifications of the expression for extended cavities in photon beams. His approach is in this respect similar to the method used by Bertilsson.[15]

Unfortunately, none of the above theories has treated the problems of the effective point of measurement or the fluence perturbation in a way consistent with their detailed discussion of the ratio of the absorbed dose in the detector to that in the medium.

4. BASIC PHYSICAL DATA

4.1. Background

A determination of the absorbed dose at a point in a phantom is generally based on a measurement of the signal from a detector, the use of one or more calibration factors (e.g., for a detector and an electrometer), the application of one or several physical constants or conversion factors, and corrections of these factors to the situation valid at the point of measurement. The purpose of this section is to analyze the presently available value of the different physical constants, like G, W, and $s_{m,i}$, while the complete absorbed dose calculation procedure is discussed in detail in Section 5.

The experimental determination of $s_{m,i}$ [for definition see Section 3.3 and Eqs. (21), (26), and (27)] is based on certain numerical values of W and G; see, e.g., Refs. 43 and 136. On the other hand, W and G are often determined using certain assumed values of $s_{m,i}$. For instance, in order to determine W from measurements with a graphite calorimeter and a graphite ionization chamber, values on $s_{\mathrm{gr,air}}$ (m = graphite, and i = air) are applied. The W value recommended by the ICRU No. 31[76] is thus mainly based on such measurements in Co-60 gamma ray beams or in beams at lower energies.

The aim in the present analysis is to avoid such a circular reasoning.

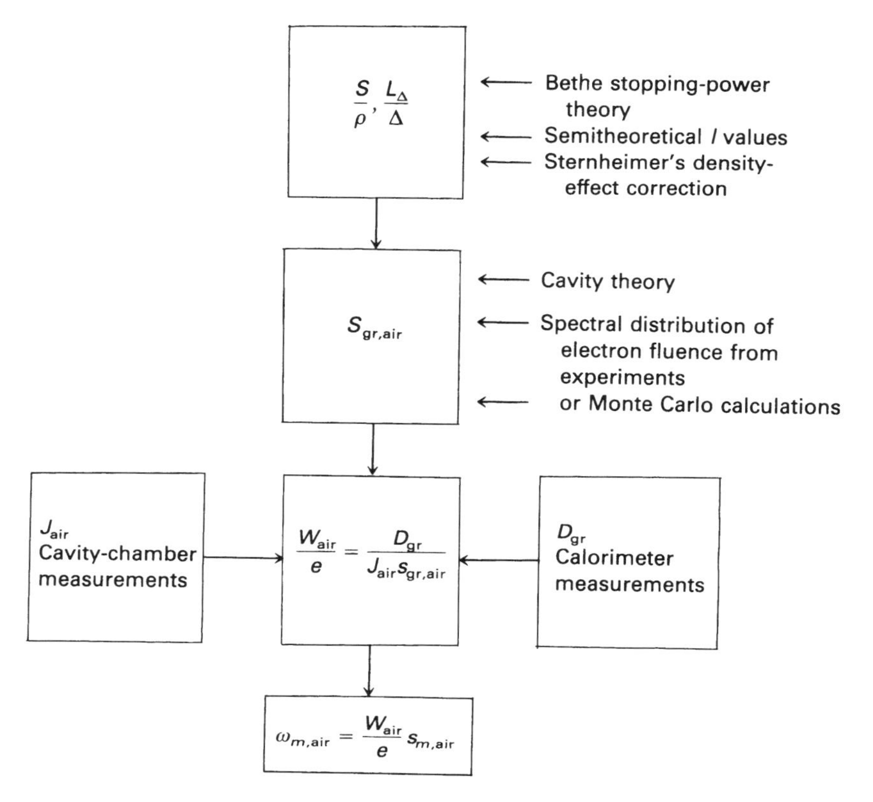

Figure 8. A flow-chart for the determination of W/e and $\omega_{m,\mathrm{air}}$. This is the method generally used for Co-60 gamma rays but is here also applied at higher energies. It is seen that W/e depends on the assumption of values for stopping power ratios. For use in dose determinations, numerical values of $\omega_{m,\mathrm{air}} = W/e.\ s_{m,\mathrm{air}}$ are needed. These values can be checked by experiments, e.g., from $\omega_{m,\mathrm{air}} = D_m/J_{\mathrm{air}}$ or $\omega_{m,\mathrm{air}} = \omega_{\mathrm{gr,air}} s_{m,\mathrm{gr}}$.

A flow chart has therefore been followed which has stopping power values as a starting point; see Fig. 8. These values are based on theories and experiments that are independent of assumptions of the values of certain dosimetric constants. The next step is the application of one of the cavity theories (see Section 3) in order to evaluate the stopping power ratios, $s_{m,i}$. Input data for $s_{m,i}$ computations are, besides the stopping powers, the spectral distribution of the electron fluence at the point of measurement, or in a simplified treatment the mean energy of the primary electrons; see Section 4.3. These $s_{m,i}$ values are then used to determine W as shown in the figure (see also Section 4.5) and G as discussed in Section 4.4.

4.2. Stopping Power, S

The collision stopping powers in most widespread use up to 1982 have been those published by Berger and Seltzer in 1964 and 1966.[9,10] They were adopted by the ICRU in the photon report No. 14[73] and electron report No. 21.[74] These stopping powers have since been reevaluated. Some of these new data from water are shown in Fig. 9. The changes are mainly due to different density effect corrections and use of different mean excitation energies (I values).

The computation by Berger and Seltzer[11] was undertaken in connection with a new ICRU report on the subject and represents the most complete data for different elements and compositions. These computations are considered to give the present state-of-the-art values and are therefore used below. It is of great interest to compare these stopping powers with those of Berger and Seltzer[9,10] as these will directly influence the determined absorbed dose value keeping other factors unchanged. For instance, the decrease in $S_{col,w}$ is 1.8% at an electron energy of

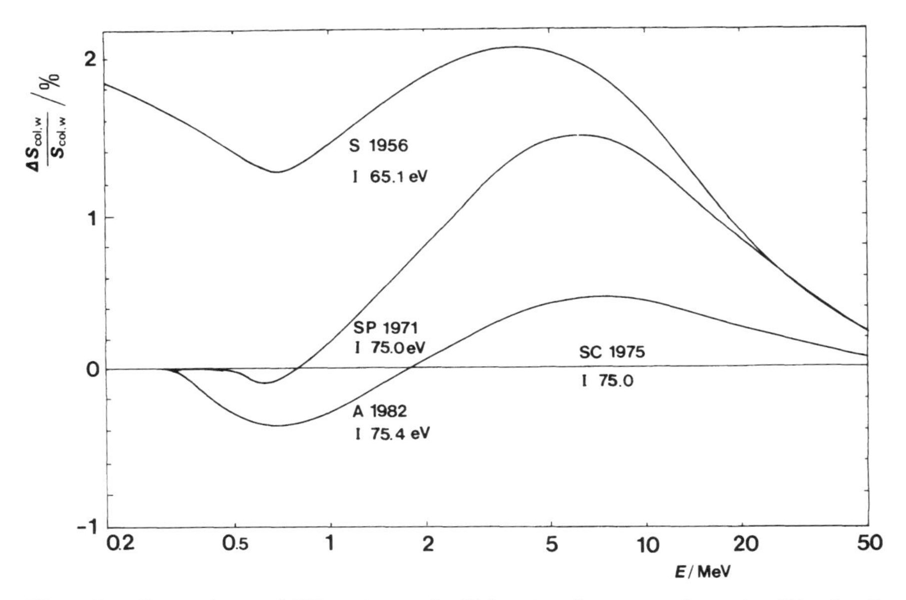

Figure 9. Comparisons of different sets of collision stopping powers for water. The density effect corrections have been performed according to Sternheimer[128] (S 1956), Sternheimer and Peirels[129] (SP 1971), Sternheimer[127] and Carlsson[33] (SC 1975), and Ashley[7] (A 1982). The computations have been carried out by Berger and Seltzer (see the text). The SC 1975) data have been recommended by Berger and Seltzer[11] and are therefore used in the present report. The (S 1956) data were used in Berger and Seltzer.[9,10]

TABLE I
The Change in I Values for Some Important Materials for Dosimetry

Material	I values by Berger and Seltzer (eV)		ΔI (eV)
	1964	1982	
Air	86.8	85.7	−1.1
Water	65.1	75.0	9.9
Graphite	78.0	78.0	0
Lucite (PMMA)	65.6	74.0	8.4
Polystyrene	63.6	68.7	5.1
Polyethylene	54.6	57.4	2.8

0.3 MeV (Fig. 9), i.e., at an energy of importance for Co-60 gamma dosimetry.

The I values from 1964–1966 and 1982 are compared for some materials of importance for clinical dosimetry in Table I. The change of the I value is particularly large for water, about 10 eV. The change in collision stopping-power values will, however, be moderate as I is included in the logarithmic term in the stopping power formula. Thus

$$S_{\text{col}} \propto \rho \frac{Z}{A} [f(E) - 2 \ln I - \delta] \tag{37}$$

where ρ is the density, Z/A is the atomic number to the atomic weight (in compounds or mixtures weighted by the mass fractions of the different elements), $f(E)$ is a function of the energy of the incident electron, and δ is the density effect correction.

The new density effect corrections differ up to about 2% from earlier calculations, a change large enough also to be of importance in clinical dosimetry. The situation is particularly complicated for graphite as it is not clear if the bulk density (about 1.7 g cm^{-3}) or the crystalline density (2.265 g cm^{-3}) should be used. The reduction in $S_{\text{col,gr}}$ as compared to the data from 1964 is shown in Fig. 10; curves are shown for the two densities. The density problem for graphite represents a special complication, since for many years graphite calorimeters have been used as the standard instrument to determine absorbed dose at national standards laboratories. The flow scheme (Fig. 8) using both these two densities is therefore discussed in Sections 4.3 and 4.4 in order to find arguments in the field of dosimetry for one or the other density.

The I value for air has been almost unchanged (Table I). The density effect correction is below 1% for energies <50 MeV. The stopping power

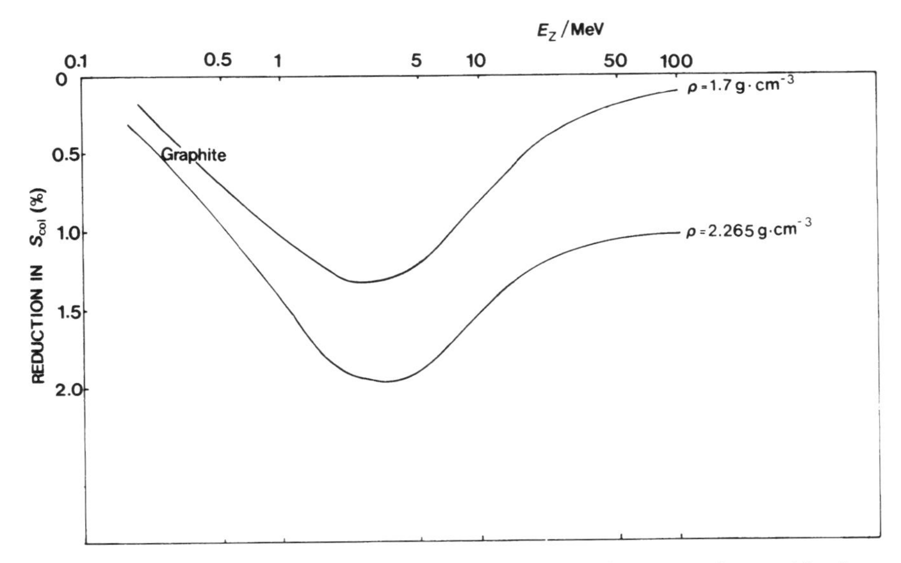

Figure 10. The percentage reduction of the collision stopping power for graphite from Berger and Seltzer 1964 (Ref. 9) to their 1982 values (Ref. 11). The graphite density 1.7 g cm^{-3} was assumed in Ref. 9.

ratios $s_{m,\text{air}}$ will therefore mainly be influenced by changes for the material m.

4.3. The Stopping Power Ratio, $s_{m,i}$

Two major irradiation geometries for the medium and the detector will be discussed in this section as shown in Fig. 2. These lead to two different sets of stopping power ratios, $s_{m,i}$, to be applied in the Bragg–Gray equation [see Eqs. (21), (26), and (27)].

In the *Bragg–Gray* (BG) approach it is assumed (see Section 3.1) that the spectral fluence of the primary electrons in the detector is the same as that in the medium, $\Phi^{\text{p}}_{E,m}$. Also, it is assumed that the energy dissipation by secondary electrons occurs "on the spot," or alternatively, that there is a secondary electron equilibrium, i.e., that the energy from secondary electrons entering the cavity is equal to the energy from such electrons leaving the cavity. Interface effects (Fig. 5d) between the medium, chamber wall, and the cavity gas are not taken into account. Under this assumption the ratio of the energy dissipation in the uniform medium and in the air inside a cavity is given by [compare Eq. (16)]

$$s_{m,\text{air}}^{\text{BG}} = \frac{\displaystyle\int_0^{E_{\max}} (\Phi_{E,m}^{p}) \left(\frac{S}{\rho}\right)_{\text{col},m} dE}{\displaystyle\int_0^{E_{\max}} (\Phi_{E,m}^{p}) \left(\frac{S}{\rho}\right)_{\text{col,air}} dE} \tag{38}$$

In order to avoid significant interface effects the medium, chamber wall, and cavity should be of very similar composition. This is not possible in many practical situations. However, many workers[19,68,62] showed that a difference in composition may be overcome by having a fairly thin lining of detector-equivalent material around the radiation sensitive material of the detector. This lining practically eliminates the interface problem. Figure 2b illustrates this geometry for an ionization chamber.

With such a lining Eq. (38) may be simplified, as suggested by Harder,[62] by using the collision stopping power at the mean energy of the primary electrons:

$$s_{m,\text{air}}^{\text{H}} = \left[\frac{S(E)}{\rho}\right]_{\text{col},m} \bigg/ \left[\frac{S(E)}{\rho}\right]_{\text{col,air}} \tag{39}$$

This simplification is valid when the collision stopping powers vary linearly with the energy of the primary electrons or when the stopping power ratio is a slowly varying function of energy over the energy range of interest.

The situation for which the Spencer–Attix (SA) stopping power ratio is valid is an air cavity with walls of the same composition as the surrounding medium, m; see Fig. 2a. This theory accounts in an approximate manner for the finite ranges of secondary electrons or δ rays. Here the spectral fluence of both primary and secondary electrons, $\Phi_{E,m}$, above a cutoff energy, Δ, must be computed. The stopping powers L_Δ/ρ are now restricted to losses smaller than the cutoff energy. According to Eq. (15) the Spencer–Attix stopping power ratio, $s_{m,\text{air}}^{\text{SA}}$ can be written

$$s_{m,\text{air}}^{\text{SA}} = \frac{\displaystyle\int_\Delta^{E} \Phi_{E,m} \left(\frac{L_\Delta}{\rho}\right)_m dE + \Phi_e(\Delta)_m \left(\frac{S(\Delta)}{\rho}\right)_m \Delta}{\displaystyle\int_\Delta^{E} \Phi_{E,m} \left(\frac{L_\Delta}{\rho}\right)_{\text{air}} dE + \Phi_E(\Delta)_m \left(\frac{S(\Delta)}{\rho}\right)_{\text{air}} \Delta} \tag{40}$$

The second term in the numerator and denominator is the energy dissipation by the track ends (see Refs. 102 and 103).

The cutoff energy is often chosen to be equal to the energy just needed for an electron to be able to cross the air cavity. The cutoff energy has thus to do with the size and shape of the chamber, but this cutoff is not very critical for the values of $s^{SA}_{m,air}$. As an example, the practical range (standard pressure and 20°C) for 10-keV electrons is about 2 mm and about 7 mm for 20 keV. The stopping power ratio, $s^{SA}_{w,air}$, at an electron energy of 10 MeV, decreases by only about 0.2% when the cutoff is increased from the lower to the higher value. The exact value is somewhat dependent on the electron energy. Berger *et al.*[13] gave an empirical formula for the relative variation of $s^{SA}_{w,air}$ with the cavity size with input parameters depending on the value of Δ, electron energy, and the depth in the phantom.

The values of the three different stopping power ratios will now be discussed. In particular, $s_{w,air}$, i.e., the stopping power ratio water to air will be analyzed, first for electron beams and then for photon beams.

Values of $s^{SA}_{w,air}$ for different electron energies at the phantom surface and different depths have been computed for electron radiation, using a Monte Carlo technique, by Berger and Seltzer[12] and by Nahum.[104] These two sets of calculations agree very well, as can be seen in Fig. 11. This agreement shows that the calculated electron spectra are sufficiently similar as the same set of input stopping power data have been used. There is also indirect experimental evidence that at least the computation of $\Phi_{E,m}$ is sufficiently accurate for the dosimetry. It is possible to find the same numerical values of $s^{SA}_{m,air}$ (E_0, z) for various incident energies, E_0, and phantom depths z. For instance, according to computations by Berger and Seltzer,[12] $s^{SA}_{gr,air}$ for $E_0 = 15$ MeV, and $z = 1$ cm will have the same value as for $E_0 = 50$ MeV and $z = 14$ cm ($s^{SA}_{gr,air} = 0.865$ using $\rho = 1.7\ \mathrm{g\ cm^{-3}}$ and $\Delta = 10$ keV). In spite of the extreme difference in parameters, experiments based on absolute measurements of D_{gr} and J_{air}[43] also show that D_{gr}/J_{air} at these two sets of energies and depths are the same within the uncertainty of measurement (a few parts of 1%), thus supporting the theoretical $\Phi_{E,m}$. More details will be found in Section 4.5, where W/e has been evaluated from experimental determinations of D_{gr}/J_{air} and the input of theoretical $s^{SA}_{gr,air}$. A consistent W/e value, independent of the combination of E_0 and z, is obtained there. The same type of experiment has been carried out in water, i.e., D_w and J_{air} were determined. Also in this case good agreement is obtained between Monte Carlo calculations and measurements. Computations of the *Bragg–Gray* stopping power ratios (Eq. (38)] have been carried out by Nahum[102,103] for various electron energies. These computations have been reevaluated using the stopping power data from Ref. 11. The

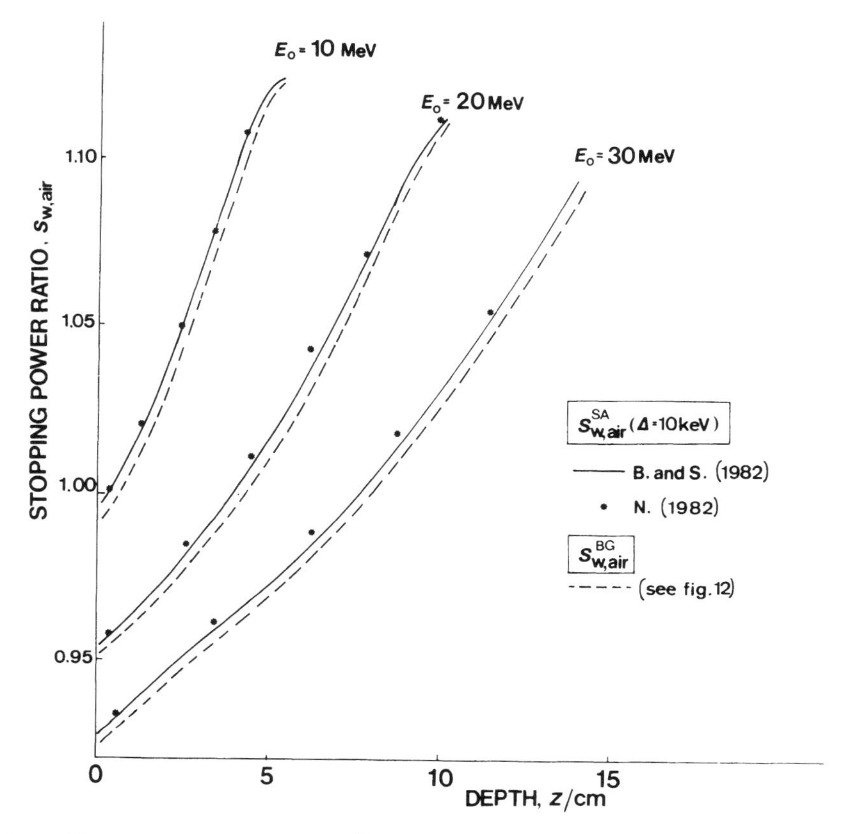

Figure 11. Monte Carlo calculated $s_{w,\text{air}}^{\text{SA}}$ from Berger and Seltzer[12] (——) and Nahum[104] (*). The $s_{w,\text{air}}^{\text{BG}}$ are taken from Fig. 12.

calculations, shown as dotted lines in Fig. 12, are made in a straightforward way using Eq. (38). In practice complete secondary electron equilibrium is never achieved. This is particularly so at small depths where the slowing down spectrum of secondary electrons is not yet built up.

A modified BG ratio has therefore also been computed (dash-dot line in Fig. 12) taking into consideration the lack of equilibrium. Here $(L/\rho)_\Delta$ replaces $(S/\rho)_{\text{col}}$. The restriction value Δ is adjusted so that the range of electrons with energy Δ is equal to z;[20] this modified $s_{w,\text{air}}^{\text{BG}}$ is included as the dashed line in Fig. 12. It is seen that the difference amounts to about 1.5% and is largest at small depths.

In Fig. 12 $s_{w,\text{air}}^{\text{H}}$ is also shown. The mean energy at the point of

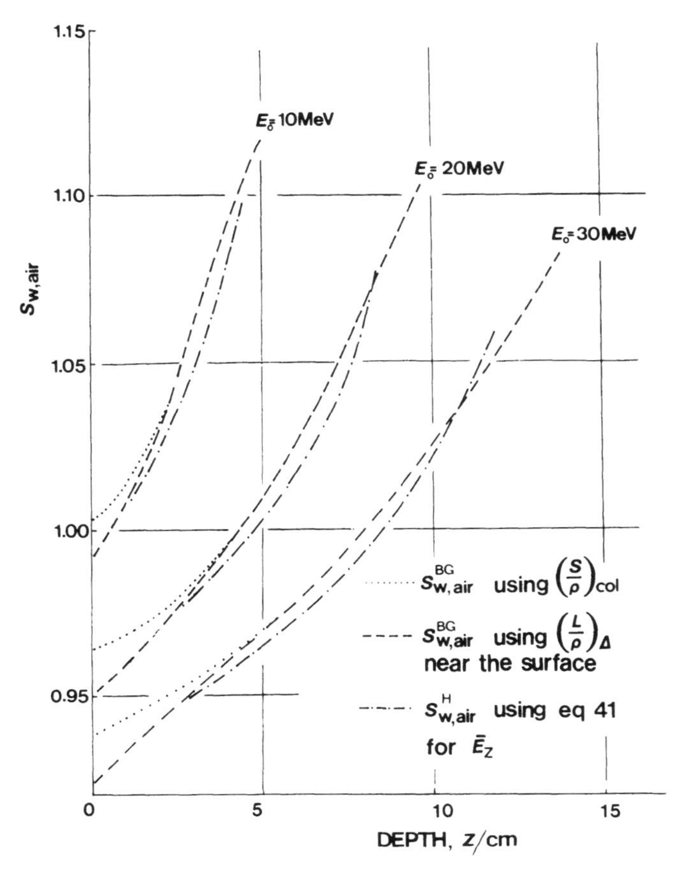

Figure 12. The straightforward Bragg–Gray stopping power ratios for water to air are shown (· · ·). Corrections for lack of secondary electron equilibrium near the surface have been performed (– – –). The result from a simplified computation of the Bragg–Gray ratios, indicated $s_{w,\text{air}}^{\text{H}}$, are also shown (– · – · –).

measurements was calculated from the relation [20,6]

$$\bar{E}(z) = \bar{E}(0)\frac{S_{\text{tot}}\exp\left[-zS_{\text{rad}}/\bar{E}(0)\right] - S_{\text{col}}}{S_{\text{rad}}} \tag{41}$$

The $s_{w,\text{air}}^{\text{BG}}$ value [Eq. (38)] agrees within about 0.5% with the $s_{w,\text{air}}^{H}$ value [Eq. (39)], at least at depths corresponding to about dose maximum.

It is also of interest to compare $s_{w,\text{air}}^{\text{BG}}$ and $s_{w,\text{air}}^{\text{SA}}$ (see Fig. 11). It is seen that the BG ratio is 0.3%–0.5% smaller than the SA ratio. Svensson and Nahum[137] computed the difference $s_{m,\text{air}}^{\text{SA}} - s_{m,\text{air}}^{\text{BG}}$ as a function of Δ

$I = I_{air} - I_m$, where I is the mean excitation energy. They found that the difference in stopping power ratios is very weakly dependent on the quality of the electron and photon beams, but increases with $|\Delta I|$. The difference is below about 0.5% for $|\Delta I| < 10$ eV.

All the stopping power ratio expressions [Eqs. (38), (39), and (40)] are equally applicable for photons and electrons. For photons the complete spectrum of secondary electrons must be known. Very few data have been reported in the literature. However, $s_{w,air}^{SA}$ ratios are now available for some qualities from Nahum[102,103] and Andreo.[5] These data agree fairly well, except at small phantom depths, i.e., before dose maximum; cf. Section 4.6.

Numerical data of $\omega_{w,air} = s_{w,air}^{SA}(W/e)$ are presented both for electron and photon beams in Section 5. They may thus also be used for electrons in the case where the ionization chamber has an air-equivalent wall, as the difference in $S_{w,air}^{SA} - s_{w,air}^{BG}$ for most practical situations is unimportant, i.e., below 0.5%. The situation is more complicated for photons because photons generate electrons differently in water and in an air-equivalent wall. In practical dosimetry, it is therefore more straightforward to use the Spencer–Attix values for photons and then correct for the non-water-equivalence of the chamber wall by means of a perturbation factor.[100]

4.4. Mass Energy Absorption Coefficient, μ_{en}/ρ

Values of μ_{en}/ρ are needed in order to compute the influence of different phantom and wall materials on the signal of the detector in photon beams. The μ_{en}/ρ for the detector material is also of interest in cases where a significant part of the electron fluence in the detector is generated in the detector material itself; see Sections 3 and 5. Values of μ_{en}/ρ have been computed by Hubbel[71,72] and they are considered to be known within a fraction of 1%.

4.5. Radiation Chemical Yield, $G(X)$

The radiation chemical yield is defined by $G(X) = n(X)/\bar{\varepsilon}$, where $n(X)$ is the mean amount of substance of a specified entity, X, produced, destroyed, or changed by the mean energy imparted, $\bar{\varepsilon}$, to the matter (see ICRU No. 33).[77] Here the radiation chemical yield of the ferrous sulfate dosimeter will be discussed.

The dose measurement technique based on the ferrous sulfate dosimeter is well established (e.g., ICRU report No. 14[73]). It is assumed here that the standard dosimeter solution is used, i.e., consisting of 1 mol m^{-3} ferrous sulfate, 400 mol m^{-3} sulfuric acid, and 1 mol m^{-3} NaCl.

The reference temperature during the irradiation and spectrophotometric measurements is assumed to be 25°C.

The absorbed dose to the dosimeter solution D_i is determined from the relationship

$$D_i = \frac{\Delta A_t}{\rho l(\varepsilon_m)_t G_{t'}} \tag{42}$$

where D_i is the absorbed dose to the dosimeter solution, ΔA_t is the increase in absorbance due to irradiation at temperature t (=25°C) during the spectrophotometric measurements, ρ is the density of the solution, l is the length of the light path in the photometer cell, $(\varepsilon_m)_t$ is the difference between the molar linear absorption coefficient for ferric ions and for ferrous ions at temperature t (=25°C), and $G_{t'}$ is the radiation chemical yield of ferric ions at the irradiation temperature t' (=25°C).

Experimental data for $\varepsilon_m G$ from different investigators will be compared rather than G values. The reason is that the experimental determination of ε_m needs great care and large systematic errors are often introduced. However, in several intercomparisons it has been shown by using a reference solution that the difference in measured absorbance with different spectrophotometers generally is very small. This seems to be the reason why a smaller spread is obtained if experimental values of $\varepsilon_m G$ are compared instead of G (see Ref. 136).

Values of $\varepsilon_m G$ determined from calorimeter measurements for electron radiation are given in Table II. The $\varepsilon_m G$ values for different types of calorimeters are given in different columns. The authors' values have been corrected in two ways:

(a) The calorimeter absorber is in some experiments replaced by a ferrous sulfate dosimeter. This dosimeter has different scattering properties than the absorber. The absorber and the dosimeter will therefore "see" a somewhat different fluence of primary electrons. The applied correction is dependent on the shape and the material of the dosimeter and on the spectral and angular distribution of the electrons [see Section 3.4.4, in particular Eq. (36)].

(b) The stopping-power data by Berger and Seltzer[11] have been applied. For the graphite calorimeter stopping powers based both the bulk density ($\rho = 1.7\,\mathrm{g\,cm^{-3}}$) and the crystalline density ($\rho = 2.265\,\mathrm{g\,cm^{-3}}$) are used. These stopping powers are considered to represent the extreme values within which the true effective density is located.

Table II includes the corrected values for electron radiation. The mean values are calculated giving the same weight to the result from each investigator. If $\rho = 1.7\,\mathrm{g\,m^{-3}}$ is used when selecting stopping powers, a

TABLE II
ε_m and G Values Determined by Different Investigators[a]

Graphite calorimeter				
		$\varepsilon_m G$ ($m^2\,kg^{-1}\,Gy^{-1}$)		
Reference	$\bar{E}_z$ (MeV)	$\rho = 1.7\,g\,cm^{-3}$ ($\times 10^{-6}$)	$\rho = 2.265\,g\,cm^{-3}$ ($\times 10^{-6}$)	ε_m ($m^2\,mol^{-1}$)
37	3.5	348.2	345.8	
	7	351.9	349.2	
				217.34
	9	352.3	349.4	
	13	352.6	349.2	
	14.5	352.8	349.4	
		351.6	348.6	
55	6	352.2	349.5	220.0
	16	348.4	245.1	
		350.3	347.3	
115	16.5	355.7	352.4	220.0
97	24	349.1	345.2	216.4
Mean:		351.7	348.4	218.4

Nongraphite calorimeter			
Reference	$\bar{E}_z$ (MeV)	$\varepsilon_m G$ ($m^2\,kg^{-1}\,Gy^{-1}$) ($\times 10^{-6}$)	ε_m ($m^2\,mol^{-1}$)
122	1	351.1	219.6
53	3	349.0	217.0
3	7	350.4	222.5
95	5	351.9	
	15	354.1	
		353.0	
114	20	354.2	219.6
Mean:		351.5	

[a] The $\varepsilon_m G$ values are corrected as described in the text.

value of $\varepsilon_m G$ is obtained ($351.7 \times 10^{-6}\,m^2\,kg^{-1}\,Gy^{-1}$), in close agreement with the determination from nongraphite calorimeters ($351.5 \times 10^{-6}\,m^2\,kg^{-1}\,Gy^{-1}$). This is one indication that the bulk density should be used at least with the most recent formulation of the density effect. However, the value using the higher density ($348.4 \times 10^{-6}\,m^2\,kg^{-1}\,Gy^{-1}$), is 0.9% lower but still within the systematic errors. The calorimetric experiments alone are therefore not sufficiently conclusive for the use of the lower density.

For Co-60 gamma-ray beams a G value of 15.5 ± 0.2 per 100 eV (i.e., $1.607 \times 10^{-6}\,mol\,kg^{-1}\,Gy^{-1}$) has been recommended by the ICRU in report No. 14.[73] The uncertainties of the individual determinations referred to in report No. 14 are fairly large due to uncertainties in spectrophotometer calibrations (ε_m value) and also in physical constants. A reevaluation of earlier measurements is complicated as the exact values of some constants used by the authors are not known. This is so in particular for G values based on the ionization chamber method as the stopping powers have undergone large changes. These changes may also influence the numerical value of W (see Section 4.5), and in some experiments the systematic errors may cancel out. A systematic error, or rather inconsistency in comparison with other methods of absorbed dose determinations, of a few percent would probably have been discovered, as ferrous sulfate dosimetry has been in considerable use for long time. The G value 15.5 per 100 eV would correspond to a value of $\varepsilon_m G = 354 \times 10^{-6}\,m^2\,kg^{-1}\,Gy^{-1}$ if the $\varepsilon_m = 220.5\,m^2\,mol^{-1}$ is used. This ε_m value was the mean found from a literature study of 83 investigators[30] and used by ICRU report No. 21.[74] There are, however, some very recent measurements that can be utilized to give a more solid base for evaluation of $\varepsilon_m G$.

Experiments that have been carried out by or in close cooperation with some of the national laboratories can be analyzed in order to determine $\varepsilon_m G$. These experiments are of particular interest as all the spectrophotometers have been intercompared.[50] A test sample of potassium dichromate in aqueous perchloric acid solution was used. Measurements of absorbance were made at the wavelengths of 313 and 350 nm, thus close to the 304 nm as used for Fricke dosimeters. The deviations between the absorbance standard and the measured values at the ten participating laboratories were fairly small, of the order of one or two tenths of one percent at an absorbance above 0.2.

The BIPM (Bureau International des Poids et Mesures) undertook irradiation of Fricke dosimeters from some of the laboratories in a water phantom. The absorbed dose rate to water, $\dot{D}_w$, at $5\,g\,cm^{-2}$ depth at a reference point 1 m from the BIPM ^{60}Co source was calculated from measurements of the dose rate, $\dot{D}_{gr}$, in a corresponding irradiated geometry using a graphite phantom. At the BIPM the measurement of absorbed dose in graphite is performed with a graphite wall cavity ionization chamber. From these ionometric measurements and also from calorimetric determinations which were carried out in cooperation with four national laboratories using their instruments at the BIPM, a weighted mean of absorbed dose rate to graphite in the BIPM beam was determined with an uncertainty (one σ) of 0.1%.[17] Calculation of $\dot{D}_w/\dot{D}_{gr}$ was performed by Boutillon. A ratio of 1.050 was derived with an uncertainty of 1% (cf. Fig. 23). The BIPM could thus report the absorbed dose to water at the position of the ferrous sulfate dosimeters.

From the paper by Ellis *et al.*[50] it is now possible to evaluate the value of $\varepsilon_m G$ that should have been used by the various laboratories in order to give the absorbed dose stated by BIPM for their irradiated dosimeters. Table III gives this information. The mean value for $\varepsilon_m G$ is $353.3 \times 10^{-6}\, m^2\, kg^{-1}\, Gy^{-1}$.

One of the centers participating in this intercomparison and reported in Table III (No. 4) also carried out a separate determination of $\varepsilon_m G$. Two liquid ionization chambers[147] were used in order to transfer absorbed dose to water determinations from the NBS and to laboratory No. 4 (i.e., University of Gothenburg). The standard deviation in a single measurement was about 0.1% using these chambers. The transfer was carried out with an uncertainty less than about 0.2%. A graphite calorimeter was used by NBS to determine absorbed dose to graphite and a thick wall graphite ionization chamber was used as a transfer instrument by the NBS in order to determine absorbed dose to water at

TABLE III

$\varepsilon_m G$ Determined from Experiment at BIPM[a]

Laboratory	$\varepsilon_m G$ ($m^2\, kg^{-1}\, Gy^{-1}$) ($\times 10^{-6}$)
2.[b] Laboratoire Metrologie des Rayonnements Ionisants, Saclay, France	354.5
3. National Institute of Radiological Sciences/ ETL., Chiba-shi, Japan	352.7
4. NIRP/University of Gothenburg, Sweden	354.5
5. National Physical Laboratory, Teddington, U.K.	351.6
7. International Atomic Energy Agency, Vienna	353.1
Mean:	353.3

[a] A recalculation from Ellis *et al.*[50]
[b] The numbers are as given in Table VIII of Ref. 50.

the point of interest in the water phantom. Stopping-power ratios are not used in the calculations (see Ref. 117). From these transfer experiments $\varepsilon_m G = 352.3 \times 10^{-6}\,\mathrm{m^2\,kg^{-1}\,Gy^{-1}}$ was determined for ^{60}Co gamma radiation[95] (14 cm × 14 cm, 5 cm depth in water, source–detector distance 100 cm). This value should be compared with $\varepsilon_m G = 354.5 \times 10^{-6}\,\mathrm{m^2\,kg^{-1}\,Gy^{-1}}$ (laboratory No. 4 in Table III) based on the BIPM comparison, a difference of 0.6%.

Mail intercomparisons using the ferrous sulfate dosimeter have been in extensive use for many years and it is well known that the storage of Fricke solution in irradiation cells may influence both the G value (generally an increase) and the precision (see, e.g., Ref. 138). Most of the participating laboratories in the experiment reported above used glass irradiation cells which should have an insignificant storage effect[49] but may instead cause interface problems between glass and liquid.[32] In the intercomparison reported above the standard deviation for the single measurements (σ for each laboratory) at the different participating laboratories varied from about 0.1% to 2%. In the transfer experiments by Mattsson *et al.*[95] based on the transfer with a liquid ionization chamber these problems do not appear, as the storage time is very short, and therefore also plastic irradiation cells can be used. The precision of the measurement was generally better than 0.1% (i.e., σ of a single measurement).

In a further experiment Mattsson *et al.*[95] compared the absorbed dose determinations with ferrous sulfate dosimeters and with a copy of the Domen water calorimeter.[40] Taking $\varepsilon_m G = 352.3 \times 10^{-6}\,\mathrm{m^2\,kg^{-1}\,Gy^{-1}}$ an exothermic chemical (or heat) defect of 3.5% has to be assumed in order to have the two methods agree. Domen[42] has reported exactly the same value, which further supports the results from Mattsson *et al.* All the analysis above results in $\varepsilon_m G$ for Co-60 gamma rays between 352 and $354 \times 10^{-6}\,\mathrm{m^2\,kg^{-1}\,Gy^{-1}}$, and a value of $353 \times 10^{-6}\,\mathrm{m^2\,kg^{-1}\,Gy^{-1}}$ is therefore recommended.

In the ICRU report No. 14 somewhat higher G values were recommended for high-energy photons from an accelerator than for Co-60 gamma rays. A very large uncertainty was, however, stated; for 11–30 MV $G = 15.7 \pm 0.6$ per 100 eV. More recent studies instead indicate that the G value may be even lower for these qualities than for Co-60 gamma radiation and high-energy electrons.[102] Conclusive experiments are not available for these qualities.

The value of $\varepsilon_m G$ for electron beams in Table II was assumed to be energy independent. However, there might be a weak energy dependence of G, as discussed by Nahum.[102] He carried out a semiempirical evaluation of G values based on Monte Carlo calculated local energy dissipation spectra ("local" meant a cutoff of 100 eV) for various

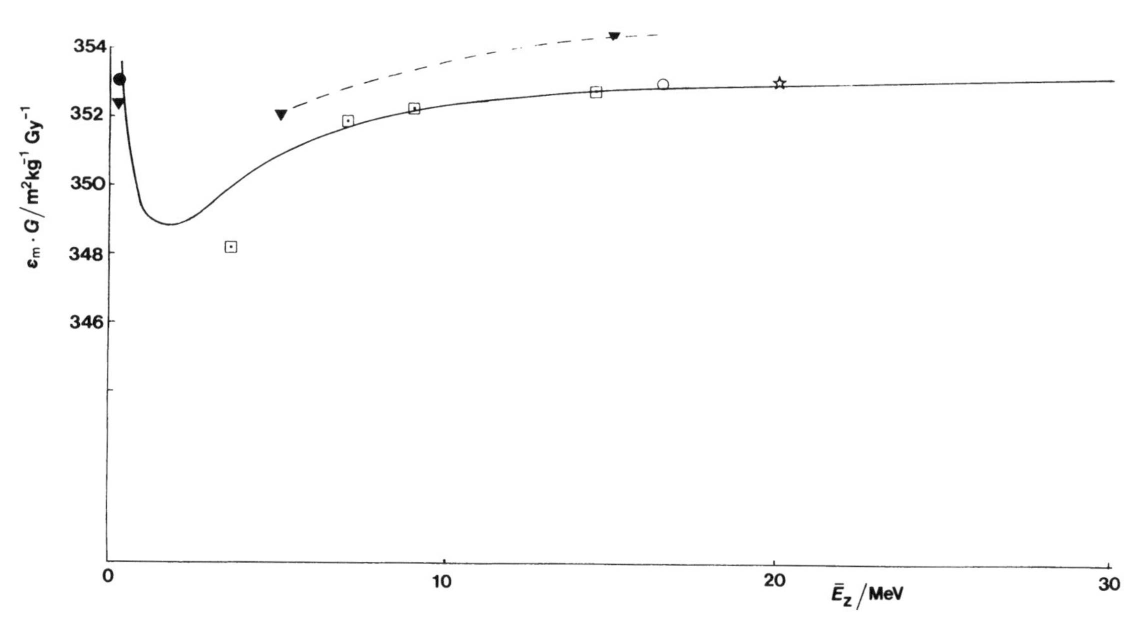

Figure 13. Semitheoretical values of $\varepsilon_m G$ ($\times 10^{-6}$) by Nahum[102] (——) normalized after the experimental value at 14.5 MeV by Cottens *et al.*[38] (□). Experimental values of $\varepsilon_m G$ from Mattsson *et al.*[95] (▼) and Ellis *et al.*[50] (●). Pettersson[114] (★) and Pinkerton[115] (○) measured G-values both for ^{60}Co gamma and electron radiation. However, in the figure a normalization to $\varepsilon_m G = 353 \times 10^{-6}\, m^2\, kg^{-1}\, Gy^{-1}$ at ^{60}Co gamma quality has been performed, as it is felt that the uncertainty in the absolute values of the product $\varepsilon_m G$ may be somewhat larger than for the recent investigations.

radiation qualities. This information, in combination with the application of some experimentally well-established G values of radiations differing widely in quality, made it possible to predict trends also for small quality changes in the case of high-energy photon and electron beams (see Fig. 13).

Also experimentally, it is indicated by the very careful measurements by Cottens *et al.*[38] (the systematic and random uncertainty, not including the stopping powers, are totally a few tenths of one percent) that G may increase with the electron energy. An opposite result was obtained by Geisseloder *et al.*[55] (see Table II). The uncertainty in these earlier measurements seems, however, to be much larger. In Fig. 13 the values by Cottens *et al.*[38] and Nahum[102] are compared, normalized to the experimental values at 14.5 MeV (see also Ref. 105). A weak energy dependence is seen.

Of special interest are calorimetric determinations that have been performed by one investigator both for electrons and Co-60 gamma beams. Three experiments of this kind will be analyzed here, namely, those by Pettersson,[114] Pinkerton,[115] and Mattsson *et al.*[95] Pettersson used a water calorimeter. His determinations do not involve stopping power ratios, except that for water to ferrous sulfate dosimeter solution, which has an insignificant uncertainty. Practically the same $\varepsilon_m G$ was obtained for Co-60 gamma and 20 MeV (mean energy, $\bar{E}_z$) electrons, 15.57 and 15.56 per 100 eV, respectively. Pettersson reported that he overcame part of the chemical defect in water by carrying out preirradiations of the water. A possible chemical effect ought, in any case, to be of the same amount both for Co-60 gamma radiations and electron beams (see Ref. 42). Mattsson *et al.* used the Domen calorimeter to determine $\varepsilon_m G$ at $\bar{E}_z = 5$ and 15 MeV. The same chemical defect was assumed as for Co-60 gamma rays, i.e., 3.5%.[42] This would result in the $\varepsilon_m G$ values given in Table II (Ref. 95) and Fig. 13. Finally, the experiment by Pinkerton was carried out using a graphite calorimeter. Pinkerton's results were 15.60 and 15.62 per 100 eV for Co-60 gamma rays and 16.5 MeV ($\bar{E}_z$) electrons, respectively. Pinkerton has not reported the exact stopping power values in use for his experiment but has referred to the table by Berger and Seltzer. In Table II corrections to the new data on stopping powers[11] were carried out. In addition a small scattering correction was applied (see Section 3.4.4). If the bulk density of graphite is used, as discussed above, then $\varepsilon_m G = 355.7 \times 10^{-6}\,\mathrm{m^2\,kg^{-1}\,Gy^{-1}}$ is obtained for $\bar{E}_z$ 16.5 MeV that is exactly the same value as for Co-60 gamma radiation (i.e., 15.60 per 100 eV, which corresponded to this $\varepsilon_m G$ value).

In conclusion: comparisons of graphite and nongraphite calorimeter determinations of G values for electron beams indicate a better agreement using the low-density $\rho = 1.7\,\mathrm{g\,cm^{-3}}$ than the high one. There are several indications that there is a small energy dependence of the G value for Co-60 gamma rays, photon and electron beams. The following $\varepsilon_m G$ values ($\times 10^{-6}$) are therefore recommended:

Co-60 gamma rays, 2–4 MV: $353 \pm 3\,\mathrm{m^2\,kg^{-1}\,Gy^{-1}}$

X rays, 5–50 MV and electrons E_z = 2–10 MeV: $352 \pm 4\,\mathrm{m^2\,kg^{-1}\,Gy^{-1}}$

electrons $\bar{E}_z$ = 10–50 MeV: $353 \pm 3\,\mathrm{m^2\,kg^{-1}\,Gy^{-1}}$

4.6. Mean Energy Expended in Gas per Ion Pair Formed, *W*

It has generally been assumed that the value of W for air (defined by $W = E/\bar{N}$, where $\bar{N}$ is the mean number of ion pairs formed when the initial kinetic energy E of a charged particle is completely dissipated in the gas, ICRU No. 33[77]) does not vary with photon or electron energy in the MeV^{-1} range. There is, however, no experimental proof of this assumption.

Several investigations have been carried out to determine W for Co-60 gamma rays; see ICRU No. 31.[76] In these measurements the Bragg–Gray relation for an ionization chamber has been made use of [Eq. (21b)]:

$$\frac{W}{e} = \frac{D_{\mathrm{gr}}}{J_{\mathrm{air}} s_{\mathrm{gr,air}}} = \frac{\omega_{\mathrm{gr,air}}}{s_{\mathrm{gr,air}}} \tag{43}$$

The absorbed dose to graphite, D_{gr}, is determined from graphite calorimeter measurements. The mass ionization, J_{air} (i.e., ionization charge of one sign per mass of air in the air cavity of an ionization chamber), is measured with a graphite walled chamber. The $s_{\mathrm{gr,air}}$ is the mean restricted mass stopping power ratio at the point of measurement for the cavity size and radiation quality in use. The ICRU (in report No. 31[76]) has recalculated the various experimental W values for the same value of $s_{\mathrm{gr,air}}$ and also included, if necessary, corrections to the dry air condition. For dry air a value of $W = 33.85 \pm 0.15$ eV was recommended. However, the $s_{\mathrm{gr,air}}$ values applied by ICRU, based on previous calculations by Berger and Seltzer, have now been revised (Ref. 11; see also Section 4.2). The change in $s_{\mathrm{gr,air}}$ will increase W by about 1.3% for Co-60 gamma; see Table IV.

Only a few experimental determinations have been carried out at higher energies.[84,88,113] They are based on a different type of measurement than for Co-60 gamma rays. A Faraday cage was used for the determination of the number of incident electrons on an ionization chamber. The charge of one sign liberated in this chamber per incident electron and per unit pathlength dJ/dX was then measured. For the

TABLE IV
Experimental *W* Values Determined by Niatel[108] using the "Domen"-Calorimeter and a BIPM Graphite Ionization Chamber[a]

Quality	$\omega_{gr,air}$ (JC^{-1})	$s^{SA}_{gr,air}$	W (eV)	$s^{SA}_{gr,air}$ parameters
Co-60 γ rays depth 10 g cm^{-2}	34.13	1.0105	33.78	Value from Ref. 108
		1.0018	34.07	Δ = 14 keV, ρ = 1.7 g cm^{-3} SC-density correction
		0.9996	34.15	Δ = 14 keV, ρ = 2.26 g cm^{-3} SC-density correction

[a] Different assumptions for $s^{SA}_{gr,air}$ are applied.

calculation of *W*, Kretschko[84] used the relation

$$\left(\frac{dE}{dX}\right)_r = W\frac{dJ}{dX} \tag{44}$$

where $(dE/dX)_r$ is the energy loss per unit pathlength including those energy transfers that are so small that the secondary electrons produced leave all their energy in the cavity chamber (i.e., a restricted stopping power). In addition, the part of the energy transfer from higher-energy secondary electrons that give ionization in the cavity is included.

The value of $(dE/dX)_r$ is theoretically derived. Its value is critically dependent on the cavity shape and an uncertainty less than a few percent seems hard to obtain. The *W* value given by Markus was 31.9 ± 1.9 eV at 14.8 MeV, by Ovadia *et al.* 34.3 ± 1 at 17.5 MeV, and by Kretschko varying from 33.3 ± 0.5 to 34.0 ± 0.5 eV from 10 to 25 MeV (these latter values should probably be increased by about 1% applying more recent stopping power data). It is very unsatisfactory to base precision dosimetry on these uncertain *W* values. Therefore, owing to the lack of reliable experimental data for high energies, it has generally been assumed that W = 33.85 eV, obtained at low energies, also applies at high energies. However, a more logical way would be to use the method for determination of *W* at Co-60 gamma also for electron beams, i.e., Eq. (43). Experimental data are available but have to be reevaluated as the purpose of the authors was to determine $s_{m,i}$ from the assumption of a value of W/e and not the other way around. The flow chart here will follow Fig. 8

The most accurate measurements which may be utilized for *W* caclulations both for Co-60 gamma rays and electrons are based on results from Niatel[108] at the BIPM and Domen and Lamperti[43] at the NBS. The NBS graphite calorimeter was used to determine D_{gr} in these two sets of measurements and plane-parallel graphite chambers were used for the measurements of J_{air}.

Niatel obtained $W_{dry\ air} = 33.78\ \mathrm{eV}$ at the depth of $10\ \mathrm{g\ cm^{-2}}$ in graphite for Co-60 irradiation (see Table IV). The $s_{gr,air}$ value applied in this experiment was higher than those from Berger and Seltzer.[11] A recalculation using the latter stopping powers would instead give $W = 34.07\ \mathrm{eV}$ assuming $\rho = 1.7\ \mathrm{g\ cm^{-3}}$ and $34.15\ \mathrm{eV}$ assuming $\rho = 2.265\ \mathrm{g\ cm^{-3}}$. Domen and Lamperti measured absolute D_{gr}/J_{air} for electron beams of energies at the phantom surface of 15, 20, 25, 30, 40, and 50 MeV. W/e is obtained by using $s^{SA}_{gr,air}$ from Berger and Seltzer.[12] Here $s^{SA}_{gr,air}$ is the restricted stopping power ratio graphite to air using a cutoff energy of 10 keV. The spectral distribution of the electron fluence was computed with the Monte Carlo method. The W values derived in this way (see also Fig. 8) were plotted as a function of the electron energy, $\bar{E}_z$, at the point of measurement, calculated from Eq. (41) (see Fig. 14). Also in this case, computations were carried out using $s_{gr,air}$ based on both the graphite density $\rho = 1.7$ and $2.265\ \mathrm{g\ cm^3}$. The experimental points are, however, only indicated for the lower density in Fig. 14. The W value for Co-60 gamma rays by Niatel[108] is also shown in this figure recalculated to new values of $s^{SA}_{gr,air}$; see Table IV. The values in Table IV are valid for dry air and have therefore been multiplied by 0.997 to yield the value for humid air at about 50% relative humidity at normal temperature and pressure. It is seen that the W-value for 60 Co γ rays constitutes a point in a realistic curve together with those points obtained for electron radiation, provided the lower density $\rho = 1.7\ \mathrm{g\ cm^{-3}}$ is accepted for graphite with electron radiation. The mean energy expended in humid air per ion pair determined in this way depends on the value of $s_{w,air}$. The fairly rapid change, that is 1.5%, in the W value from the Co-60 gamma quality ($\bar{E}_z \approx 0.3\ \mathrm{MeV}$) to 10-MeV electrons may simply depend on an error in $s_{gr,air}$. It can be argued that the largest differences in the various sets of stopping power data are in this energy region (see Figs. 9 and 10) and therefore the largest uncertainty should be expected here. However, an energy dependence in W cannot be excluded. In order to point out that the experimental W-values plotted in Fig. 14 are based on such assumptions a subscript has been used, i.e., W_{ex}.

A similar analysis of D_w/J_{air} determined from Svensson[133] and Mattsson *et al.*[94] may be performed. In these investigations D_w was determined from ferrous sulfate dosimeter measurements. Recalculations to $\varepsilon_m G = 352 \times 10^{-6}\ \mathrm{m^2\ kg^{-1}\ Gy^{-1}}$ were used. A probable energy dependence of $\varepsilon_m G$ was disregarded. Furthermore, recent data, to correct for the perturbation effect in the ionization chamber dosimetry, were applied (see Ref. 135). W/e may then be determined from $D_w/J_{air}s^{SA}_{w,air} = W/e$, taking $s^{SA}_{w,air}$ from Berger and Seltzer.[12]

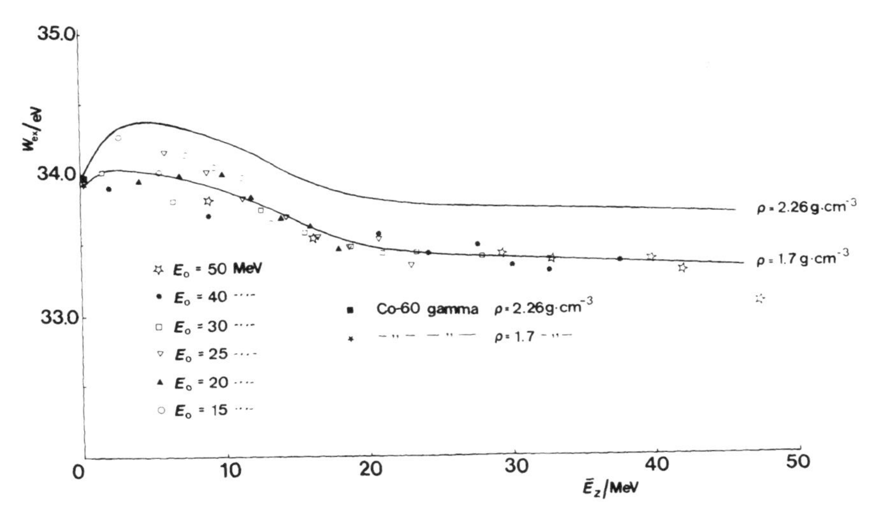

Figure 14. The mean energy expended in humid air per ion pair formed. The experimental points are obtained from the relation $W/e = D_{gr}/J_{air}s_{gr,air}$, where D_{gr}/J_{air} are taken from Ref. 43 for electrons and from Ref. 108 for Co-60 γ rays. The stopping power ratio, $s_{gr,air}$, is taken from Ref. 12, Different sets of data are obtained for the various incident energies, E_0. The mean energy at a depth z, $\bar{E}_z$ is computed using Eq. (41). The W has been given an index ex in order to stress the semiempirical basis of the values. The lower solid line is a fit to the experimental points assuming $s_{gr,air}$ is calculated for $\rho = 1.7\,\mathrm{g\,cm^{-3}}$. The upper solid line would have been obtained for $\rho = 2.26\,\mathrm{g\,cm^{-3}}$.

The data from Svensson and Mattsson *et al.* agree closely with the data in Fig. 14 (see Fig. 15). The W_{ex} "lines" from Fig. 14 are included as a comparison for the two assumptions about the density of graphite (solid lines). The values based on measurements on D_w/J_{air} are scattered between the two solid lines, except at the highest energies. The data from Svensson and Mattsson *et al.* show a somewhat larger spread than those from Domen and Lamperti. One reason is that several different types of medical accelerators have been used as radiation sources in the experiments. The values at the highest energies are from a betatron. It is known that energy-degraded electrons from the collimator appear in this beam, resulting in an underestimate of $s^{SA}_{w,air}$ by possibly 0.5%–1.5%.[82] Furthermore, an energy dependence of $\varepsilon_m G$ as discussed in Section 4.3 would decrease W_{ex} at Co-60 quality and electrons above $E_z \approx 10\,\mathrm{MeV}$; cf. Fig. 13. Considering these facts the best experimental values should be along the broken line, which also falls between tne two "graphite experiment" lines.

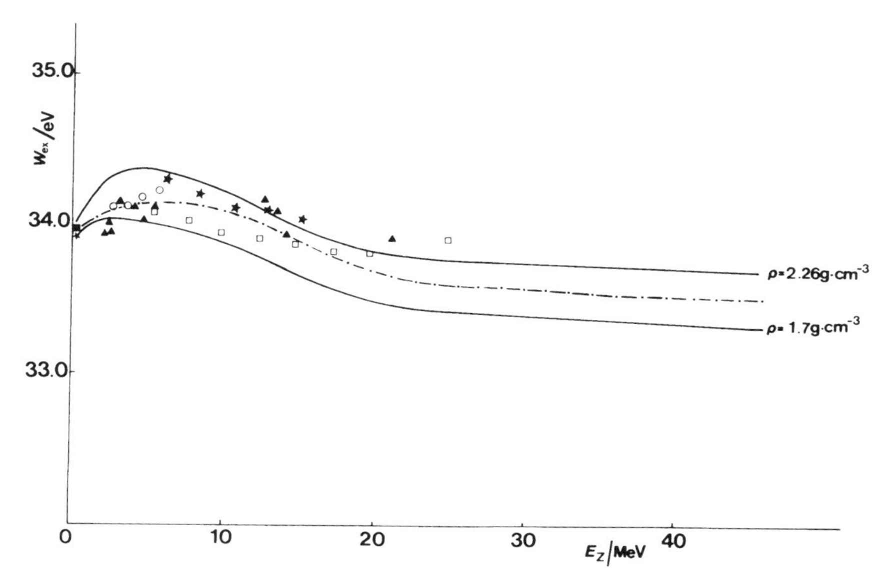

Figure 15. Experimental W values based on $W/e = D_w/J_{\text{air}}s_{w,\text{air}}$, where D_w/J_{air} has been measured by Mattsson *et al.*[94] (▲) at the reference depth for various beam energies and by Svensson[133] at different depths using $\bar{E}_0 = 30$, 20, and 10 MeV. The $s_{w,\text{air}}^{\text{SA}}$ are taken from Ref. 12. The solid lines are explained in Fig. 14 as well as the two experimental data for Co-60 gamma rays (at the lowest energy in the figure).

To avoid problems with different stopping powers used over the years, it would be preferable to investigate the variation of the quantity $\omega_{m,\text{air}} = (W/e)s_{m,\text{air}}$ with energy in a similar way, as $\varepsilon_m G$ is more accurately known than ε_m or G alone. This has the additional advantage that $\omega_{m,\text{air}}$ is exactly the quantity needed for determination of absorbed dose D_m from measurements of J_{air} using the relation $D_m = \omega_{m,\text{air}}J_{\text{air}}$ [cf. Eq. (21c)].

4.7. Absorbed Dose to Mass Ionization Factor $\omega_{m,\text{air}}$

Numerical values of $\omega_{w,\text{air}}$ are given in Figs. 16a and 16b for electron beams between 4 and 50 MeV. These data are based on the W_{ex} from Fig. 15 (broken line), and $s_{w,\text{air}}^{\text{SA}}$ from Berger and Seltzer.[12] A comparison of the present $\omega_{w,\text{air}}$ values with $(W/e)s_{w,\text{air}}$ data from ICRU is shown in Fig. 17. Here a constant value of W/e is assumed for humid air, 33.73 J C^{-1}. Different sets of stopping powers as well as fluence spectra calculations are used. The present $\omega_{w,\text{air}}$ values should be used for a fairly "clean" electron beam, while the data indicated as the dash-dotted line (in Fig. 17) probably apply better for clinical beams, as these beams often are somewhat contaminated by low-energy collimator electrons at small phantom depths (Ref. 82 see also Section 5.1).

A similar analysis was carried out for photon beams. The result is given in Fig. 18 where the $\omega_{w,\text{air}}$ values are computed using the Monte Carlo method.[4,5] The Berger and Seltzer[11] stopping power data were used as input. The solid lines are $\omega_{w,\text{air}}$ (i.e., $s_{w,\text{air}}^{\text{BG}} W_{\text{ex}}/e$) for different monoenergetic photon beams given as a function of the phantom depth. The broken lines represent ^{60}Co gamma rays, and 6- and 21-MV X rays. The spectral distributions of the Bremsstrahlung were taken from Nilsson and Brahme.[110] Andreo's results agree fairly well with those by Nahum[102,103] (recalculated to stopping powers from Berger and Seltzer[11]) given as dotted lines for the monoenergetic photon beams. The deviations seen at low depths are probably due to the larger energy cutoff used by Nahum[102] for the transport of secondary electrons. In the figure experimental values for ^{60}Co gamma rays are included from Niatel.[108] These measurements gave $\omega_{\text{gr,air}}$ (i.e., $D_{\text{gr}}/J_{\text{air}}$) and are corrected to yield $\omega_{w,\text{air}}$ by multiplication by $(S_{\text{col}}/\rho)_{w,\text{gr}}$, taking the collision stopping powers for 0.3 MeV from Berger and Seltzer.[11] This experiment agrees very well with the computed values.

To allow calculation for other materials, the mean energies of the primary electrons at a few depths of interest are given in Table V. The corresponding stopping power ratios are also included. The stopping power ratio at the mean electron energy (i.e., $s_{w,\text{air}}^{\text{H}}$) is about 0.5%–1% lower than these values

5. THE CALIBRATION PROCEDURE

5.1. Background

Several steps are needed in the calibration of a medical electron accelerator. The measuring program after installation of a new accelerator and during its servicable life often includes

1. a survey of different safety aspects (radiation, mechanical, and electrical);
2. investigation of the radiation source, e.g., in terms of some of the parameters tabulated in Fig. 19;
3. investigation of the therapeutic beam, with regard to factors like energy, perturbation, uniformity, and penumbra;
4. absorbed dose calibration often based on the ionization chamber method as indicated in the figure.

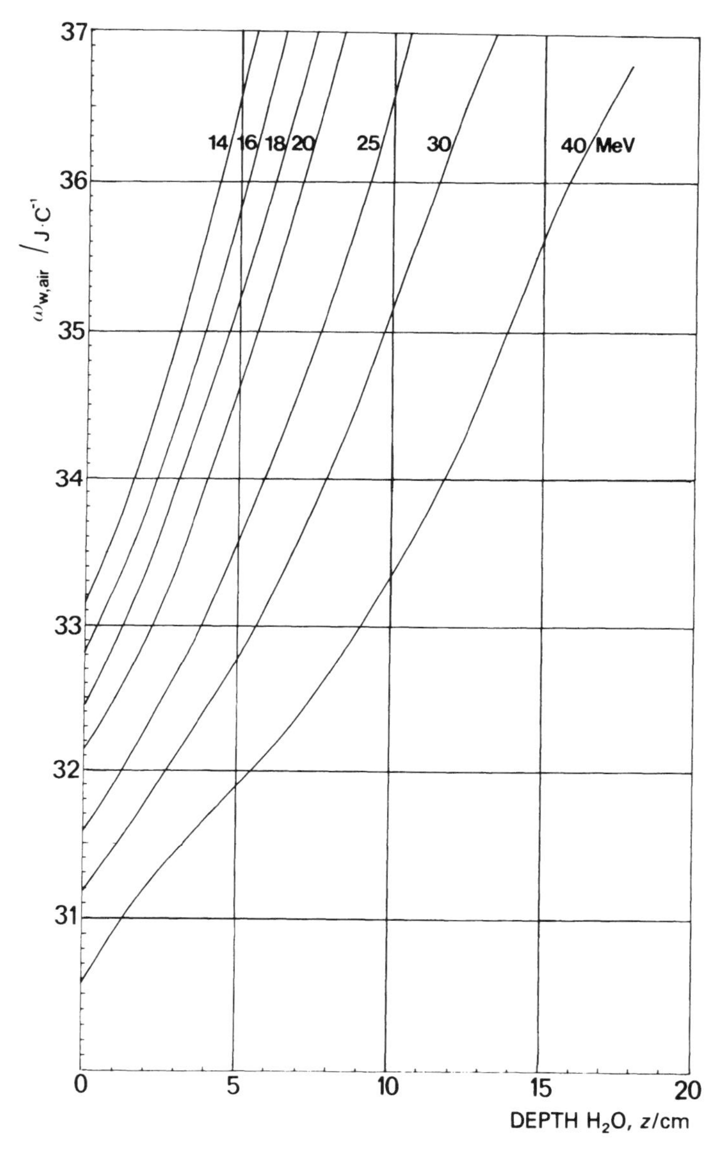

Figure 16. Absorbed dose to mass ionization values for water to air, $\omega_{w,\mathrm{air}}$, as a function of depth for different electron energies at the phantom surface. The data are based on W_{ex} from Fig. 15 (broken line) and $s_{w,\mathrm{air}}^{\mathrm{SA}}$ from Ref. 12.

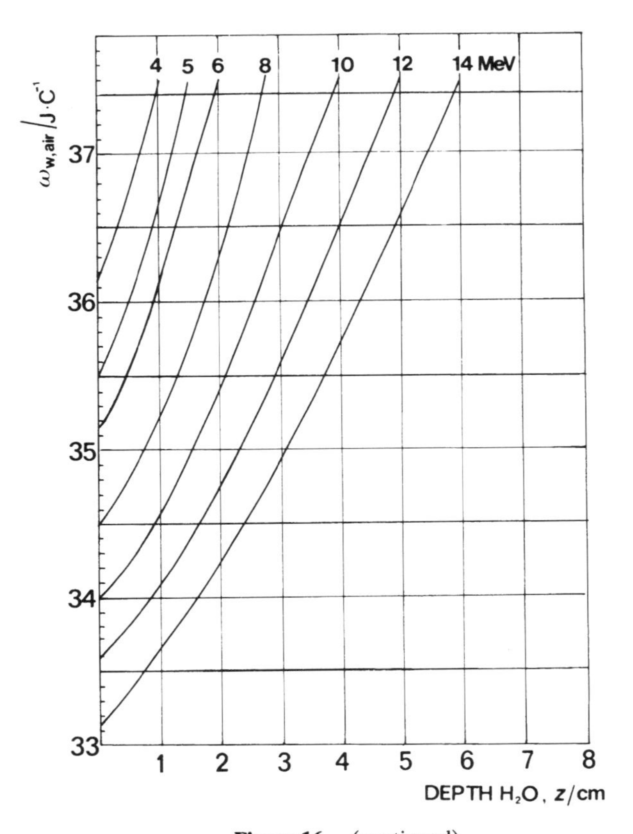

Figure 16. (continued)

Here only those parts of the procedure that are necessary in order to carry out accurate absorbed dose determinations will be discussed.

5.2. Characteristics of Electron Beams for Dosimetric Purposes

The dosimetrical properties of a clinical electron beam depend significantly on the spectral distribution of the electrons. Essential parameters are the mean energy $\bar{E}$ and the most probable energy E_p of the incident electrons. Indices a, i, 0, and z are used to indicate whether the energy is specified for the intrinsic accelerator beam (a), the initial electron beam at the inner side of the vacuum window (i), the beam at the phantom surface (0), and the beam at a depth z inside a phantom (z).

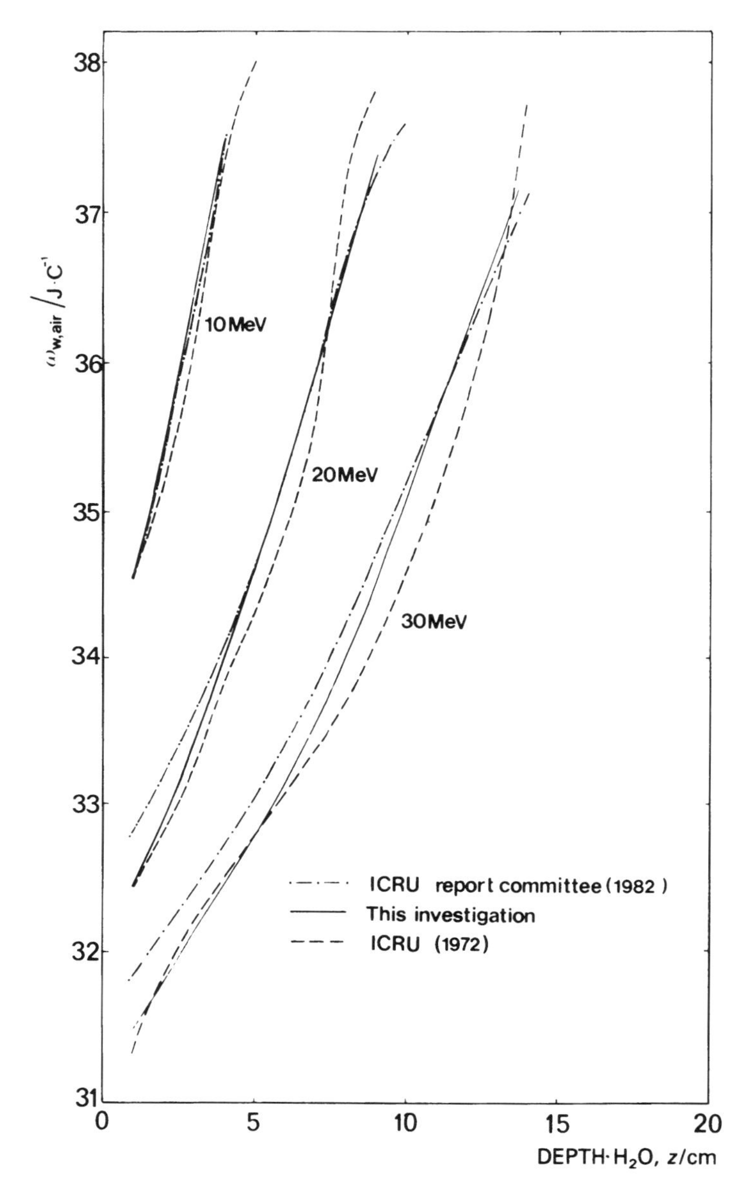

Figure 17. Comparison of $\omega_{w,air}$ as obtained in this report with data based on other assumptions.

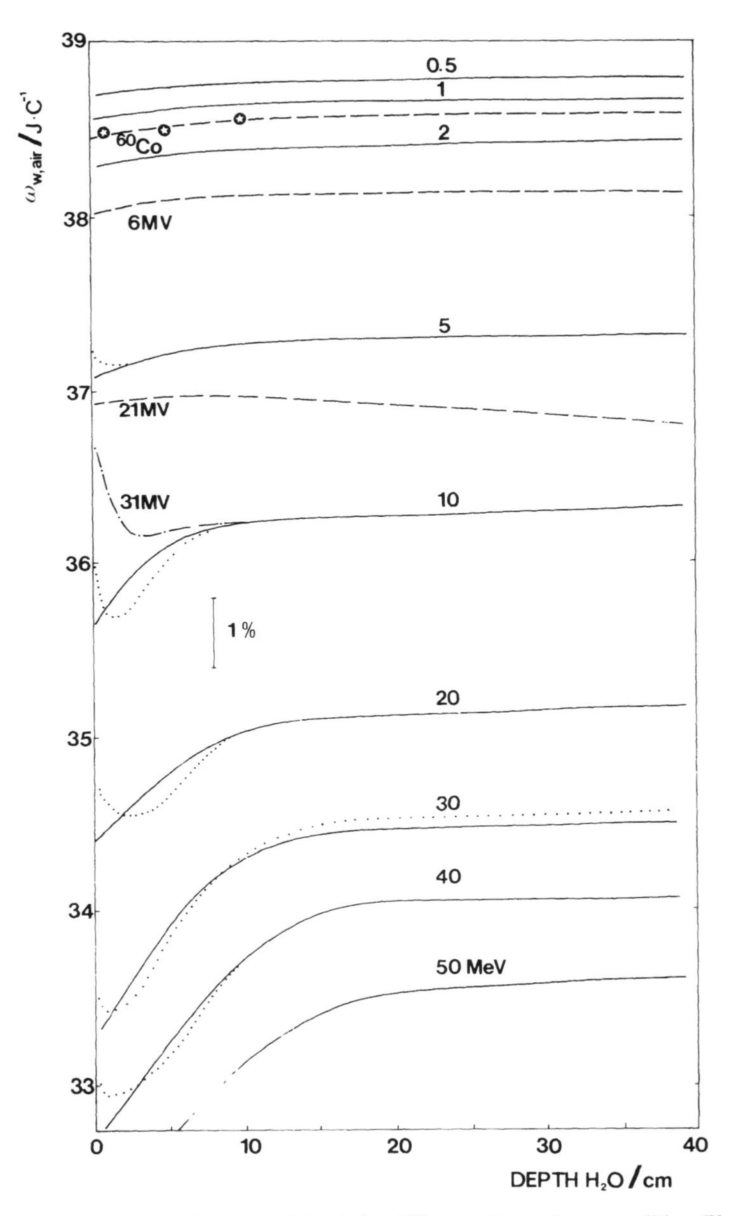

Figure 18. $\omega_{w,\mathrm{air}}$ as a function of depth for different photon beam qualities. The notations are as follows: ——, monoenergetic photons from Ref. 5; · · ·, monoeneregetic photons from Ref. 104; – – –, 6- and 21-MV X rays from Ref. 5; –·–·–, 31-MV X rays from Ref. 102; o o o, experimental points from Ref. 108.

TABLE V
Mean Energies of Primary Electrons, Stopping Power Ratios, and Linear Attenuation Coefficients in Photon Beams

Energy or accelerating potential	Depth (cm)	$\bar{E}_{prim}$ (MeV)	$s^{BG}_{w,air}$	μ_{MC} (m^{-1})
0.5 MeV	0.5	0.142	1.135	5.24
	10	0.125	1.137	
	40	0.102	1.138	
1.0 MeV	0.5	0.351	1.131	3.94
	10	0.312	1.133	
	40	0.265	1.134	
2 MeV	0.5	0.774	1.119	2.75
	1.5	0.759	1.120	
	10	0.695	1.122	
	40	0.620	1.125	
6 MV	0.5	1.08	1.112	2.87
	1.5	1.08	1.112	
	10	0.95	1.116	
	40	0.95	1.117	
21 MV	0.5	2.10	1.092	2.29
	3.5	1.90	1.086	
	10	1.85	1.090	
	40	1.67	1.085	
5 MeV	0.5	2.65	1.087	2.05
	3.5	2.45	1.091	
	10	2.50	1.093	
	40	2.80	1.098	
10 MeV	0.5	5.02	1.047	1.66
	4.5	3.60	1.062	
	10	3.60	1.064	
	40	3.40	1.067	
20 MeV	1.0	9.26	1.016	1.52
	10	6.85	1.033	
	40	6.30	1.040	
30 MeV	1.0	14.7	0.990	1.18
	10	10.3	1.012	
	40	9.2	1.020	
40 MeV	1.5	18.7	0.979	1.14
	10	13.8	0.996	
	20	12.4	1.005	
	40	12.1	1.007	
50 MeV	1.5	23.0	0.970	1.4
	10	17.6	0.985	
	25	15.0	0.997	
	40	14.5	0.999	

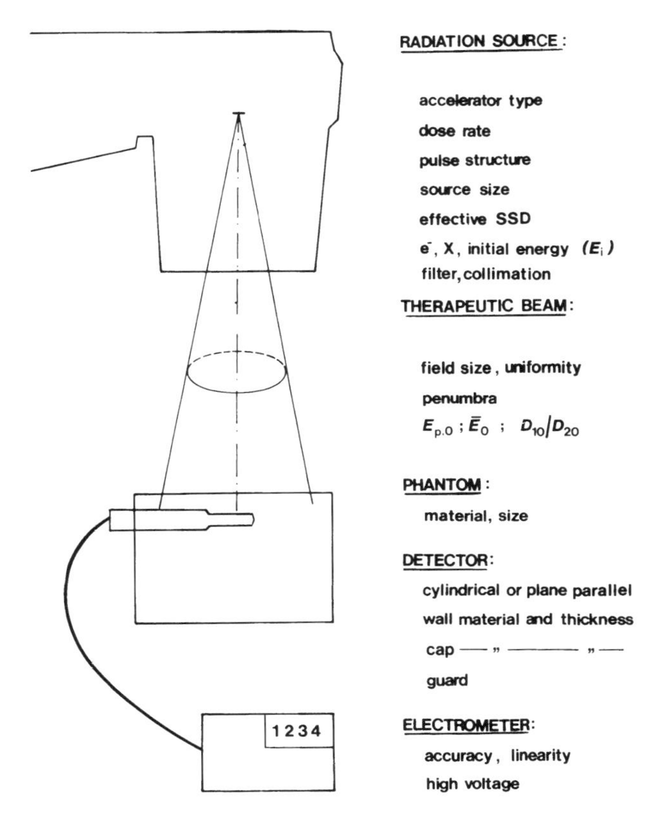

Figure 19. The different factors that affect an absorbed dose calibration.

In theoretical calculations of dose distributions and stopping power ratios it is often assumed that the electron beam is monoenergetic. This is, however, not the case with a clinical beam. For example, the energy spread of the intrinsic accelerator beam is sometimes significant. This energy distribution will often be changed in the beam transport system (i.e., in the bending magnets and associated slits), and further in the scattering material from the inner side of the accelerator window to the patient (phantom).

The most probable energy at the phantom surface, $E_{p,0}$, has in the past often been used as an input parameter in the determination of absorbed dose conversion factors for electron dosimetry, i.e., C_E or $s_{w,\mathrm{air}}$ values (see, e.g., Ref. 74). The recommended values of C_E or $s_{w,\mathrm{air}}$ have, however, been computed assuming a broad monoenergetic beam perpendicularly incident on a phantom. In Ref. 100 the use of $\bar{E}_0$ as an input parameter was recommended as this was considered to give more correct values of $s_{w,\text{air}}$, at least at the reference depth.

It has been shown that $\bar{E}_0$ is correlated to the half value depth, R_{50}, of the central axis depth dose curve[26] according to

$$\bar{E}_0 = k_1 R_{50} \tag{45}$$

where $k_1 = 2.33\,\text{MeV}\,\text{cm}^2\,\text{g}^{-1}$. The NACP[100] has described a simple procedure for the measurement of R_{50} from a depth ionization curve, which can be recommended.

The most probable energy at the phantom surface $E_{p,0}$ can be obtained from the practical range R_p:

$$E_{p,0} = k_2 + k_3 R_p + k_4 R_p^2 \tag{45}$$

where $k_2 = 0.22\,\text{MeV}$, $k_3 = 1.98\,\text{MeV}\,\text{cm}^{-1}$, and $k_4 = 0.0025\,\text{MeV}\,\text{cm}^{-2}$.[29,100]

Johansson and Svensson[82] investigated six different types of accelerators and found that $E_{p,0} - \bar{E}_0$, as determined with the range formulas, was as large as 2 MeV in the energy range 17–22 MeV. Using $E_{p,0}$ instead of $\bar{E}_0$ would therefore have given too low $s_{w,\text{air}}$, in some cases by about 1%. Furthermore, they found that $\bar{E}_0$ values determined from R_{50}, were probably overestimated, the reason being that the relation between R_{50} and $\bar{E}_0$ holds only if the electrons have been degraded in energy when passing through materials in the beam (see Ref. 26), like scattering foils and vacuum windows, but does not fully apply if there is a large energy degradation and scattering of electrons in the collimators. Such electrons will reach the phantom and may be fairly important, at least up to the depth of dose maximum, as shown by Lax and Brahme.[85] Their presence has therefore only minor influence on the 50% depth.

In order to investigate the consequence for absorbed dose determinations Johansson and Svensson[82] measured $\omega_{w,\text{air}}$, i.e., D_w/J_{air}. D_w was measured with the ferrous sulfate dosimeter and J_{air} with a plane-parallel chamber. They found that $\omega_{w,\text{air}}$ at the reference depth for a given value of $\bar{E}_0$ (about 20 MeV) determined from R_{50} varied by up to 4% for different accelerators. The highest values were obtained using accelerators with cone types of collimators, since cones give a large contamination of low-energy electrons.

From the results of these experiments it is obvious that the methods applied today often overestimate the mean energy, resulting in an overestimation of $\omega_{w,\text{air}}$ ($s_{w,\text{air}}$) of 0.5%–1.5%, with most accelerators at about $\bar{E}_0 = 20\,\text{MeV}$. This figure can even be larger for some beams. A further increase by about 1% results if $E_{p,0}$ is used instead of $\bar{E}_0$ as an input parameter.

It could thus be concluded that the ionization chamber dosimetry is rather uncertain for beams contaminated with a large contribution of low-energy electrons and that absorbed dose calibrations for such beams should be carried out with a less energy-dependent method, e.g., ferrous sulfate dosimeter or a liquid ionization chamber. It is, therefore, important to be able to analyze the beam quality to find out if there is a large contamination of low-energy electrons. This may be done by studying the shape of the central axis depth absorbed dose curve. The NACP[100] has recommended some minimum values of the dose gradient. These recommendations have been set in order to assure an acceptable beam performance for radiation treatment. It seems, however, that if these recommendations are fulfilled then also an acceptable accuracy is obtained in the dosimetric constants, i.e., an uncertainty below about 3% (see Ref. 134). This is, of course, a fairly large uncertainty and simple methods other than those based on ionization dosimetry ought to be further developed.

5.3. Characteristics of Photon Beams for Dosimetric Purposes

It is not simple to define a unique radiation quality parameter to use in dosimetry of photon beams from therapy accelerators. For instance, Rawlinson and Johns[118] found that the central axis depth–dose curve from a linear accelerator at 25-MV X rays was very similar to that from a betatron for 16-MV X rays. The reasons were attributed to differences in the target and flattening filter.

The main features influencing the quality of photon beams are as follows:

1. E_i (the electron energy at the target), and possible energy spread Γ_i. Generally, the intrinsic distribution is very narrow for betatrons, $\Gamma_a < 20$ keV,[57] and microtrons, $\Gamma_a < 40$ keV,[28] and broader for linear accelerators, particularly of the standing wave type, where Γ_a is around 10% (i.e., of $E_{p,a}$) and the traveling wave type, where Γ_a is about half this value.[39,140,143]

2. For a betatron, the target could be fairly thin as electrons penetrating the target are turned away by the magnetic field. For other medical accelerators, the electrons are instead completely stopped in a thick target. It is clear that the mean energy of the generated photons is greatest in a thin target. A composite target, with a thin layer of high atomic number (Z) material followed by a low-Z electron stop, falls between the thin and thick target situation (see Ref. 27). The high-energy electrons here mainly produce X rays in the first layer. The remaining electron energy is dissipated in collision losses in the low-Z material without stopping the high-energy bremsstrahlung produced in the first layer.

3. Depending on the relative location of the bremsstrahlung spectrum and the absorption minimum of the flattening filter, the spectral distribution of the beam can be influenced considerably. Prodgoršak *et al.*[116] showed that a substantial hardening of the photon spectrum was obtained by replacing a lead flattening filter by one of aluminum at 25 MV. However, this will certainly introduce problems if uniform dose distribution is the aim at all phantom depths. The reason is that the mean photon energy decreases with the distance from the central ray, in an unflattened photon beam. This effect will be magnified with a filter that increases the mean photon energy (i.e., filter more low-energy photons than high-energy photons, particularly in the central part of the field, where the filter is thickest). The penetration of the photons is therefore greatest at the beam axis. Even if an optimization of the uniformity is achieved at one phantom depth, poor uniformity may be obtained at other depths. The manufacturers make different types of compromises to achieve both an acceptable uniformity and depth–dose distribution. This fact means that the spectral distributions may differ considerably for accelerators using the same accelerating potential.

4. The beam may be contaminated by secondary photons and electrons from the flattening filter, the collimator system, and the air. This influence is particularly large in the dose buildup region of broad low- and high-energy photon beams and is at low energies mainly due to Compton electrons from the air.[109] The influence from scattered photons is generally of the order of 3% of the primary photon dose.[6,110]

It appears from the discussion above that the maximum acceleration potential, U, or maximum photon energy $h\nu_{\max}$, in the X-ray spectrum gives an insufficient description of the beam quality. This is also the case from a dosimetric point of view. The increase in C_λ by replacing an aluminum filter by a lead filter was, for instance, calculated by Nahum[102] to 1% at 31 MV.

It was, therefore, suggested by NACP[100] to base the choice of stopping power ratio on the slope of the exponential portion of the depth–dose curve as characterized by the D_{10}/D_{20} dose ratio rather than the acceleration potential U. This dose ratio reflects the attenuation coefficient of the photon spectrum in the beam and should therefore be better related to its dosimetric properties that depends on μ_{en} (cf. Eq. (18)]. In fact, if both D_{10} and D_{20}, or more generally D_{z_1} and D_{z_2}, are located on the pure exponential falloff portion of the depth–dose curve, the practical attenuation coefficients (cf. Refs. 27 and 61) can be obtained from

$$\mu_p = \frac{\ln(D_{z_1}/D_{z_2})}{z_2 - z_1} \tag{47}$$

Of course, this practical attenuation coefficient includes the inverse square decrease in photon dose with distance from the effective source, but this can easily be removed by elementary calculations:

$$\bar{\mu} \simeq \mu_p - \frac{2}{s_{\text{vir}} + (z_1 + z_2/2)} \tag{48}$$

where $\bar{\mu}$ is the mean attenuation coefficient for the field size and photon spectrum at hand (cf. Refs. 121 and 27). If D_{10}/D_{20} data are used, the negative factor in Eq. (48) becomes $1.74\,\text{m}^{-1}$ assuming $s_{\text{vir}} = 100\,\text{cm}$.

To illustrate the interrelation of various types at attenuation coefficients in water, the most important ones are plotted in a single diagram as a function of photon energy in Fig. 20. The total (μ) and the energy (μ_{en}) absorption coefficients are taken from data by Hubbel.[72] As discussed by Nilsson and Brahme,[111] the mean attenuation coefficient for various field sizes should fall within these bounds. This is indeed the case with the various experimental data also included in the figure.

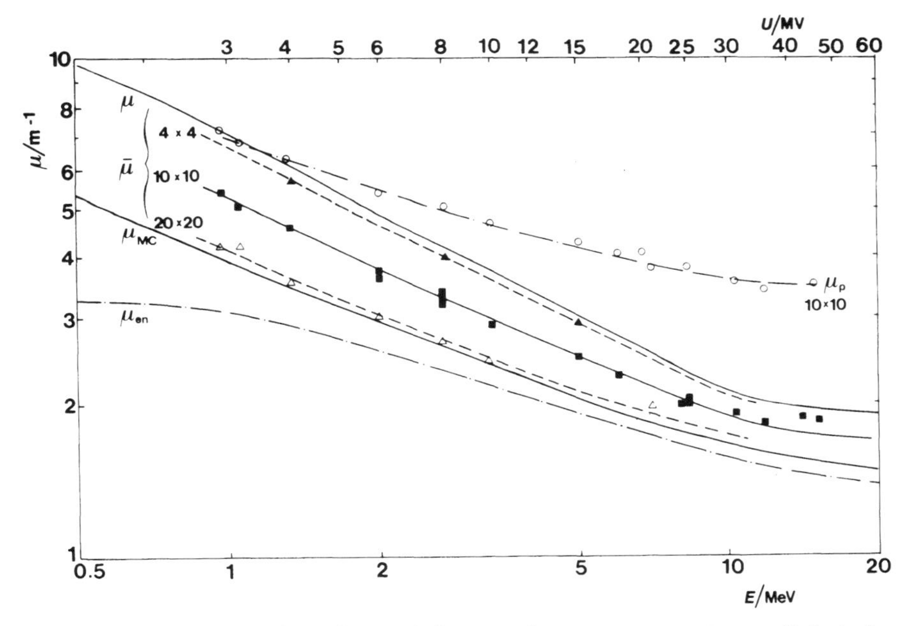

Figure 20. The energy dependence of the most important attenuation coefficients for photon beams in water. The experimental data points for μ_p and $\bar{\mu}$ are mainly taken from Ref. 27, μ_{MC} is from Ref. 5, and μ_{en} and μ from Ref. 72. The lower scale is for monoenergetic photons (μ_{MC}, μ_{en}, and μ). The upper scale for bremsstrahlung spectra is shifted a factor of 3 relative to the monoenergetic scale.

It is seen that, for small fields (<4 cm across), the total attenuation coefficient is approached, whereas for large fields (>20 cm across) the curve is close to the energy absorption coefficient curve. For comparison, the attenuation coefficient in infinite parallel beams calculated by the Monte Carlo method (μ_{MC}) is included. As should be expected, it practically coincides with the experimental broad beam attenuation data. The increasing difference between μ_{MC} and μ_{en} at low photon energies is due to the increased multiple scatter and "local" absorption of the scattered photons that are disregarded in μ_{en}. It is also seen from Fig. 20 that $\bar{\mu}$ for $10 \times 10\,cm^2$ fields falls approximately halfway between μ and μ_{MC}.

The experimental bremsstrahlung points are plotted at a monochromatic energy equal to one third of the acceleration potential U in MV (see the upper scales). This potential corresponds precisely to the energy of the electrons incident on the target and thus the energy of most energetic photons $h\nu_{max}$. To illustrate the influence of beam divergence, the μ_p values for a $10 \times 10\,cm^2$ field are also included in the diagram. Naturally this curve represents the primary experimental data and the $\bar{\mu}$ values are obtained after application of Eq. (48).

Based on this information, it would be most natural to relate the dosimetric properties of photon beams to $\bar{\mu}$ in broad beams, as this parameter is closest to the dosimetrically important quantity μ_{en}. However, owing to the well-behaved relation between $\bar{\mu}$ for large and small fields and μ_p, it might be more convenient to continue the use of standard depth–dose data at the standard field size of $10 \times 10\,cm^2$. It is therefore recommended to use μ_p determined for a field size of $10 \times 10\,cm^2$ and an SSD of 100 cm as a beam quality parameter for choice of dosimetric quantities. This has the advantage that previously used D_{10}/D_{20} values can be used simply by using Eq (45).

For this purpose the $\omega_{w,\mathrm{air}}$ value for different photon beams is plotted in Fig. 21 as a function of μ_p. The dashed line pertains to monoenergetic photon beams, whereas the solid line represents the typical values for ordinary bremsstrahlung spectra. Both curves are based on Monte Carlo calculations by Andreo.[5] Because the data are plotted as a function of μ_p, the dependence of $\omega_{w,\mathrm{air}}$ on the spectral shape and filtration is considerably reduced. It is natural that the solid line falls below the dashed line for monoenergetic beams, as the low-energy photon component of the bremsstrahlung spectrum decreases the beam penetration more than it increases the $\omega_{w,\mathrm{air}}$ value owing to the slow variation of $\omega_{w,\mathrm{air}}$ and $s_{w,\mathrm{air}}$ with energy when the density effect becomes negligible at low energies. The dashed line thus represents the maximum possible deviation for extremely hard photon spectra. The solid line could thus be recommended for use over a wide range of bremsstrahlung

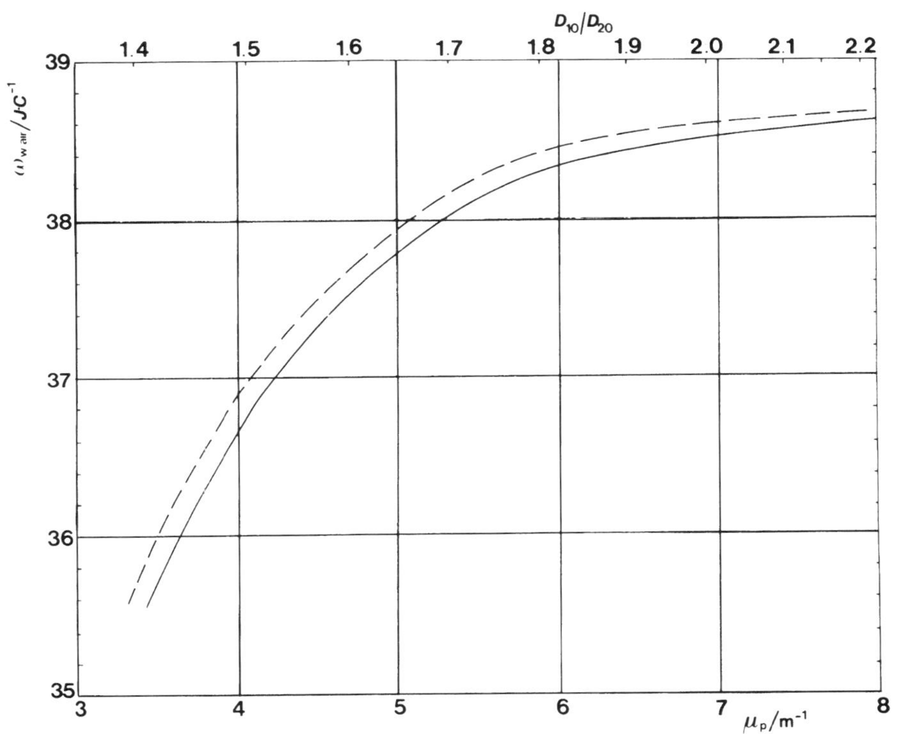

Figure 21. The relation between $\omega_{w,\mathrm{air}}$ and the practical attenuation coefficient μ_p. The dashed line holds for monoenergetic photons, whereas the solid line is representative for most common bremsstrahlung spectra.

spectra. Only when the spectrum is unusually soft, e.g., after thick high atomic number filters at high photon energies, or when it is unusually hard after a thick low atomic number filter, will the true values fall slightly below and above the solid line, respectively. The differences should rarely be more than 0.5%.

5.4. Phantom Materials

5.4.1. Background

The general practice in radiation therapy with high-energy photon and electron beams is to state the absorbed dose to the target volume or an organ of interest in units of absorbed dose to water. This is done even though it is well known that the absorbed dose to the tissue in question may differ considerably from the absorbed dose to water.

With the increased use of advanced diagnostic equipment, such as computerized tomography (CT) and to some extent also ultrasound and magnetic resonance imaging, the accurate delineation of the patient geometry has been greatly improved. In addition the CT scanner and also the Compton scatter tomograph give the possibility of determining the density and even the atomic composition of the different tissues. In some cases it may therefore be desirable to reconsider the old practice of stating absorbed dose to water, especially since failure to do so may add to the total uncertainty in dosimetry.

In a recent analysis,[25] it was shown that both muscle and adipose tissues are water equivalent to within 1% from 2 to 40 MeV for electrons and from 0.2 to 20 MV for photons. Deviations of 5% or larger are generally obtained for the absorbed dose to bone as compared to water. Water is thus a very good soft tissue substitute and the natural phantom material for accurate absolute dosimetry in therapy beams. However, in some cases it is more convenient to use solid phantom materials owing to the ease of handling and high geometric reproducibility. When a solid phantom is used to calibrate in units of absorbed dose to water, it is essential that all necessary corrections factors are known, as both the particle fluence and the absorbed dose at dose maximum or the reference point generally are different. When the phantom is made of a solid insulator like polystyrene and polymethylmethacrylate, care should be taken when it is used in electron beams owing to the buildup of long-lasting electrostatic fields inside the insulator that may influence the detector reading. Large errors in dosimetry based on a cylindrical chamber placed in such a preirradiated phantom have been demonstrated.[54,92]

5.4.2. Electrons

With a given electron fluence incident on a uniform medium, the absorbed dose at the reference point (dose maximum) in the material will generally differ from that in water. This is due to two principal effects. First, the mass stopping power of the medium will generally be different, resulting in different absorbed doses for a given electron fluence. Secondly, the mass scattering power may be different, resulting in a difference in the buildup of the fluence of primary electrons.

The ratio of the absorbed doses in a medium to that in water may under assumption of equilibrium in the transport of secondary electrons be written

$$\frac{D_m}{D_w} = \frac{\int (\Phi_E)_m \left(\frac{S_{\text{col}}}{\rho}\right)_m dE}{\int (\Phi_E)_w \left(\frac{S_{\text{col}}}{\rho}\right)_w dE} = \frac{\Phi_m \left(\frac{\bar{S}_{\text{col}}}{\rho}\right)_m}{\Phi_w \left(\frac{\bar{S}_{\text{col}}}{\rho}\right)_w} \tag{49}$$

where the integrals are taken over the spectrum of primary electrons, Φ_E, and the mean collision stopping powers are defined according to

$$\left(\frac{\bar{S}_{\text{col}}}{\rho}\right)_m = \frac{\int (\Phi_E)_m \left(\frac{S_{\text{col}}}{\rho}\right)_m dE}{\int (\Phi_E)_m \, dE} \tag{50}$$

There are two main sources for the difference in fluence at dose maximum. First, because different depths of dose maximum in different media, the inverse square law will produce different fluence reductions at dose maximum in divergent beams. Secondly, because of differences in linear scattering powers the fluence at dose maximum will differ owing to differences in the angular distribution of the electrons.

According to Brahme and Lax,[25] the ratio of the primary electron fluences at dose maximum may be written

$$\frac{\Phi_m}{\Phi_w} = 1 + \tfrac{1}{2}(T_m R_{100,m} - T_w R_{100,w}) - 2\left(\frac{R_{100,m} - R_{100,w}}{s_{\text{vir}}}\right) \tag{51}$$

where T is the linear scattering power, R_{100} the depth of dose maximum, and s_{vir} the distance to the viritual point source.[24]

The fluence factor given by Eq. (51) is plotted in Fig. 22 as a function of the most probable energy at the surface for characteristic tissues and some phantom materials of interest. Also included in the diagram are experimental data points from measurements by Mattsson *et al.*[93] It is seen that for therapeutic electron beams at least in the range 3–30 MeV, PMMA (polymethylmethacrylate) has the interesting property that the response of a small Bragg–Gray detector at dose maximum agrees within half a percent with its response in water. This is essentially because the linear scattering power of PMMA and water are very close and thus the electron fluence at dose maximum is practically the same even though the absolute dose level is about 3% lower in PMMA due to its lower mass stopping power. For graphite (C) and polystyrene (PS) over a wide range of energies, the fluence is about 1% higher and lower, respectively, as compared to water. It should be pointed out that small high atomic number additions may change the properties of the phantom material. This is the case with TiO_2 in white polystyrene, which therefore is more water equivalent than clear polystyrene.

5.4.3. Photons

For a given incident photon energy fluence differential in energy, Ψ_E, the ratio of the absorbed dose at secondary electron equilibrium in a

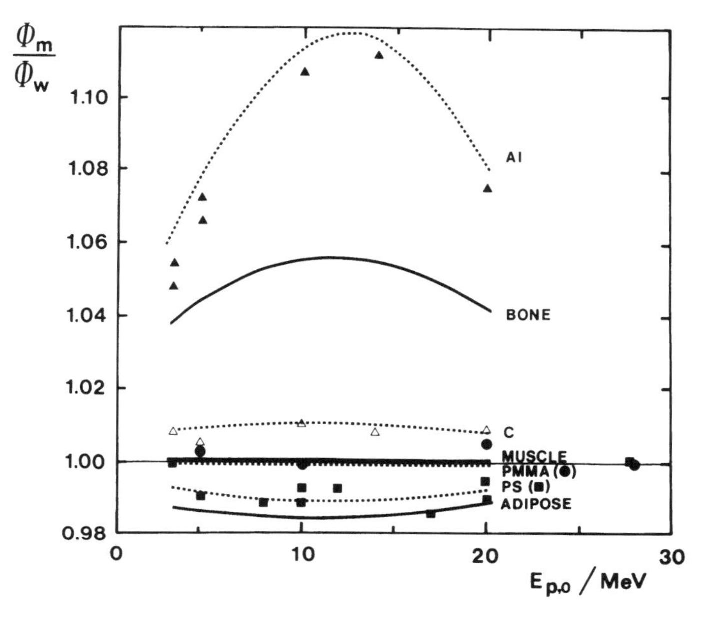

Figure 22. The ratio of the fluence at dose maximum in a medium to that in water as a function of the most probable energy at the surface $E_{p,0}$ (dotted and solid lines from Ref. 25). The experimental points are taken from Ref. 93.

uniform medium to the equilibrium dose in water is given by

$$\frac{D_m}{D_w} = \frac{\int (\Psi_E)_m \left(\frac{\mu_{\text{en}}}{\rho}\right)_m dE}{\int (\Psi_E)_w \left(\frac{\mu_{\text{en}}}{\rho}\right)_w dE} = \frac{\Psi_m}{\Psi_w} \frac{\left(\frac{\bar{\mu}_{\text{en}}}{\rho}\right)_m}{\left(\frac{\bar{\mu}_{\text{en}}}{\rho}\right)_w} \tag{52}$$

where the bar indicates a mean mass energy absorption coefficient averaged over the energy fluence differential in energy:

$$\left(\frac{\bar{\mu}_{\text{en}}}{\rho}\right)_m = \frac{\int (\Psi_E)_m \left(\frac{\bar{\mu}_{\text{en}}}{\rho}\right)_m dE}{\int (\Psi_E)_m \, dE} \tag{53}$$

The ratio of the signal measured with a small Bragg–Gray detector, that is, the ratio of the electron fluence in the two media, is obtained by

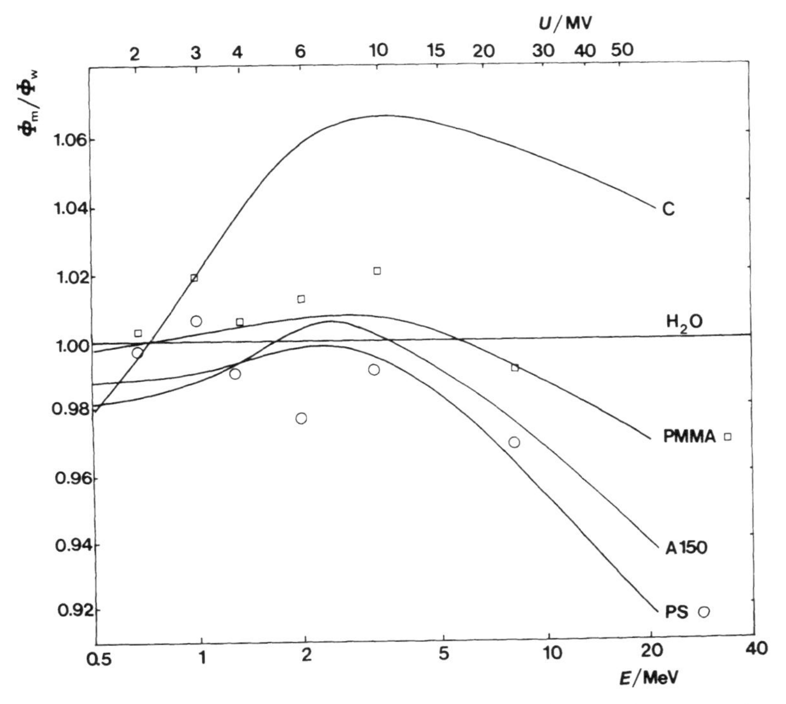

Figure 23. The ratio of the electron fluence at dose maximum in a medium to that in water for different photon beam energies. The experimental data points for PMMA and PS for a $15 \times 15\,\text{cm}^2$ field[107] represent the signal ratio for a Farmer chamber at $5\,\text{g/cm}^2$ depth.

dividing by the respective mean stopping powers according to

$$\frac{\Phi_m}{\Phi_w} = \frac{\Psi_m}{\Psi_w} \frac{\left(\frac{\bar{\mu}_{\text{en}}}{\rho}\right)_m \left(\frac{\bar{S}_{\text{col}}}{\rho}\right)_w}{\left(\frac{\bar{\mu}_{\text{en}}}{\rho}\right)_w \left(\frac{\bar{S}_{\text{col}}}{\rho}\right)_m} \tag{54}$$

This ratio is plotted in Fig. 23 using stopping powers determined at the mean energy of primary electrons [cf. Eq. (39)] taken from Table III. The fluence ratio was evaluated at dose maximum with correction for the inverse square attenuation according to

$$\frac{\Psi_m}{\Psi_w} \approx 1 + 2\left(\frac{R_{100,w} - R_{100,m}}{s_{\text{vir}}}\right) \approx 1 + \frac{2R_{100,w}}{s_{\text{vir}}}\left[1 + \frac{(r_0/\rho)_m}{(r_0/\rho)_w}\right] \tag{55}$$

where r_0 is the continuous slowing down range in g cm^{-2} [cf. Eq. (51) and Ref. 25]. The last step in Eq. (55) is based on the assumption that the buildup depth is a constant fraction of r_0, independent of the material.

As seen by comparing Figs 23 and 22 PMMA is almost as water equivalent for photons as for electrons, at least from ^{60}Co to about 20 MV. Nath *et al.*[107] showed that the response of a Farmer-type chamber at a depth of 5 g cm^{-2} in PMMA is the same as in water within 1% in the same energy range. Nilsson and Brahme[111] showed by electron transport calculations based on Eq. (6) that the ratio between the absorbed dose at dose maximum and the collision kerma at the phantom surface for ^{60}Co gamma rays was the same within 0.1% for water, air, and aluminum, at infinite SSD. This indicates that in photon beams, energy fluence differences are mainly due to differences in the inverse square corrections at the depth of dose maximum.

5.5. Absorbed Dose Determination by the Ionization Method

5.5.1. Background

A determination of the absorbed dose in high-energy photon and electron beams by the user is, in principle, carried out as a two-step procedure. First the ionization chamber is calibrated at a reference quality by the user at a standards laboratory, and then the chamber is used to determine the absorbed dose with the user's beam. A number of conversion and correction factors have to be applied. Different sets of factors are needed depending on what physical quantity the calibration refers to (X, K_{air}, etc.), on the calibration geometry (free in air, in phantom, etc.), and on the design of the chamber (size, shape, wall material). Factors may be needed to convert from one physical quantity (X or K_{air}) at the calibration quality to a different quantity of interest for the user, i.e., $s_{w,air}$; also, these are dependent on the irradiation geometry and chamber design.

The ICRU reports No. 14[73] and No. 21[74] have given some sets of factors to be applied under special conditions (see Fig. 2). Similar factors were recommended by the national protocols, i.e., SCRAD,[123] HPA,[70] and NACP.[99] The uncertainty in using these procedures seems, however, to be fairly large, as none of the protocols gives a clear description of under what conditions these factors are to be applied. For instance, the wall material in the chamber is seldom water equivalent, as assumed, but not stated, in the ICRU photon report (No. 14),[73] or air equivalent, as assumed, but not stated, in the electron report (ICRU No. 21).[74]

A discussion of the influence of the wall material on the calibration and on the measurement of absorbed dose was given by Mutsuzawa,[98]

Greening,[60] and Almond and Svensson.[2] It was shown that just an incorrect assumption regarding the wall materials often gave errors of 3% or more.

A very careful analysis of all the factors involved was undertaken by Loevinger.[86] His analysis covered different calibration assumptions, i.e., exposure, air-kerma, and absorbed dose to water calibration for ^{60}Co gamma beams. It appears from this analysis that the dosimetrical procedure, if not restricted to certain well-defined geometries (e.g., chamber design), will be very complicated and that some simplifications are needed.

The NACP[100] described a procedure to be applied for specially designed ionization chambers. Cylindrical chambers, with a cavity diameter between 4 and 6 mm, were recommended for all high-energy photon beams and electron beams above 10 MeV. The chamber wall and buildup caps should be of the same material. Factors were given for walls of graphite, and air equivlent and water (≈tissue) equivalent materials. For electron beams with $\bar{E}_0$ below 10 MeV, a specific plane-parallel chamber was recommended, in order to reduce the uncertainty in the perturbation correction at these qualities.[101]

It was shown that very good agreement was obtained in the absorbed dose values as determined from measurements according to the NACP[99,100] and from ferrous sulfate measurements using $\varepsilon_m G = 352 \times 10^{-6}\,\mathrm{m^2\,kg^{-1}\,Gy^{-1}}$, generally within 1% (see Ref. 82). However, it should be pointed out that some of the factors recommended in the NACP (e.g., $s_{w,\mathrm{air}}$) are in error by several percent if newer data from Berger and Seltzer[11] are adopted. In most cases the error is canceled by the product of mutually compensating factors in the dose equation. It is therefore important that the stopping power ratios are not changed, if more correct values are made available, without also changing other factors in the protocol such as k_m (see Table VI) and W/e. These factors will be discussed below in order to identify their interrelations.

5.5.2. Calibration Quality

At the standards laboratory much experience has been gained in determining exposure in a ^{60}Co gamma beam. The exposure standards from different laboratories generally agree within one or two tenths of one percent. In order to obtain a uniform dosimetry throughout the world this very good agreement should be used.

The exposure is determined from the following relationship (see, e.g., Ref. 87):

$$X_c = J_{\mathrm{air}} \prod_i K_i \tag{56}$$

TABLE VI
The Calibration Procedure by NACP[100,101] and That Recommended in the Text

	NACP[100,101]	This paper
Calibration at calibration quality (index c)	*Geometry* Co-60 γ rays SSD = 100 cm, to chamber center; field size 10 cm × 10 cm; free in air *Quantity calibrated* $K_{\text{air},c}$; air-kerma derived from $K_{\text{air},c}(1-g) = X_c W/e$ *Calibration factor* $N_K = \dfrac{K_{\text{air},c}}{M_c}$ Derived factor: $N_D = \dfrac{\bar{D}_{\text{air},c}}{M_c} = N_K(1-g)k_m k_{\text{att}}$	*Geometry* As NACP[100,101] *Quantity calibrated* X_c *Calibration factor* $N_x = \dfrac{X_c}{M_c}$ Derived factor: $N_J = \dfrac{J_{\text{air}}}{M_c} = N_x k_m k_{\text{att}}$
Assumption	$N_D = \dfrac{\bar{D}_{\text{air},c}}{M_c} = \dfrac{D_{\text{air},u}}{M_u}$	No assumptions
	Geometry Water phantom; chamber center at the reference point; chamber without build-up cap	*Geometry* As NACP (1980)
Absorbed dose measurement at users' quality (index u)	*Dose determination* $D_{w,u} = \bar{D}_{\text{air},u}(s_{w,\text{air}})_u p_u$ $= N_D M_u (s_{w,\text{air}})_u p_u$	*Dose determination* $D_{w,u} = J_{\text{air}}(\omega_{w,\text{air}})_u (p_{w,\text{air}})_u$ $= N_J M_u (\omega_{w,\text{air}})_u (p_{w,\text{air}})_u$

where X_c is the exposure at the calibration radiation quality; J_{air} is the mass ionization, i.e., charge of one sign liberated in the air cavity per unit mass of air [see Eqs. (4b) and (4c)]; and $\prod_i K_i$ is the product of the various correction and conversion factors. The factors, K_i, convert J_{air} into exposure. Some of these factors are trivial, e.g., the correction for recombination losses, and are therefore not discussed here. Other factors are of more fundamental character. Corrections thus have to be applied for the non-air-equivalence of the chamber wall, given the symbol K_m, and the attenuation of the photons in the buildup cap and chamber wall, given the symbol K_{att}. The material in the standard chamber is generally graphite, as this material is fairly "air equivalent." K_m is calculated from [see Eq. (55) and Refs. 18 and 87]:

$$K_m = (\mu_{\text{en}}/\rho)_{\text{air,gr}} s_{\text{gr,air}}^{\text{SA}} \tag{57}$$

K_{att}, as applied below, does not include all the attenuation in all the wall as the electrons giving ionization are produced up-stream in the chamber wall and buildup cap. K_{att} depends on the thickness of the wall and cap and also on its shape. K_{att} can be computed, e.g., using the Monte Carlo method,[106] electron transport calculations,[111] or measurements using an extrapolation method.

The exposure X_c, for a given irradiation time, is in this way determined for a specified point in the ^{60}Co gamma beam. The very good agreement in exposure, as determined by different standards laboratories, relies on the fact that K_m is computed in the same way.

However, the user is not really interested in a calibration of the ionization chamber in exposure. The main interest is instead to make it possible to determine the relation between the detector signal and the mass ionization, J_{air} [Eqs. (4a–c)]. In practice, dose determinations are based on a knowledge of $\bar{D}_{air}$. The Bragg–Gray relation Eq. (21) or its generalized form Eqs. (26) and (27) are then applied to determine absorbed dose to water at the user's quality (see Table VI).

It may be possible for the user to measure, Q_{air}, directly, but the mass of air, m_{air}, is more complicated to determine. In effect the "effective mass," $m_{air,eff}$, is what is determined during calibration. However, if a graphite chamber of design similar to that used by the standards laboratory is available, then it should be very simple to make an exposure calibration of the chamber at the standards laboratory and applying Eq. (56) to determine mass ionization per meter reading J_{air}/M_c $(=N_J)$.

As the chamber often is of a very different design from that at the standards laboratory, other correction factors must often be supplied. In the NACP the inverse of the K_m and K_{att}, given the symbols k_m and k_{att}, respectively, are reported for some types of chambers. It is thus obvious that k_m for a graphite chamber is the inverse value used by the standards laboratory. The cavity size dependence of $s^{SA}_{air,gr}$ can generally be neglected, as the variation for ordinary cavity sizes is below 0.2%, cf Boutillon and Niatel.[18] An error in K_m is therefore canceled out provided that the same physical data are used both by the standards laboratory and the user.

In the Nordic protocol an absorbed dose to air ionization chamber factor

$$N_D = \frac{\bar{D}_{air,c}}{M_c} = \frac{J_{air}(W/e)}{M_c}$$

was defined. Here $\bar{D}_{air,c}$ is the mean absorbed dose to air inside the air cavity of the chamber [see Eq. (4b)]. J_{air} and therefore also $\bar{D}_{air}$ may be determined from Eq. (55), i.e., $J_{air} = k_m k_{att} X_c$, again leaving out the normal corrections. The Nordic standards laboratories give the ionization chamber calibration in air-kerma rather than exposure. The relation between these two quantities is

$$K_{\text{air},c}(1 - g) = X_c \frac{W}{e} \tag{58}$$

where g is the fraction of the charged particle energy lost to bremsstrahlung in air ($g = 0.004$ at ^{60}Co; see Boutillon[16]). Thus N_D may be written

$$N_D = \frac{K_{\text{air},c}(1 - g)k_m k_{\text{att}}}{M_c} \tag{59}$$

It is seen [Eq. (58)] that the numerical value of W/e is needed at the calibration quality in order to determine N_D. As the analysis in Section 4 indicates that this value still is fairly uncertain, a more direct way would be to determine a mass ionization factor $N_J = J_{\text{air}}/M_c = X_c k_m k_{\text{att}}/M_c$ at the calibration quality. Furthermore, $(\omega_{w,\text{air}})_u$ for the users quality can be used which takes a possible energy dependence of W/e into consideration (see Section 4.5 and Table VI).

5.5.3. Absorbed Dose to Water Determination at the User's Quality

The calibrated ionization chamber can now be used to determine the absorbed dose to water utilizing the Bragg–Gray equation (see Table VI). The NACP[100,101] used a total perturbation factor, p_u, that corrects both for the fluence perturbation of the wall and cavity as shown in Fig. 5 (Section 3.4.4) and for the shift of effective point of measurement as the center of the chamber is placed at the reference point. Experimental values of p_u have been recommended by NACP[100] and are given for a cylindrical chamber that has an inner diameter of 5 mm with a 0.5-mm-thick graphite or water equivalent wall. The modified procedure, described in Table VI, uses the equation

$$D_{w,u} = N_j M_u (\omega_{w,\text{air}})_u (p_{m,\text{air}})_u \tag{60}$$

Values of $(\omega_{w,\text{air}})_u$ are given in Fig. 16 for electron radiation and in Fig. 18 for photon radiation.

$(p_{m,\text{air}})_u$ is the fluence perturbation correction correcting for these effects indicated in Fig. 5. However, it is here assumed that the ionization chamber is placed with its effective point of measurement at the reference depth. In most practical cases with electron and photon beams the effective point of measurement is located at $0.55r$ in front of the chamber center for cylindrical chambers (see Section 3.4.3.1).

The experimental $(p_{m,\text{air}})_u$ values from Fig. 7 are recommended for electron beams (the broken line). The values are plotted as a function of the mean energy of the primary electrons at the point of measurements. This energy could be derived from Eq. (41). A more simple estimate could, however, be obtained by assuming that the electron energy decrease with 2 MeV per cm penetration in the water phantom. These values can also be used for chambers having thin walls of low atomic number materials other than water as the primary electrons mainly are responsible for the absorbed dose to the air in the cavity and the difference between in- and outflow from the cavity of secondary electrons should be fairly small.

The $(p_{m,\text{air}})_u$ for photon beams are generally assumed to be equal to 1.000 if the chamber wall is made of water equivalent material ($m = w$) and if the chamber is placed with the effective point of measurement at the depth of interest. The wall material (m) is, however, more critical than for electron radiation as the photon generated electrons in the wall may give a large contribution to J_{air}. Almond and Svensson[2] used a simple two-component theory, assuming that a fraction β of the ionization comes from electrons generated in the water and a fraction $(1 - \beta)$ from electrons generated in the wall. Using their approach $p_{m,\text{air}}$ would be approximated by

$$(p_{m,\text{air}})_u = \frac{\beta(s_{w,\text{air}})_u + (1 - \beta)(s_{m,\text{air}})_u[(\mu_{\text{en}}/\rho)_{w,m}]_u}{(s_{w,\text{air}})_u} \tag{61}$$

As an example, in Table VII, $p_{\text{gr,air}}$ values are shown for $\beta = 0.2$ and $\beta = 0.8$. The lower value, $\beta = 0.2$, is relevant for a ^{60}Co gamma beam using a common thickness of the graphite wall, i.e., 0.5–1 mm, while for acceleration potentials >20 MV values of $\beta > 0.8$ should be applied instead. The agreement between theory and experiment is reasonable. A minor systematic error may, however, be introduced in the experiment as it was assumed that the reference chamber made of A-150 was completely water equivalent, which may not be the case (see Fig. 21), particularly if the recent stopping power data from Berger and Seltzer[11] are trusted.

5.6. Calibration Procedures Based on Solid Dosimeters

The ionization chamber is today the base for most dose measurements at a hospital. It is therefore very unfortunate that some fundamental features of this dosimeter always will cause problems resulting in fairly large uncertainties, e.g., the quality dependence and perturbation correction. The ferrous sulfate dosimeter does not have these drawbacks and

TABLE VII
Values of $(p_{gr,air})_u$ Given as a Function of the Beam Quality[a]

Radiation quality	$(p_{gr,air})$ Experimental 0.5 mm wall	Theoretical $\beta = 0.2$	Theoretical $\beta = 0.8$
^{60}Co γ rays	0.975	0.985	0.996
4 MeV	0.978	0.983	0.996
6 MeV	0.980	0.978	0.995
8 MeV	0.982	0.980	0.995
10 MeV	0.984	0.981	0.996
15 MeV	0.989	0.986	0.996
20 MeV	0.992	0.990	0.997
25 MeV	0.993	0.996	0.998
30 MeV	0.995	0.999	1.000
40 MeV	0.997	1.005	1.001

[a] The experimental values have been measured as the ratio of the response from a cylindrical chamber made of A-150 and from a graphite chamber at 10 cm depth in a water phantom. The responses of the chambers were normalized in a high-energy electron beam. The theoretical values were obtained from Eq. (60). The stopping power ratios $s_{m,i}^{BG}$ were calculated for 10 cm depth using data from Berger and Seltzer.[11] The values of μ_{en}/ρ were taken from Hubbell.[71]

seems at present to be the best choice when accuracy is most important. However, this dosimeter system has been in use for many years and has still not become very popular owing to the laborious technique. Therefore there is still a need for an accurate method that is attractive to the hospital physicist.

Recently Domen[40] developed a new type of calorimeter which made it possible to measure the temperature increase at a point of interest in water after it had been irradiated at common dose rates. The method is based on the use of a microthermistor sandwiched between two very thin insulating polyethylene films in a water phantom. With a dose rate of 1 Gy min^{-1} and an exposure time of 3 min it was possible to achieve a standard deviation of about 0.6%. This technique is surely possible to improve. Also it needs to be explained why this calorimeter measure an absorbed dose value 3.5% higher for ^{60}Co gamma rays than if the dosimetry is based on a graphite calorimeter measurement transferred to water with the ionization chamber method; see Domen.[41] This difference is overcome by using the microthermistor in a polystyrene phantom,[42] but then absorbed dose to water is not directly measured.

Still, the transfer from an absorbed dose determination at the standards laboratory to the user's quality must be carried out. If an ionization chamber method is applied, then again most of the uncertainties discussed in Sections 4 and 5 will be introduced. Thus, there is a great need for a precision dosimeter with small energy dependence. Also, the dosimeter should not disturb the electron and photon fluence at the point of interest in the water phantom and should have a high spatial resolution. It seems that the liquid ionization chamber would satisfy these criteria; cf. the transfer of dosimetry carried out in the experiment reported by Mattsson *et al.*[95] and Wickman and Svensson[147] to determine $\varepsilon_m G$ (see Section 4.3).

The liquid ionization chamber was described already by Strahel.[131] However, a more detailed analysis of the physical properties was first carried out by a group in Toulouse (Refs. 34, 90, 91, etc.). A practical chamber design was developed by Wickman,[144,145] who used 2.2.4-trimethylpentane as the radiation sensitive liquid. The chambers by Wickman made high-resolution and precision measurements possible. The corrections for general recombination and temperature were of about the same magnitude as for an air ionization chamber, that is, generally below a few percent. Furthermore, the chambers seemed to have an insignificant quality dependence for electron beams, that is, a constant response per unit absorbed dose independent of the energy at the phantom surface and the depth in the phantom. Measurements with this type of chamber by Svensson *et al.*[139] and Johansson and Svensson[82] confirmed these results and showed, in addition, that the quality dependence for photon beams was fairly small (2% between 6 and 40 MV).

The liquid ionization chamber thus has many advantages for use as a transfer instrument. The chamber has until now been in use in only a few departments. The reason seems to be that the chamber needs some special care in handling. However, recently, Wickman has constructed a chamber that also seems to be sufficiently rugged for use in most departments.[146]

REFERENCES

1. M. Abou-Mandour and D. Harder, Scheinbare Reflexion schneller Elektronen bei streifendem Einfall, *Z. Naturforsch.* **30a,** 265 (1975).
2. P. R. Almond and H. Svensson, Ionization chamber dosimetry for photon and electron beams, *Acta Radiol. Ther. Phys. Biol.* **16,** 177 (1977).
3. A. R. Anderson, A calorimetric determination of the oxidation yield of the Fricke dosimeter at high dose rates of electrons, *J. Phys. Chem.* **66,** 180 (1962).

4. P. Andreo, Monte-Carlo simulation of electron transport in water. FANZ 80–3. Department of Nuclear Physics, University of Zaragoza, Spain (1980).
5. P. Andreo and A. Brahme, Mean energy in electron beams, *Med. Phys.* **8,** 682 (1981).
6. P. Andreo and A. Brahme, Dosimetry and quality specifications of high energy photon beams: 1. Theoretical background, submitted to *Phys. Med. Biol.* (1985).
7. J. C. Ashley, Density effect in liquid water, *Radiat. Res.* **89,** 32 (1982).
8. F. M. Attix, The partition of kerma to account for bremsstrahlung, *Health Phys.* **36,** 347 (1979).
9. M. J. Berger and S. M. Seltzer, Tables of energy losses and ranges of electrons and positrons, in NAS-NRC publication 1133, National Academy of Sciences—National Research Council, Washington, pp. 205–268; also NASA Publication No. SP-3012, National Aeronautics and Space Administration (1964).
10. M. J. Berger and S. M. Seltzer, Additional stopping power and range tables for protons, mesons and electrons, NASA SP-3036, National Aeronautics and Space Administration (1966).
11. M. J. Berger and S. M. Seltzer, Stopping powers and ranges of electrons and positrons, NBS IR 82–2550 (1982a).
12. M. J. Berger and S. M. Seltzer, private communication (1982b).
13. M. J. Berger, S. M. Seltzer, S. R. Domen, and P. J. Lamperti, Stopping-power ratios for electron dosimetry with ionization chamber, in *Biological dosimetry,* IAEA-SM-193139, p. 594 (1975).
14. M. Bernard, Etude de l'ionisation au voisinge des interfaces planes situées entre deux milieux de nombres atomiques differents soumis aux rayons gamma de Cobalt 60 et aux rayons x d'un betatron de 20 MV, thesis, Université de Paris (1964).
15. G. Bertilsson, Electron scattering effects on absorbed dose measurements with LiF-dosemeters, thesis, University of Lund (1975).
16. M. Boutillon, Some remarks concerning the measurement of kerma with a cavity ionization chamber, Bureau International des Poids et Mesures, CCEMRI (I)/77–114 (1977).
17. M. Boutillon, Determination of absorbed dose in a water phantom from the measurement of absorbed dose in a graphite phantom, Bureau International des Poids et Mesures, CCEMRI (I)/81-6, Rapport BIPM-81/2 (1981).
18. M. Boutillon and M.-T. Niatel, A study of graphite cavity chamber for absolute exposure measurements of ^{60}Co gamma rays, *Metrologia* **9,** 139 (1973).
19. W. H. Bragg, *Studies in radioactivity,* Macmillan, New York, p. 94 (1912).
20. A. Brahme, Simple relations for the penetration of high energy electron beams in matter, SSI 1975-11, National Institute of Radiation Protection, Stockholm (1975).
21. A. Brahme, Restricted energy absorption coefficients for use in dosimetry. *Rad. Res.* **73,** 420 (1978).
22. A. Brahme, Correction of measured distribution for finite extension of the detector, *Strahlentherapie* **157,** 258 (1981).
23. A. Brahme, Physics of electron beam penetration: Fluence and absorbed dose, in *Proc. Symp. Elec. Dosimetry and Therapy,* AAPM (B. Paliwal, ed.), p. 45 (1982).
24. A. Brahme, Correction for the angular dependence of a detector in electron and photon beams, accepted for publication in *Acta Radiol. Oncol* **24,** 301 (1983).
25. A. Brahme and I. Lax, Absorbed dose distribution of electron beams in uniform and inhomogeneous media, *Acta Radiol. Suppl.* **364,** 26 (1983).
26. A. Brahme and H. Svensson, Specification of electron beam quality from central axis depth absorbed dose distribution, *Med. Phys.* **3,** 96 (1975).
27. A. Brahme and H. Svensson, Radiation beam characteristics of a 22 MeV microtron, *Acta Radiol. Oncol.* **18,** 244 (1979).
28. A. Brahme, G. Hultén, and H. Svensson, Electron depth absorbed dose distribution for a 10 MeV clinical microtron, *Phys. Med. Biol.* **20,** 39 (1975).

29. A. Brahme, T. Kraepelien, and H. Svensson, Electron and photon beams, from a 50 MeV racetrack microtron, *Acta Radiol. Oncol.* **19,** 305 (1980).
30. R. K. Broszkiewicz and Z. Bulhak, Errors in ferrous sulphate dosimetry, *Phys. Med. Biol.* **15,** 549 (1970).
31. P. R. J. Burch, Cavity ion chamber theory, *Rad. Res.* **3,** 361 (1955).
32. T. E. Burlin, Cavity-chamber theory, in *Radiation dosimetry,* Vol. 1 (F. H. Attix and W. C. Roesch, eds.), p. 331, Academic, New York (1968).
33. T. A. Carlsson, *Photoelectron and Auger Spectroscopy,* Plenum, New York (1975).
34. J. Casanovas, J. P. Patau, U. J. Mathieu, and D. Blanc, Comparaison entre l'ionisation produite dans le volume sensible d'une chambre d'ionisation à diélectrique et l'énergie qui y est déposée, calculée par la méthode de Monte-Carlo, dans le cas d'irradiation par des électrons monocinétiques de 1.5 et 1.9 MeV, in *Third symposium on microdosimetry, Stresa, Italy, EUR 4810 d-f-e,* p. 571 (1971).
35. J. Casanovas, Etude de la conduction induite par des rayonnements à transferts lineiques d'énergie très different dans certains liquides organiques non polaires, thesis, Université Paul Sabatier, Toulouse, France (1975).
36. R. S. Caswell, Deposition of energy by neutrons in spherical cavities, *Rad. Res.* **27,** 92 (1966).
37. E. Cottens, Geabsorbeerde dosis kalorimetrie bij hog energie elektronenbundels en onderzoek van de ijzersulfaat dosimeter, thesis, University of Gent (1979).
38. E. Cottens, A. Janssens, G. Eggermont, and R. Jacobs, Absorbed dose calorimetry with a graphite calorimeter and *G*-value determinations for the Fricke dosemeter in high energy electron beams, IAEA-SM-249/32 (1980), p. 488.
39. G. W. Dolphin, N. H. Gale, and A. L. Bradshaw, Investigations of high energy electron beams for use in therapy, *Br. J. Radiol.* **32,** 13 (1959).
40. S. R. Domen, Absorbed dose water calorimeter, *Med. Phys.* **7,** 157 (1980).
41. S. R. Domen, An absorbed dose water calorimeter: Theory, design and performance, *J. Res. N.B.S.* **87,** 211 (1982).
42. S. R. Domen, A polystyrene-water calorimeter, *J. Res. N.B.S.* **88,** 373 (1983).
43. S. R. Domen and P. J. Lamperti, Comparisons of calorimetric and ionometric measurements in graphite irradiated with electrons from 15 to 50 MeV, *Med. Phys.* **3,** 294 (1976).
44. J. J. Duderstadt and W. R. Martin, *Transport Theory,* Wiley, New York (1979).
45. A. Dutreix, Problems of high energy x-ray beam dosimetry, in *HIgh Energy Photons and Electrons* (S. Kramer, ed.), Wiley, New York (1976).
46. J. Dutreix and A. Dutreix, Etude comparée d'une série de chambres d'ionisation dans des faisceaux d'électron de 20 et 10 MeV, *Biophysik* **3,** 249 (1966).
47. P. J. Ebert, A. F. Luzon, and E. M. Lent, Transmission and backscattering of 4 to 12 MeV electrons, *Phys. Rev.* **183,** 422 (1969).
48. M. Ehrlich and C. G. Soares, Measurement assurance studies of high-energy electron and photon dosimetry in radiation-therapy applications, in *IAEA Technical Report Series. Intercomparison Procedures in the Dosimetry of High-Energy X-Ray and Electron Beams* (1979).
49. S. C. Ellis, The dissemination of absorbed dose standards by chemical dosimetry, *Rad. Sci.* **30,** (1974). National Physical Laboratory.
50. S. C. Ellis, J. H. Barrett, P. H. G. Sharpe, and M.-T. Niatel, Preliminary report on an intercomparison of Fricke dosimetry systems. Bureau International des Poids et Mesures, CCEMRI (1)/31-18 (1981).
51. G. Failla, Dosimetry of ionizing radiation, *Progr. Nucl. Energy Ser. VII,* 147 (1956).
52. U. Fano, Introductory remarks on the dosimetry of ionizing radiations, *Rad. Res.* **1,** 3 (1954).
53. H. Feist, private communication (1980).

54. D. M. Galbraith, J. A. Rawlinson, and P. Munro, Dose errors due to charge storage in electron irradiated plastic phantom, *Med. Phys.*, **11,** 253 (1984).
55. J. Geisseloder, K. Koepke, and J. S. Laughlin, Calorimetric determination of absorbed dose and $G_{Fe^{+++}}$ of the Fricke dosimeter with 10 MeV and 20 MeV electrons, *Rad. Res.* **20,** 423 (1963).
56. M. Goitein, A technique for calculating the influence of thin inhomogeneities on charged particle beams, *Med. Phys.* **5,** 258 (1978).
57. E. L. Goldwasser, F. E. Mills, and A. O. Hanson, Ionization loss and straggling of fast electrons, *Phys. Rev.* **88,** 1137 (1952).
58. L. H. Gray, The absorption of penetrating radiation, *Proc. R. Soc. (London) Ser. A* **122,** 647 (1929).
59. L. H. Gray, Ionization method for the absolute measurement of gamma-ray energy, *Proc. R. Soc. (London) Ser. A* **156,** 578 (1936).
60. J. R. Greening, Dose conversion factors for electrons, *Phys. Med. Biol.* **19,** 746 (1974).
61. J. R. Greening, Fundamentals of radiation dosimetry, *Medical Physics Handbooks 6,* Adam Hilger Ltd., Bristol, in collaboration with the HPA (1981).
62. D. Harder, Berechnung der Energiedosis aus Ionisationsmessungen bei Sekundärelektronen-Gleichgewicht, *Symposium on High-Energy Electrons 1964.* (Zuppinger and Poretti, eds.) p. 40, Springer, Berlin (1965).
63. D. Harder, Einfluss der Vielfachstreeung von Elektronen auf der Ionisations im gasgefüllten Hohlräumen, *Biophysik* **5,** 157 (1968).
64. D. Harder, Some general results from the transport theory of electron absorption, in *Second symposium on microdosimetry, Stresa, Italy, EUR 4452 d-f-e,* p. 567 (1970).
65. D. Harder, Fano's theorem and the multiple scattering correction, *Proceedings of the 4th Symposium on microdosimetry, Pallanza,* Vol II (J. Booz, H. G. Ebert, R. Eickel, and A. Waker, eds.), Euroatom, Brussels (1973).
66. D. Harder, Present status of electron beam dosimetry, invited paper at *XIVth International Congress of Radiology,* Rio de Janeiro (1977).
67. G. Hettinger, C. Pettersson and H. Svensson, Displacement effect of thimble chambers exposed to a photon or electron beam from a betatron, *Acta Radiol. Ther. Phys. Biol.* **6,** 61 (1967).
68. H. Holthausen, Über die Bedingungen der Röntgenstrahlenenergiemessung bei verschiedenen Impulsbreiten auf luftelektrischen Wege, *Fortschr. Röntgenstr.* **26,** 211 (1919).
69. Y. S. Horowitz and A. Dubi, A proposed modification of Burlin's general cavity theory for photons, *Phys. Med. Biol.* **27,** 867 (1982).
70. HPA, The Hospital Physicists' Association: A practical guide to electron dosimetry 5–35 MeV, HPA Report Series No 4, HPA, London (1971).
71. J. H. Hubbel, Photon mass attenuation and mass energy-absorption coefficients for H, C, N, O, Ar, and seven mixtures from 0.1 keV to 20 MeV, *Rad. Res.* **70,** 58 (1977).
72. J. H. Hubbel, Photon mass attenuation and energy absorption coefficients from 1 keV to 20 MeV, *Int. J. Appl. Rad. Isot.* **33,** 1269 (1982).
73. ICRU, International Commission on Radiation Units and Measurements, *Radiation Dosimetry: X-Ray and Gamma Rays with Maximum Photon Energies between 0.6 and 50 MeV,* ICRU report 14, International Commission on Radiation Units and Measurements, Bethesda, Maryland (1969).
74. ICRU, International Commission on Radiation Units and Measurements, *Radiation Dosimetry: Electron Beams with Energies Between 1 and 50 MeV,* ICRU report 1351, International Commission on Radiation Units and Measurements, Bethesda, Maryland (1984).

75. ICRU, International Commission on Radiation Units and Measurements, *Determination of Absorbed Dose in a Patient Irradiated by Beams of X or Gamma Rays in Radiotherapy Procedures,* ICRU report 24, International Commission on Radiation Units and Measurements, Bethesda, Maryland (1976).
76. ICRU, International Commission on Radiation Units and Measurements, *Average Energy Required to Produce an Ion Pair,* ICRU report 31, International Commission on Radiation Units and Measurements, Bethesda, Maryland (1979).
77. ICRU, International Commission on Radiation Units and Measurements, *Radiation Quantities and Units,* ICRU report 33, International Commission on Radiation Units and Measurements, Bethesda, Maryland (1980).
78. A. C. A. Janssens, G. Eggermont, R. Jacobs, and G. Thielens, Spectrum perturbation and energy deposition models for stopping power ratio calculations in general cavity theory, *Phys. Med. Biol.* **19,** 619 (1974).
79. A. C. A. Janssens, G. Eggermont, and R. Jacobs, Cavity theory, a general formulation of the problem and a proposal for practical application, *Proceedings of the 8th Long. Int. Soc. Fr. de Radioprotection 65* (1976).
80. A. C. A. Janssens, Modified energy deposition model for the computation of the stopping power ratio for small cavity sizes, *Phys. Rev. A* **23,** 1164 (1981).
81. K.-A. Johansson, L. O. Mattsson, L. Lindborg, and H. Svensson, Absorbed dose determination with ionization chambers in electron and photon beams with energies between 9 and 50 MeV, in International symposium on national and international standardization of radiation dosimetry, Atlanta, IAEA-SM-222/35 (1978).
82. K.-A. Johansson and H. Svensson, Liquid ionization chamber for absorbed dose determinations in photon and electron beams, *Acta Radiol. Oncol.* **21,** 359 (1982).
83. E. E. Kearsley, General cavity theory for photon and neutron dosimetry, thesis, University of Wisconsin-Madison (1981).
84. J. Kretschko, Absolutmessungen an schnellen Elektronen mit einem Faraday-Käfig, thesis, University of Frankfurt (1960).
85. I. Lax and A. Brahme, On the collimination of high energy electron beams, *Acta Radiol. Oncol.* **19,** 199 (1980).
86. R. Loevinger, A formalism for calculations of absorbed dose to the medium from photon and electron beams, *Med. Phys.* **8,** 1 (1981).
87. T. P. Loftus and J. T. Weaver, Standardization of ^{60}Co and ^{137}Cs gamma ray beams in terms of exposure, *J. Res. Nat. Bur. Stand.* **78A,** 465 (1975).
88. B. Markus, Spezifische totale Ionisation und Ionisierungsaufwand von 15-MeV-Elektronen in Luft und einigen anderen Gasen, *Naturwissenschaften* **46,** 1 (1959).
89. B. Markus, Eine Parallelplatten-Kleinkammer zur Dosimetrie schneller Elektronen und ihre Anwendung, *Strahlentherapie* **152,** 517 (1976).
90. J. Mathieu, Thése d'Etat ès Sciences Physique, No. 313, Toulouse (1968).
91. J. Mathieu, D. Blanc, J. Casanovas, A. Dutreix, A. Wambersie, and M. Prignot, Mesure de la réparation de la dose déposée en profondeur dans un fantôme de plexiglas irradié par un faisceau d'électrons moncinétiques de 10, 15, 20 ou 30 MeV, in *Second Symposium on Microdosimetry, Stresa, Italy, EUR 4452 d-f-e,* p. 437 (1969).
92. L. O. Mattsson and H. Svensson, Charge build-up effects in insulating phantom materials, *Acta Radiol. Oncol.* **23,** 393 (1984).
93. L. O. Mattsson, K.-A., Johansson, and H. Svensson, Calibration and use of plane-parallel ionization chambers for the determination of absorbed dose in electron beams, *Acta Radiol. Oncol.* **20,** 385 (1981).
94. L. O. Mattsson, K. A. Johansson, and H. Svensson, Ferrous sulphate dosimeter for control of ionization chamber dosimetry of electron and ^{60}Co gamma beams, *Acta Radiol. Oncol.* **21,** 139 (1982).

95. L. O. Mattsson, Application of the water calorimeter, Fricke dosimeter and ionization chamber in clinical dosimetry. Paper I, p. 12. Thesis, University of Gothenborg, JSBN 91-7222-7 29-X (1984).
96. W. T. Morris and B. Owen, An ionization chamber for therapy level dosimetry of electron beams, *Phys. Med. Biol.* **20,** 718 (1975).
97. D. Mosse, M. Cance, K. Steinschaden, M. Chartier, A. Ostrowsky, and J. P. Simoen, Détermination du rendement du dosimètre au sulfate ferreux dans un faisceau d'électrons de 35 MeV, *Phys. Med. Biol.* **27,** 583 (1982).
98. H. Mutsuzawa, K. Kawashima, and T. Hiaoka, Dose conversion factors for electrons, *Phys. Med. Biol.* **19,** 744 (1974).
99. Nordic Association of Clinical Physics (NACP), Procedures in radiation therapy dosimetry with 5 to 50 MeV electrons and roentgen and gamma rays with maximum photon energies between 1 and 50 MeV, *Acta Radiol. Ther. Phys. Biol.* **11,** 603 (1972).
100. NACP, Procedures in external radiation therapy dosimetry with electron and photon beams with maximum energies between 1 and 50 MeV, *Acta Radiol. Oncol.* **19,** 55 (1980).
101. NACP, Electron beams with mean energies at the phantom surface below 15 MeV, Supplement to the recommendations by the Nordic Association of Clinical Physics (NACP) 1980, *Acta Radiol. Oncol.* **20,** 401 (1981).
102. A. E. Nahum, Calculations of electron flux spectra in water irradiated with megavoltage electron and photon beams with applications to dosimetry, thesis, University of Edinburgh, U.K. (1976).
103. A. E. Nahum, Water/air mass stopping power ratios for megavoltage photon and electron beams, *Phys. Med. Biol.* **23,** 24 (1978).
104. A. E. Nahum, private communication (1982).
105. A. E. Nahum, H. Svensson, and A. Brahme, The ferrous sulphate G-value for electron and photon beams: A semi-empirical analysis and its experimental support, *Proceedings of the 7th symposium on microdosimetry,* Vol. II, Harward Academic Publishers, Chur, Switzerland, p. 841 (1981).
106. R. Nath and R. J. Schulz, Calculated response and wall correction factors of practical ionization chambers for Co-60 gamma rays, Annual AAPM meeting Atlanta (1979).
107. R. Nath, L. Friedman, and R. J. Schulz, A comparison of plastic and water phantoms for absorbed dose calibration of high energy x-rays, *Phys. Med. Biol.* **23,** 1093 (1978).
108. M.-T. Niatel, Détermination de W_{air} (énergie moyenne nécessaire pour produire une paire d'ions dans l'air) basée sur des comparaisons d'étalons de dose absorbée effectuéesau BIPM, CCEMRI (I)/77-113, p. R(1) 56 (1977).
109. B. Nilsson and A. Brahme, Absorbed dose from secondary electrons in high energy photon beams, *Phys. Med. Biol.* **24,** 901 (1979).
110. B. Nilsson and A. Brahme, Contamination of high energy photon beams, by scattered photons, *Strahlentherapie* **157,** 181 (1981).
111. B. Nilsson and A. Brahme, Relation between kerma and absorbed dose in photon beams, *Acta Radiol. Oncol.* **22,** 77 (1983).
112. O. T. Ogunleye, F. H. Attix, and B. R. Paliwal, Comparison of Burlin cavity theory with LiF TLD measurements for Cobalt 60 gamma rays, *Phys. Med. Biol.* **25,** 203 (1980).
113. J. Ovadia, M. Danzker, J. W. Beattie, and J. S. Laughlin, Ionization of 9 to 17.5 MeV electrons in air, *Rad. Res.* **3,** 430 (1955).
114. C. Pettersson, Calorimetric determination of the G-value of the ferrous sulphate dosimeter with high energy electrons and ^{60}Co gamma-rays, *Ark. Fys.* **34,** 385 (1967).
115. A. P. Pinkerton, Comparison of calorimetric and other methods for the determination of absorbed dose, *Ann. Acad. Sci.* **161,** 63 (1969).

116. E. B. Prodgoršak, J. A. Rawlinson, M. I. Glavinvic, and H. E. Hohns, Design of x-ray targets for high energy linear accelerators in radiotherapy, *Am. J. Roentgenol.* **121,** 873 (1974).
117. J. S. Pruitt, S. R. Domen, and R. Loevinger, The graphite calorimeter as a standard of absorbed dose for Cobalt-60 gamma radiation, *J. Res. NBS* **86,** 495 (1981).
118. J. A. Rawlinson and H. E. Johns, Percentage depth dose for high energy x-ray beams in radiotherapy, *Am. J. Roentgenol. Rad. Ther. Nucl. Med.* **118,** 919 (1973).
119. H. Roos, P. Drepper, and D. Harder, The transition from multiple scattering to complete diffusion of high-energy electrons, in *4th symposium on microdosimetry, EUR 5122 d-e-f,* p. 779 (1974).
120. B.-I. Rudén and L. B. Bengtsson, Accuracy of megavoltage radiation dosimetry using thermoluminescent lithium fluoride, *Acta Radiol. Ther. Phys. Biol.* **16,** 157 (1977).
121. C. Samuelsson, Influence of air cavities on central depth dose curves for 33 MV roentgen rays, *Acta Radiol. Ther. Phys. Biol.* **16,** 465 (1977).
122. R. H. Schuler and A. O. Allen, Yield of the ferrous sulphate radiation dosimeter: An improved cathode-ray determination, *J. Chem. Phys.* **24,** 56 (1956).
123. SCRAD, The Sub-Committee on Radiation Dosimetry of the American Association of Physicists in Medicine: Protocol for the dosimetry of high energy electrons, *Phys. Med. Biol.* **11,** 505 (1966).
124. R. J. Shalek, P. Kennedy, M. Stovall, J. H. Cundiff, W. F. Gagnon, W. Grant, and W. F. Hanson, Quality assurance for measurements in therapy, National Bureau of Standards SP 456 111 (1976).
125. L. S. Skaggs, Depth dose of electrons from the betatron, *Radiology* **53,** 868 (1949).
126. L. V. Spencer and F. H. Attix, A theory of cavity ionization, *Rad. Res.* **3,** 239 (1955).
127. R. M. Sternheimer, The density effect for the ionization loss in various materials, *Phys. Rev.* **88,** 851 (1952).
128. R. M. Sternheimer, Density effect for the ionization loss in various materials, *Phys. Rev.* **103,** 511 (1956).
129. R. M. Sternheimer and R. F. Peierls, General expression for the density effect for the ionization loss of charged particles, *Phys. Rev. B* **3,** 3681 (1971).
130. R. M. Sternheimer, S. M. Seltzer, and M. J. Berger, Density effect for the ionization loss of charged particles, *Phys. Rev. B* **26,** 6067 (1982).
131. Strahel, *Strahelentherapie* **31,** 582 (1929).
132. H. Svensson, Influence of scattering foils, transmission monitors and collimating system on the absorbed dose distribution from 10 to 35 MeV electron radiation, *Acta Radiol. Ther. Phys. Biol.* **10,** 443 (1971a).
133. H. Svensson, Dosimetric measurements at the Nordic medical accelerators. II. Absorbed dose measurements, *Acta Radiol. Ther. Phys. Biol.* **10,** 631 (1971b).
134. H. Svensson, Quality assurance in radiation therapy; physical aspects, in *Supplement to International Journal of Radiation Oncology, Biology and Physics,* **10** (7), (1984) pp. 59–65.
135. H. Svensson and A. Brahme, Ferrous sulphate dosimetry for electrons. A re-evaluation, *Acta Radiol. Oncol.* **18,** 326 (1979).
136. H. Svensson and A. Brahme, Fundamentals of electron beam dosimetry, p. 17 of *Proceedings of the symposium on electron beam therapy* (F. C. H. Chu and J. S. Laughlin, eds.), Memorial Sloan Kettering Cancer Center, New York (1981).
137. H. Svensson and A. E. Nahum, Present knowledge of stopping-power ratios for ionization chambers, Invited paper presented at the World Congress on Medical Physics and Biomedical Engineering, Hamburg (1982).
138. H. Svensson, C. Pettersson, and G. Hettinger, Effects on ferrous sulphate dosimeter solution stored in small polystyrene cells, p. 251 in *Solid State and Chemical Radiation Dosimetry in Medicine and Biology,* IAEA, Vienna (1967).

139. H. Svensson, G. Hultén, G. Hettinger and G. Wickman, Determination of absorbed dose conversion factors of an air ionization chamber for 10–33 MeV electron and photon beams with the aid of a liquid ionization chamber, Paper presented at the *XIII Int. Congress of Radiology,* Madrid (1973).
140. W. M. Telford, J. E. Crawford, H. H. Zwick, and L. G. Stephens-Newsham, Linear electron accelerator for medical purposes, *J. Can. Assoc. Radiol.* **28,** 298 (1967).
141. H. Weatherburn and B. Stedeford, Effective measuring position for cylindrical ionization chambers when used for electron beam dosimetry, *Br. J. Radiol.* **50,** 921 (1977).
142. H. Weatherburn, A. D. Welsh, and B. Stedeford, Re-calculation of perturbation correction factors for thimble ionization chambers when used for electron dosimetry, in Digest of the *12th International Conference on Medical and Biological Engineering, Jerusalem,* p. 516 (1979).
143. B. W. Wessels, B. R. Paliwal, M. J. Parrot, and M. C. Choi, Characterization of Clinac-18 electron-beam energy using a magnetic analysis method, *Med. Phys.* **6,** 45 (1979).
144. G. Wickman, A liquid ionization chamber with high spatial resolution, *Phys. Med. Biol.* **19,** 66 (1974a).
145. G. Wickman, Radiation quality independent liquid ionization chamber for dosimetry of electron radiation from medical accelerators, *Acta Radiol. Ther. Phys. Biol.* **13,** 37 (1974b).
146. G. Wickman, personal communication (1983).
147. G. Wickman and H. Svensson, Personal communication (1983).
148. J. F. Wochos, L. A. De Werd, R. Hilko, J. A. Meyer, M. Stovall, D. Spearman, C. Thomason, and G. L. Dubuque, Mailed thermoluminescent dosimetry reviews in radiation therapy, *Med. Phys.* **9,** 920 (1982).
149. L. Zheng-Ming, A general theory of the cavity ionization, in *Collected papers in Atomic Energy Science and Technology* (1976), p. 155 (in Chinese).
150. L. Zheng-Ming, An electron transport theory of cavity ionization, *Rad. Res.* **84,** 1 (1980).

Chapter 4

Microdosimetry and Its Application to Biological Processes

Marco Zaider and Harald H. Rossi

PART I. PHYSICS

1. INTRODUCTION: THE RATIONALE OF MICRODOSIMETRY

Microdosimetry is the study of energy deposition processes in biological media with particular accent on phenomena correlated with the physical aspects of the radiation action on living systems. As such, microdosimetry is geared toward understanding the basic mechanisms in the initiation stage of radiation action and, as a corollary, defining a set of quantities characterizing the radiation fields that are most directly related to the biological effect, all other conditions being the same.

The interaction of ionizing radiation with matter involves basically two processes: ionizations and excitations. Since the number of such elementary alterations per unit of absorbed energy is approximately constant for any type of radiation and energy, it was logical to assume that the mere concentration of such events in the sensitive site of the biological object should correlate with the biological effect (end point) observed. This basic assumption has marked the whole line of thinking in radiation physics since its inception.

Marco Zaider and Harald H. Rossi • Radiological Research Laboratory, Department of Radiology, Cancer Center/Institute of Cancer Research, Columbia University College of Physicians and Surgeons, New York, New York 10032.

Radiation effects are usually measured as average values over a large number of cells or organisms exposed to a homogeneous radiation field. The *amount* of radiation producing the effect is conveniently specified as the energy deposited per unit mass in the irradiated system, i.e., the absorbed dose, D. Although defined at each point, the absorbed dose can be considered to be a macroscopic quantity because its value is unaffected by microscopic fluctuations of energy deposition. However, these fluctuations are important if only because they are the reason why equal doses of different radiations produce effects of different magnitude. While the absorbed dose determines the average number of energy deposition events, each cell as an individual entity will react in fact to the *actual* energy deposited in it. The average response of a system of cells should therefore depend on the energy distribution on a scale that is at least as small as the dimensions of the cell. In addition, the pattern of energy deposition at the subcellular level is also of importance because radiosensitive components occupy only a portion of the cell. Such energy distributions vary greatly among different types of radiation, which raises the possibility that by correlating them with biological effectiveness, information might be obtained on radiobiological mechanisms. This realization constituted the starting point of microdosimetry, some two decades ago.

The quantity *specific energy, z,* defined as the energy imparted to finite volumes per unit mass and measured in the same units as absorbed dose, was introduced[1] in order to quantify the stochastic nature of energy deposition in cellular and subcellular objects. The specific energy, as a random variable, is characterized by the distribution function $f(z; D)\,dz$, representing the probability of depositing in a site a specific energy between z and $z + dz$. Simple physical considerations show that this distribution depends, among other things, on the dimensions of the volume under consideration and the dose D (i.e., the average value of z). The statistical fluctuations of z about its mean value are larger for smaller volumes, small doses, and high-LET. A few distributions $f(z; D)$ for the same dose but different radiations are shown in Fig. 1.

The significance of the distributions $f(z; D)$ for an understanding of biological effects can be illustrated by a simple example. Consider experiments in which a dichotomous response is observed (e.g., cell survival or cell transformation). Since the cellular effect must be initiated by some energy deposition event it is obvious that the fraction of cells unaffected must be at least equal to the probability of depositing zero energy in a cell; unless each particle traversal causes the effect, the fraction is larger. As will be shown later this probability is explicitly contained in $f(z; D)$ as a separate term, $\exp(-n)$, where n is the average number of events (particle traversals with concomitant energy deposi-

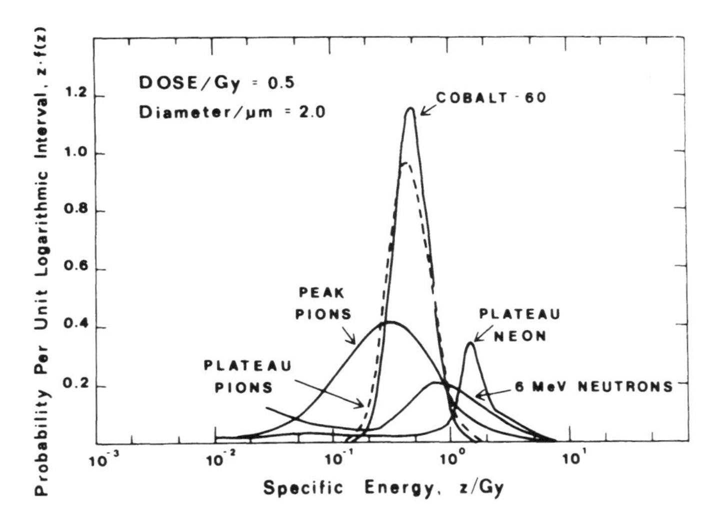

Figure 1. Distributions $f(z; D)$ of specific energy z in a spherical tissue cavity of 2.0 μm diameter at 0.5 Gy. In this representation the area under the curve $f(z; D)$ delineated by two values of z is proportional to the probability of depositing a specific energy z in that interval. The distributions are calculated from single-event microdosimetric spectra for ^{60}Co,[2] negative pions,[3] neutrons,[4] and neon ions,[5] respectively, using the computer code KFOLD.[6]

tions) per cell. This number is proportional to the absorbed dose, D, and increases generally as the square of the diameter of the volume considered. A dose of 1 cGy of 250-kV X rays, for instance, results in about 37% cells with no particle traversal, if the cell diameter is 5 μm. For the same dose delivered by 1.5-MeV neutrons, the corresponding fraction is 97%. Because of the known dependence between n and the site diameter one can use such a simple analysis to actually calculate the minimum size of the sensitive site compatible with the requirement that the fraction of cells showing no effect should be equal to or greater than the fraction of cells experiencing no event. When such an analysis is performed one finds that the sensitive region of a mammalian cell is spread over at least a few micrometers in diameter. An immediate consequence of this is the requirement that measurements and calculations of microdosimetric distributions should be carried out for site sizes of the order of micrometers. A second interesting consequence occurs when an analysis of the type described before results in a minimum diameter which is actually *greater* than the cellular dimensions. In such a case one can conclude that intercellular processes must be involved in the effect observed.

Considerations based on the absence of events are relatively simple.

The greater challenge to theories based on microdosimetry is to account for effect probability in terms of energy depositions in cells which *have* experienced events. Perhaps the simplest assumption that can be made is that the cell contains sensitive sites which are subcellular regions wherein events cause the effect in question with a probability that depends on the energy deposited. If $E(D)$ is the average effect observed per cell after irradiation with the dose, D, this model amounts to postulating the existence of an effect function $E(z)$ such that

$$E(D) = \int_0^\infty E(z) f(z; D)\, dz \qquad (1)$$

where $f(z; D)$ refers to the sites under consideration. A substantial body of radiobiological evidence can be accounted for using relatively simple effect functions, $E(z)$ (some of these will be discussed later). A significant feature of these theories is that they indeed predict site sizes consistent with the previous conclusions based on event frequency only.

Recent developments in microdosimetry have resulted in new concepts and quantities capable of describing situations where the above assumptions are actually not applicable, should this be necessary. These developments will also be described in detail.

2. MICRODOSIMETRIC QUANTITIES AND DISTRIBUTIONS

2.1. Quantities and Terminology

A unique terminology has been developed in connection with microdosimetry. Most of the basic microdosimetric terms are defined in the ICRU Report 33, Radiation Quantities and Units[7] which was recently published. The reader is referred to this publication for additional details.

The energy dissipation by a charged particle in a medium consists primarily of a series of discrete (quantal) interactions resulting in ionizations and excitations. The term *transfer point* refers to the geometrical position of an interaction point. In the transfer, the kinetic energy of the particle is reduced and additional energy may be imparted to secondary ionizing particles. The remainder is deposited locally at a transfer point and is called *energy transfer*. The interactions of an ionizing particle and its secondaries are completely characterized if all the transfer points and energy transfers are specified. Such an entity will in the following be termed a *track*.

The term *event* is used in microdosimetry to denote traversal of a region by a particle and/or its secondaries with concomitant energy deposition. In different words, an event occurs whenever at least one energy transfer associated with a track is situated inside the region of interest. The traversal of a cell by a proton, for instance, is an event. But even if the proton passes actually outside the cell and only a secondary electron (delta ray) is injected in the cell, one still speaks of an event associated with the proton track.

It is important to understand that an event usually consists of several energy transfers which are statistically *correlated,* while different events are statistically *independent.* This is the justification for introducing the notion of event. Since, as will be seen, experimental microdosimetry deals mostly with *single* events while in radiobiology one frequently encounters situations where at least several random events occur in a cell, the property of statistical independence of the events (on which their definition is actually based) becomes essential.

The *energy imparted,* ε, by ionizing radiation to the matter in a volume is[8]

$$\varepsilon = R_{\text{in}} - R_{\text{out}} + \sum \mathrm{Q} \tag{2}$$

where R_{in} is the sum of energies (excluding rest energies) of all those ionizing particles entering the volume; R_{out} is the sum of energies (excluding rest energies) of all those ionizing particles which leave the volume; $\sum Q$ is the sum of all changes of the rest mass energy of nuclei and elementary particles in any nuclear transformations which occur in the volume. For any volume, the energy imparted is equal to the sum of all energy transfers, ε_i, in that volume:

$$\varepsilon = \sum_i \varepsilon_i \tag{3}$$

Because of the statistical nature of the energy deposition processes, ε is a stochastic quantity. Several additional quantities are defined in connection with the energy imparted.

The *lineal energy, y,* is the quotient of ε by the mean chord length of the volume of reference, $\bar{l}$:

$$y = \varepsilon/\bar{l} \tag{4}$$

Here ε refers to a single event.

For any convex body, according to the well-known Cauchy theorem,[9] the mean chord length is given by

$$\bar{l} = 4V/S \tag{5}$$

where V is the volume and S is the surface area. In the rest of this presentation, and unless otherwise stated, only spherical volumes will be considered. For a sphere

$$\bar{l} = 2a/3 \tag{6}$$

where a is the diameter.

The term *specific energy,* z, denotes the quotient of ε, the energy imparted, by the mass of the reference volume:

$$z = \varepsilon/m \tag{7}$$

In contrast to the lineal energy, z refers to energy imparted by any number of events. If z is restricted to a single event only, then the following simple relation holds:

$$y = \rho S z/4 \tag{8}$$

where ρ is the mass density of the medium. The units employed for the quantities defined before are generally SI units. Historically, though, a number of special, nonstandard units have been introduced. Since these units are very frequently used in the scientific literature it is worth mentioning them. Thus, the imparted energy is measured in keV, the lineal energy in keV/μm, and the specific energy in grays. With these units a useful formulation of Eq. (8) for unit density spheres is

$$z_{\mathrm{Gy}} = \frac{0.204}{a^2_{\mu\mathrm{m}}} y_{\mathrm{keV}/\mu\mathrm{m}} \tag{9}$$

2.2. Probability Distribution Functions and Mean Values for the Microdosimetric Quantities

The probability of finding a value of specific energy between z and $z + dz$ is commonly designated as $f(z)\,dz$. The probability density, $f(z)$, is also called *frequency distribution* in z. The mean value of the specific energy, $\bar{z}_F$, is

$$\bar{z}_F = \int_0^\infty z f(z)\,dz \tag{10}$$

The absorbed dose, D (a nonstochastic quantity), is defined as the limit of $\bar{z}_F$ when the dimensions of the reference volume approach zero:

$$D = \lim_{a\to 0} \bar{z}_F \tag{11}$$

Since microdosimetry deals with very small volumes, $\bar{z}_F$ is usually identified with D. Consequently $\bar{z}_F$ and D will be used interchangeably in the following.

The specific energy may be due to one or more energy deposition events. Specific energies associated with *single* events play a special role in microdosimetry: $f_1(z)$ designates the probability distribution of specific energy for these events. The mean value of the distribution $f_1(z)$ is

$$\bar{z}_{1F} = \int_0^\infty z f_1(z)\, dz \tag{12}$$

$f_1(z)$ is known as the *single-event spectrum* in z. It should be noted that, while $f(z)$ has a finite term for $z = 0$ (i.e., the probability that no event occurred), $f_1(z)$ is not defined for $z = 0$. This latter distribution measures the probability of z *conditioned* by the fact that one event (and only one) took place. A second important observation is that the distribution $f(z)$ depends on the dose delivered, whereas the single-event distribution does not.

Because of the statistical independence property of energy absorption events one can establish a direct relation between the distributions $f(z)$ and $f_1(z)$. Indeed, if $f(z)$ represents the distribution corresponding to a dose D, the *average number of events* occurring in the reference volume is

$$n = D/\bar{z}_{1F} \tag{13}$$

The probability for exactly ν events is given by the Poisson distribution:

$$p_\nu = e^{-n} n^\nu / \nu! \tag{14}$$

and with this

$$f(z) = \sum_{\nu=0}^{\infty} p_\nu f_\nu(z) \tag{15}$$

where $f_\nu(z)$ is the distribution of z obtained in exactly ν events. The spectra $f_\nu(z)$ can be obtained by successive convolutions of the single-event spectrum $f_1(z)$:

$$\begin{aligned}
f_0(z) &= \delta(z) \\
f_1(z) &= f_1(z) \\
f_2(z) &= \int_0^z f_1(z') f_1(z - z')\, dz' \\
&\vdots \\
f_\nu(z) &= \int_0^z f_1(z') f_{\nu-1}(z - z')\, dz
\end{aligned} \tag{16}$$

Let us rewrite Eq. (15):

$$f(z) = e^{-D/\bar{z}_{1F}}\,\delta(z) + e^{-D/\bar{z}_{1F}} \sum_{\nu=1}^{\infty} \frac{1}{\nu!}\left(\frac{D}{\bar{z}_{1F}}\right)^{\nu} f_\nu(z) \tag{17}$$

The essential feature of the expression (17), and as mentioned previously, the reason for introducing the term "event", consists in the clear separation of the dose dependence of $f(z)$ (in the Poisson terms) from the dose-independent factors. By measuring $f_1(z)$ one can calculate, using Eqs. (16) and (17), the distributions $f(z)$ for any given dose. One should also remark the first term in Eq. (17) representing the probability that no event occurred in the reference volume; this term has been discussed in the Introduction.

For doses that are very small compared with $\bar{z}_{1F}$, one obtains a very simple approximation for Eq. (17):

$$f(z) = (1 - D/\bar{z}_{1F})\,\delta(z) + \frac{D}{\bar{z}_{1F}} f_1(z) \tag{18}$$

It is instructive to define $f(z)$ in a somewhat different manner, as the solution of a transport equation. Let us rewrite $f(z)$ as $f(z;n)$ where the mean number of events is explicitly stated. If n is increased by a small amount dn (i.e., a small dose is added) one has

$$f(z; n + dn) = f(z; n) + \frac{\partial f(z; n)}{\partial n}\,dn \tag{19}$$

The new distribution, $f(z; n + dn)$ can be expressed as before as a convolution between the distributions corresponding to n and dn, respectively:

$$f(z; n + dn) = \int_0^z f(z'; n) f(z - z'; dn)\,dz' \tag{20}$$

But, from Eq. (18),

$$f(z - z'; dn) = (1 - dn)\,\delta(z - z') + f_1(z - z')dn \tag{21}$$

Replacing Eqs. (20) and (21) in Eq. (19) one obtains

$$\frac{\partial f(z; n)}{\partial n} = -f(z; n) + \int_0^z f(z'; n) f_1(z - z')\,dz' \tag{22}$$

which is the transport equation defining $f(z;n)$. The interpretation of Eq. (22) is simple: $f(z;n)$ represents the fractional number of events with specific energy between z and $z + dz$. By adding an additional mean number of events, dn, two changes will occur in $f(z;n)$: (a) a fraction proportional to dn of the events in dz at z will be removed from this interval because of the additional specific energy deposited [first term in Eq. (22)], and (b) events which were not previously in the interval dz at z will be added, owing to the addition of dn events (second term).

The equation (22) is analogous to the transport equation describing the statistical fluctuations in the energy loss of charged particles by collision.[10,11] This similarity has been pointed out by Kellerer,[6] and a number of solutions of this equation, applicable to both phenomena, have been worked out. A formulation for the solution of Eq. (22) is described in the following.

Let $\phi(\omega)$ and $\phi_1(\omega)$ be the Fourier transforms of $f(z)$ and $f_1(z)$, respectively. According to the well-known convolution theorem,[12] the convolution of two functions [second term in Eq. (22)] corresponds in frequency space to the product of their transforms. By performing the Fourier transforms, Eq. (22) becomes

$$\frac{\partial\phi(\omega;n)}{\partial n} = -\phi(\omega;n) + \phi(\omega;n)\phi_1(\omega) \tag{23}$$

Therefore

$$\phi(\omega;n) = e^{n[\phi_1(\omega)-1]} \tag{24}$$

or, transforming back to the original functions:

$$f(z;n) = \int_{-\infty}^{+\infty} \phi(\omega;n)e^{2\pi i\omega z}\, d\omega \tag{25}$$

$$f(z;n) = \int_{-\infty}^{+\infty} \exp\left\{2\pi i\omega z + n\left[\int_0^{\infty} f_1(z)e^{-2\pi iz'\omega}\, dz' - 1\right]\right\} d\omega \tag{26}$$

The equation (26) is the solution of Eq. (22).

A number of useful relations between the moments of the distributions $f(z)$ and $f_1(z)$ could be obtained from Eq. (24).

For any pair of Fourier transforms, it can be easily verified from Eq. (25) that

$$\phi^{(\nu)}(0) = (-2\pi i)^{\nu} \int_0^{\infty} z^{\nu} f(z)\, dz \tag{27}$$

where $\phi^{(\nu)}(\omega)$ is the νth derivative of this function. In particular

$$\begin{aligned}\phi(0) &= 1\\ \phi'(0) &= -2\pi i\bar{z}\\ \phi''(0) &= -4\pi^2\overline{z^2}\end{aligned} \tag{28}$$

where $\overline{z^\nu}$ is the νth moment of $f(z)$.

Similar relations hold for $f_1(z)$ and $\phi_1(\omega)$. By actually performing the derivatives on Eq. (24) one obtains

$$\bar{z} = n\bar{z}_{1F} \tag{29}$$

$$\overline{z^2} = \bar{z}(\bar{z} + \bar{z}_{1D}) \tag{30}$$

where in Eq. (30) $\bar{z}_{1D}$ denotes

$$\bar{z}_{1D} = \frac{1}{\bar{z}_{1F}}\int_0^\infty z^2 f_1(z)\,dz \tag{31}$$

The equation (30) will be frequently used later on.

A second distribution is obtained if, instead of considering the *frequency* of events as a function of z [i.e., $f(z)$], one is interested in the *fraction of the total dose* delivered at a specific energy z. Since the absorbed dose, D, is simply

$$D = \int_0^\infty zf(z)\,dz \tag{32}$$

the fraction delivered in the interval dz centered at z is

$$d(z) = \frac{zf(z)}{D} \tag{33}$$

$d(z)$ is called the *dose distribution* in z. The average value of this distribution is

$$\bar{z}_D = \int_0^\infty z\,d(z)\,dz = \frac{1}{D}\int_0^\infty z^2 f(z)\,dz \tag{34}$$

A similar distribution can be defined in connection with the single-event spectrum:

$$d_1(z) = zf_1(z)/\bar{z}_{1F} \tag{35}$$

$$\bar{z}_{1D} = \int_0^\infty z\, d_1(z)\, dz = \frac{1}{\bar{z}_{1F}} \int_0^\infty z^2 f_1(z)\, dz \tag{36}$$

Finally, since for single event spectra the specific energy z is directly related to the lineal energy, y [see Eq. (8)], one can define distribution functions for y analogous to the distributions $f_1(z)$ and $d_1(z)$. Thus, if $f(y)$ is the frequency distribution in y, one has

$$\bar{y}_F = \int_0^\infty y f(y)\, dy \tag{37}$$

The dose distribution in y, i.e., the fraction of dose delivered in the interval dy at y, is

$$d(y) = \frac{1}{\bar{y}_F} y f(y) \tag{38}$$

and the dose-averaged lineal energy is

$$\bar{y}_D = \int_0^\infty y\, d(y)\, dy = \frac{1}{\bar{y}_F} \int_0^\infty y^2 f(y)\, dy \tag{39}$$

The subscript 1 is not necessary for the distributions in y and their mean values since the lineal energy corresponds, by definition, to single events only. The measured single event spectra are usually displayed as a function of lineal energy.

2.3. The Kellerer–Chmelevsky Equation and the Proximity Function

An important relation for the single-event dose-averaged specific energy, $\bar{z}_{1D}$, or the corresponding quantity in lineal energy, was obtained by Kellerer and Chmelevsky[13–15]; this relation has the advantage that the physical aspects involved in the energy absorption events are clearly separated from the geometrical properties of the reference volume. The demonstration of this relation is also useful in introducing some of the concepts and terminology related to the Monte Carlo simulation of microdosimetric distributions.

Consider a spherical volume of diameter a. Let the center of this sphere be $\vec{x}_i$. This sphere is the reference volume with respect to which the microdosimetric spectra are measured. If an event occurs in this volume then the specific energy is

$$z(\vec{x}_i) = \frac{1}{m} \sum_{k=1}^{n_i} \varepsilon_k \tag{40}$$

Here m is the mass of the spherical volume and ε_k ($k = 1, 2, \ldots, n_i$) are the energy transfers inside the volume. If particle tracks are known (i.e., energy transfers and transfer points associated with each particle) one can calculate the dose average of the single-event distribution in z following either one of the two following equivalent procedures: (a) have n tracks intersect the volume randomly; then

$$\bar{z}_{1D} \underset{n\to\infty}{=} \sum_{i=1}^{n} z^2(\vec{x}_i) \Big/ \sum_{i=1}^{n} z(\vec{x}_i) \tag{41}$$

or (b) place randomly n spheres in the associated volume of one representative track (the *associated volume*[16] of a track is the geometrical space around the track defined such that any reference volume with the center in it contrains at least one energy transfer; this notion is obviously dependent upon the diameter of the reference volume, see Fig. 2); then Eq. (41) applies again, with n being the number of spheres considered. In an actual Monte Carlo simulation scheme the latter procedure is generally preferred.

The equation (41) can be written

$$\bar{z}_{1D} = \frac{1}{m} \sum_{i=1}^{n} \left(\sum_{j,k=1}^{n_i} \varepsilon_j \varepsilon_k \right) \Big/ \sum_{i=1}^{n} \left(\sum_{l=1}^{n_i} \varepsilon_l \right) \tag{42}$$

Consider first the summation over ε_l in Eq. (42). Each energy transfer, ε_l, may appear more than once in this summation. Actually, for large n values, the *average* number of times each transfer point will appear is equal to nV_a/V, where V_a is the volume of the sphere and V is the total associated volume [this statement is easy to understand if the procedure (b) above is followed]. One has

$$\sum_{i=1}^{n} \left[\sum_{l=1}^{n_i} \varepsilon_l \right] = n \frac{V_a}{V} \sum_{k=1}^{N} \varepsilon_k \tag{43}$$

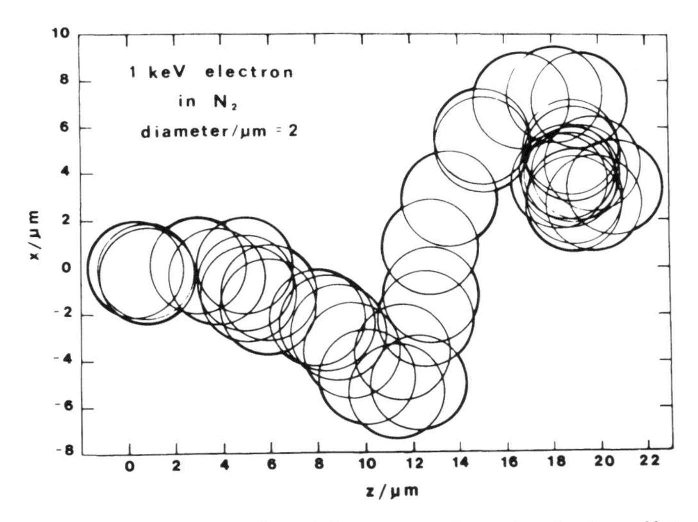

Figure 2. Schematic representation of the concept of associated volume.[16] Transfer points corresponding to ionizations and excitations have been calculated by simulating with Monte Carlo techniques a 1-keV electron track in N_2.[17] Consider spheres, of the same diameter as the reference volume, centered at each transfer point (the transfer points have been omitted in the figure for clarity). The total volume occupied by spheres is the associated volume of this particular track. It is easy to see that this construction corresponds to the definition given in the text: any reference volume with the center in the associated volume contains at least one transfer point.

where N is the total number of transfer points in the associated volume V.

Consider now the summation over cross products in Eq. (42). Each term, $\varepsilon_j \varepsilon_k$, will appear whenever one of the sampling spheres has its center in the intersecting volume of two spheres centered at ε_j and ε_k. This volume is

$$V_1 = V_a\left(1 - \frac{3x}{2a} + \frac{x^3}{2a^3}\right) \tag{44}$$

with $x = |\vec{x}_j - \vec{x}_k|$, $x \leqslant a$. Then, similar to Eq. (43),

$$\overline{\sum_{i=1}^{n}\left[\sum_{k,j=1}^{n_i} \varepsilon_j \varepsilon_k\right]} = n\frac{V_a}{V}\sum_{\substack{j,k=1\\x\leqslant a}}^{N} \varepsilon_j \varepsilon_k\left(1 - \frac{3x}{2a} + \frac{x^3}{2a^3}\right) \tag{45}$$

It is important to understand that in the right-hand sides of Eqs. (43) and (45) the summations are made such that each transfer point [for Eq. (43)]

or each pair of transfer points [for Eq. (45)] in the associated volume appears only once. With this

$$\bar{z}_{1D} = \frac{1}{m}\sum_{j,k=1}^{N} \varepsilon_j\varepsilon_k\left(1 - \frac{3x}{2a} + \frac{x^3}{2a^3}\right)\Big/\sum_{k=1}^{N}\varepsilon_k, \qquad x \leqslant a \tag{46}$$

If now k and j are restricted to transfer points separated by distances between x and $x + dx$ one can write

$$\bar{z}_{1D} = \frac{1}{m}\sum_{x}\frac{V_1(x)}{V_a}\left[\frac{1}{\sum_k \varepsilon_k}\sum_{(k,j)\in[x,x+dx]}\varepsilon_j\varepsilon_k\right] \tag{47}$$

The following notations have been introduced:

$$u(x) = \frac{V_1(x)}{V_a} = 1 - \frac{3x}{2a} + \frac{x^3}{2a^3} \tag{48}$$

$$t(x)\,dx = \sum_{(k,j)\in[x,x+dx]}\varepsilon_j\varepsilon_k\Big/\sum_k \varepsilon_k \tag{49}$$

Then

$$\bar{z}_{1D} = \frac{1}{m}\int_0^a u(x)t(x)\,dx \tag{50}$$

The equation (50) is the Kellerer–Chmelevsky equation. In this equation $u(x)$ is related to the unnormalized distribution, $\phi(x)$, of distances between two random points in the spherical volume[18]:

$$\phi(x) = 4\pi x^2 u(x) \tag{51}$$

$u(x)$ characterizes the geometry of the reference volume. The function $t(x)$ is related to the distribution of energy transfers in a particle track; it is called[15] the *proximity function of energy transfers*. A simple interpretation of the proximity function is obtained if one writes

$$t(x)\,dx = \sum_k \frac{\varepsilon_k}{E}\left(\sum_j \varepsilon_j\right), \qquad (k, j) \in [x, x + dx] \tag{52}$$

with

$$E = \sum_{k=1}^{N}\varepsilon_k \tag{53}$$

Then $t(x)$ can be defined as follows[19]:

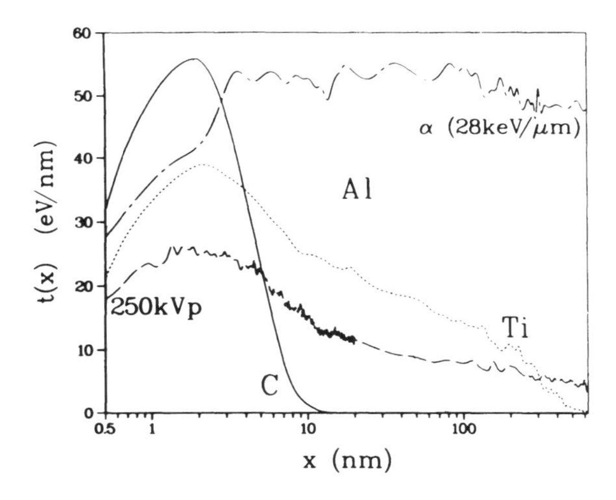

Figure 3. Calculated proximity functions for soft X-ray photoelectrons: C (0.27 keV), Al (1.5 keV); Ti (4.5 keV); and 250-kVp X rays and 28-keV/μm alpha particles.

Select energy transfers, ε, in the irradiated medium with a probability proportional to ε [i.e., the factor ε/E in Eq. (52)]. Then $t(x)\,dx$ is the expected sum of energy transfers contained in a shell of radius x and thickness dx centered at the transfer point where the energy transferred ε occurred.

As it will be seen later on, the proximity function $t(x)$ has an important biophysical significance in the context of the dual radiation action theory.[19]

Figure 3 shows several examples of proximity functions, $t(x)$.

Proximity functions could be calculated for radiation fields involving a spectrum of initial energies. If $p(E)\,dE$ is the fraction of particles with energies between E and $E + dE$ then particle tracks of energy E should be selected with a probability $p(E)$ in Eq. (52). A simple expression can be obtained for the situation where the *whole* track of the particle contributes. This situation occurs for indirectly ionizing particles (neutrons and photons) and/or for very large reference volumes. Indeed, since the selection probability for a transfer point is proportional to the energy transfer, then particles of initial energy E will contribute to the total proximity function, $t(x)$, proportional to $Ep(E)$. Therefore, if $t_E(x)$ is the proximity function for a monoenergetic particle, one has

$$t(x) = \int_0^\infty Ep(E)t_E(x)\,dx \Big/ \int_0^\infty Ep(E)\,dE \tag{54}$$

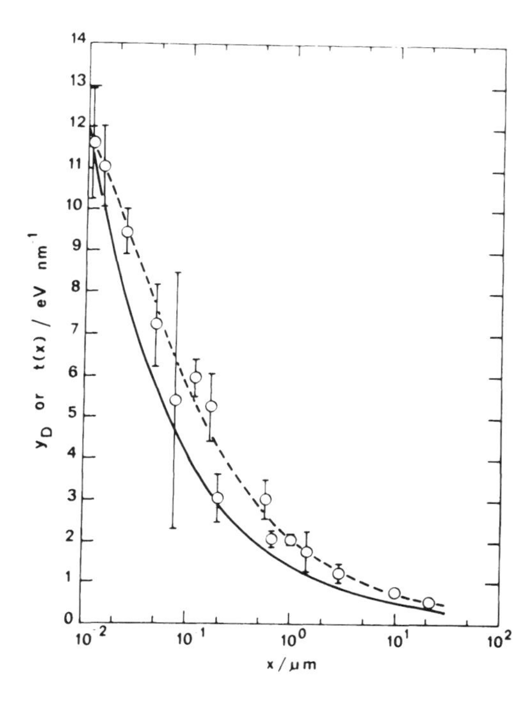

Figure 4. Circles: experimental values of y_D for Co-60 redrawn from Ref. 22. The proximity function (solid line) was calculated using Eq. (55) and a quartic B-spline fit (dashed curve) to y_D.

Finally, an inversion expression could be obtained[21] for the proximity function $t(x)$ of Eq. (50). If the dose-average single-event specific energy is known as a function of the site diameter, a, one has

$$t(a) = \frac{m}{3}\frac{d^2}{da^2}\left\{\frac{1}{a^2}\left[\frac{d}{da}a^4\bar{z}_{1D}(a)\right] - a\bar{z}_{1D}(a)\right\} \tag{55}$$

Experimental methods to measure directly $\bar{z}_{1D}$ over a wide range of site sizes are available[22] and the expression (55) could be used to "measure" the proximity functions (see Fig. 4). The equations (50) and (55) show the complete equivalence between the proximity functions $t(x)$ and the corresponding values $\bar{z}_{1D}(a)$.

3. EXPERIMENTAL TECHNIQUES

The experimental techniques used in microdosimetry are quite complex, and a complete description is beyond the scope of the present paper (see, however, Ref. 1). The goal of this section is to allow the reader to understand the main features of the microdosimetric methods used to obtain the data and to evaluate critically the available results.

3.1. General Principles

The present state of the art of experimental sciences does not provide reliable methods for the direct measurement of energy deposition in objects with dimensions of the order of 1 μm or less. Several alternate methods have been developed over the last 20 years which are capable of producing equivalent information. The basic idea behind these methods is quite simple. The mean energy imparted by a particular type of charged particle which traverses a thickness x of material of density ρ depends only on the product $x\rho$. One can obtain equal energy depositions by manipulating conveniently x and ρ such that their product remains constant. In practice, a microscopic tissue volume ($x \sim 1\ \mu$m, $\rho = 1\ \text{g/cm}^3$) is simulated with a larger cavity (x a few centimeters) by correspondingly reducing the density. Since most detectors use a gas in their active volume one can vary the density by simply changing the gas pressure. Thus, for instance, a 2.5-cm diameter sphere filled with propane-based TE gas at 17 Torr is equivalent to a 1-μm diameter sphere of unit density material.

A second aspect on which the simulation of microscopic volumes is based is the result known as Fano's theorem.[23] According to this the flux or the distribution of flux of secondary particles in an infinite and homogeneous medium is independent of the density of the medium (or density changes) provided that the mass attenuation coefficients are independent of density and that the strength density of any particle source (i.e., particles created in unit volume) is proportional to the density of the medium. In practice the conditions for Fano's theorem are realized whenever the composition of the gas cavity is identical to that of the surrounding medium. It should also be noted that the requirement that mass attenuation coefficients are independent of density is not always fulfilled owing to polarization effects in solids.[24]

The actual measurement of energy deposition spectra rests upon ionization phenomena. A typical counter consists of a spherical or cylindrical cavity (called sensitive volume) filled with gas and delineated by a solid wall material (see Fig. 5), and a central wire placed along the counter diameter or central axis and isolated electrically from the solid wall. The traversal of the sensitive volume by a charged particle results in the ejection of electrons from gas atoms. Depending on the probability of combination of these electrons with the neutral atoms (the attachment coefficient), they tend to remain free or (in the case of electronegative gases) to form negative ions. In either case one speaks generally of the production of ion pairs. The ions would normally have a random thermal motion in the gas. However, if a voltage is applied between the conductive wall of the counter and the central wire the ions will drift

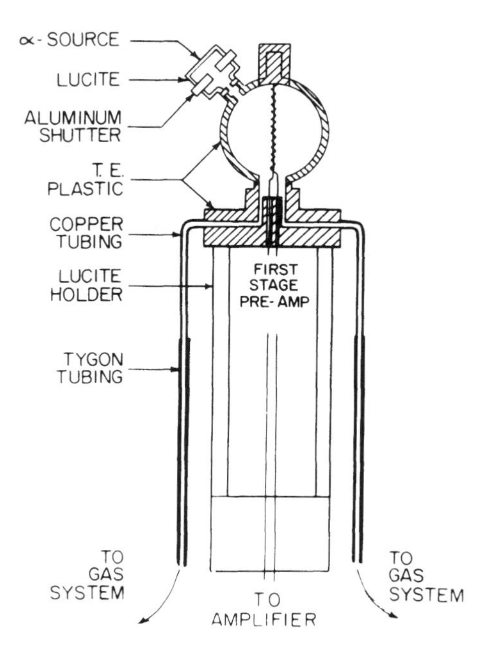

Figure 5. Schematic representation of a microdosimetric spherical proportional counter. Also shown is the first stage of the preamplifier connected directly to the counter in order to reduce electronic noise. For further details see text.

along the electrical field lines and induce a voltage change at the collecting electrode (center wire). The amount of charge collected is then a measure of the number of ion pairs produced in the counter. This number can be further converted to energy deposited if the mean energy required to produce an ion pair (i.e., the W value) is known for that particular type of particle and energy.

The size of the pulse appearing at the central wire depends on the applied voltage. For relatively low voltages (a few volts) the ion collection process competes with ion recombination phenomena since the movement of ions is relatively slow. If the voltage is further increased one reaches a region where practically all the ions produced are collected: this is called the ion chamber region since ionization chambers operate at such voltages. As the voltage is further increased gas amplification occurs: each primary ion acquires enough energy to produce second and higher-generation ions through collisions with the gas molecules. This process is especially important in gases which do not exhibit electron attachment because multiplication by electron collision is a much more efficient process. One obtains an amplified pulse, proportional to the energy deposited. A counter operating in this mode is called a proportional detector. This is the type of detector that is most commonly used in microdosimetry.

The gas gain (or amplification) in a proportional counter is governed, among other things, by the field strength across the sensitive volume. For the simple case of a cylindrical counter the field strength at a radius r is given by[25]

$$E = \frac{V}{r \ln (r_2/r_1)} \tag{56}$$

where V is the voltage applied, and r_1 and r_2 are the radii of the central wire and collection volume, respectively. It is obvious from the functional form of Eq. (56) that E decreases very fast with r and that, consequently, the gas amplification starts only in a region very close to the central wire (typically a few wire diameters). Under these conditions the gas multiplication is nearly independent of the position where the primary ion was produced and the measured pulse is proportional to the number of primary ions. A second major factor which determines the multiplication gain is the gas pressure. Detailed theoretical treatments[26,27] show that for very low pressures the multiplication region extends over a much larger radius such that the counter output ceases to be independent of the spatial distribution of primary ion pairs in the sensitive volume. This feature sets a lower limit for the minimum site size which can be simulated (typically about 0.25 μm).

3.2. Experimental Methods for Measuring Microdosimetric Spectra

The most commonly used detector in microdosimetric measurements is the spherical proportional counter. The spherical shape, although less convenient from a constructional viewpoint as compared with a cylindrical one, is preferred because of its complete directional symmetry in any radiation field. A spherical walled detector is represented schematically in Fig. 5. The sensitive volume of this detector is defined by a solid spherical shell made of tissue equivalent (TE) plastic (A-150).[28] The collector consists of a stainless steel wire about 0.0006 mm in diameter stretched along the counter diameter and surrounded by an additional electrode shaped as a helix. Because of the shape of the outer wall, the field is more intense toward the end of the collecting wire than it is in the center. By applying a voltage on the helix (about 80% of the shell voltage) one obtains a relatively uniform multiplication region along the center wire.

The sensitive volume is filled with a tissue equivalent gas mixture[29] at a pressure corresponding to the desired simulated site size. Gas purity and pressure are maintained by a flow system connected to the counter.[30]

The central wire (anode) is connected to a low-noise preamplifier which integrates the input charge signal and produces a voltage output of an amplitude that is proportional to the amount of charge deposited in the detector. After further amplification, the signals are fed into an analog-to-digital converter and stored for final processing in a memory unit. In modern microdosimetric systems the whole data taking process is computerized such that the desired distributions are obtained on-line.

As explained in the previous section, a microdosimetric measurement yields the distribution of ion pairs (charge) produced by the radiation field in the detector. One can convert this distribution into an energy deposition spectrum by using the W values for the radiation measured. This procedure might be in fact quite involved owing to two factors: (a) although for a given charged particle the W values are nearly constant over a wide energy range, strong variations occur at very low energies, and (b) for some radiations (neutrons, pions, heavy ions) microdosimetric distributions contain contributions from a variety of particles with slightly different W values. Since the measurement itself does not discriminate between particle types, the conversion from ion pairs to energy distributions has to be made based on an average W value which is representative for a range of energies and particle types. In practice this conversion is performed by calibrating the microdosimetric system with radioactive sources of known energies (such a calibration accounts automatically for the counter gain also). The detector is provided with a special opening where a small case containing the source is placed (see Fig. 5). Two types of calibrating sources are normally used. An example of the first type is a miniature X-ray tube emitting the 1.5-keV Al photons. The photoelectrons are completely absorbed in the detector and the ratio between the output pulse and the energy imparted (i.e., 1.5 keV) provides the calibration factor. A second type of source consists of a collimated beam of alpha particles (e.g., the 5.5-MeV α from Am^{241}) directed diametrically across the counter. For the pressures normally used in microdosimetry, the alpha particles traverse the sensitive volume completely and the energy deposited can be either calculated or directly measured with a solid-state device. In contrast to the first calibration method, this latter one requires in addition an accurate knowledge of the path length (counter diameter) along which the energy was imparted.

The two calibration procedures described above correspond roughly to W values for low- and high-LET particles, respectively. For each particular microdosimetric measurement the use of either procedure (or both) is dictated by the range of particle types and energy depositions under consideration.

3.3. Experimental Methods for Measuring $\bar{y}_D$

For a number of applications in radiobiology, as well as for the determination of the proximity function (see Section 2.3) the dose mean lineal energy $\bar{y}_D$, is of interest.[22] This quantity can be determined from an analysis of the measured $f(y)$ spectrum. However, an experimental procedure, known as the *variance method,* has been developed for the purpose of measuring $\bar{y}_D$ directly, without the usual intermediate step of determining a complete microdosimetric distribution. As a result, the range of site sizes which could be simulated can be extended considerably.

Consider the dose-dependent distribution of specific energy, $f(z; D)$. The variance, σ^2, of this distribution can be calculated using the result of Eq. (30):

$$\sigma^2 = \overline{z^2} - D^2 = \bar{z}_{1D} D \tag{57}$$

Here the average value of z is replaced with the dose, D. It is immediately apparent that if the relative variance, $V_r = \sigma^2/D^2$ of the distribution in specific energy, z, is measured one can obtain the single-event dose average of z, i.e., $\bar{z}_{1D}$:

$$\bar{z}_{1D} = V_r D \tag{58}$$

One should note that the measurement of V_r refers to distributions in z and *not* to the single-event distributions in z which are normally measured in microdosimetry. This is the essential point in the variance method.

The experimental procedure consists of measuring specific energy distributions of z, for a constant dose D, using proportional or ionization counters. No (or very little) gas multiplication is necessary since the size of the events measured depends on the average number of single-event contributions to z, and this can be controlled by changing D (or, equivalently, the irradiation time). Consequently, very small volumes can be simulated without many of the problems associated with proportional counters requiring a substantial gas gain. As an example, in a recent publication,[31] values for $\bar{z}_{1D}$ for Co^{60} and X rays have been reported for site sizes as small as 10 nm in diameter (unit density material).

An obvious requirement for the application of the variance method is that the relative variance associated with energy deposition should be much larger than any other spurious effects (i.e., variations in beam intensity, gas pressures, electronic noise, etc.).*

* A recent development—the so-called variance–covariance technique—makes some of the requirements unnecessary. For further details see A. M. Kellerer and H. H. Rossi, *Radiat. Res.* **97,** 237–245 (1984).

3.4. Wall Effects and Wall-less Counters

The simulation of microscopic volumes using macroscopic gas-filled cavities surrounded by solid material results in a number of spurious phenomena termed wall effects (a more appropriate name is probably density effects).

Consider a particle track in a medium of density ρ (Fig. 6). Energy depositions can be calculated by placing randomly reference spherical volumes (of diameter a) with their centers in the associated volume of the track. If the reference volume is now replaced with an equivalent gas cavity with (for instance) density $\rho_1 = \rho/2$ and diameter $a_1 = 2a$, the geometrical pattern of the energy transfer points *inside* the cavity will be simply "magnified" by a factor of 2 (see Fig. 6). The calculated energy spectrum will be, however, the same. The same result is obtained for any distribution of tracks because of Fano's theorem.[23]

Consider now a different kind of track where the lateral distribution in space of transfer points has dimensions similar to, or greater than, the diameter of the gas cavity. Also let this track deviate strongly from a quasilinear shape over portions of its length. Such deviations may be due to multiple Coulomb scattering. Two reference volumes of density ρ are shown in Fig. 7. If, again, these volumes are replaced by equivalent gas cavities (ρ_1, a_1) the "magnification" effect results in *different* energy depositions: in addition to the transfer points which were originally in the spherical volume, one now has contributions from the branch B, in one

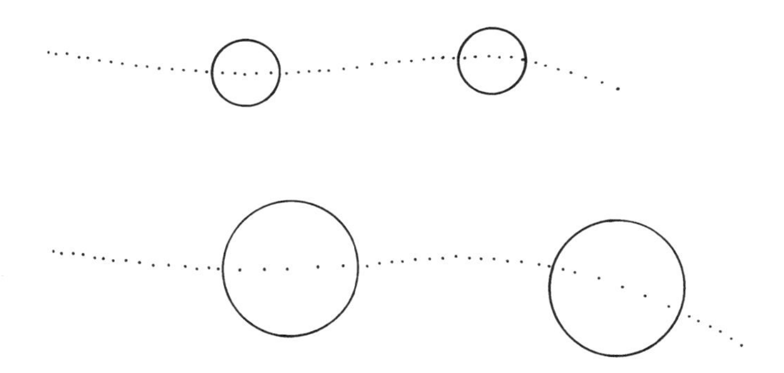

Figure 6. Illustration of the wall effect, I: a "simple" track (represented as ionization transfer points) penetrates a medium of density ρ and two cavities (density ρ, diameter a). The energy deposited in each cavity is proportional to the number of transfer points inside its volume. The same energy depositions will be recorded if the diameter of the cavity is increased ($2a$, for instance) and the density is decreased ($\rho/2$) by the same factor: transfer points inside the cavity are separated by larger distances inversely proportional to the relative reduction in density. This situation illustrates the basic idea behind simulating *micro*scopic sites (of unit density) with *macro*scopic volumes (of lower density).

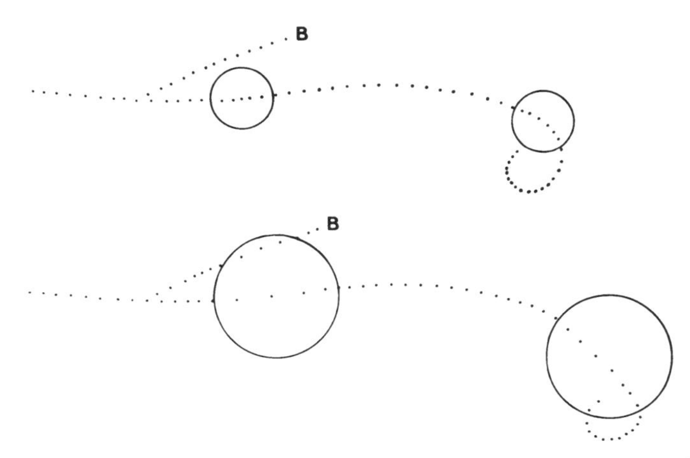

Figure 7. Illustration of the wall effect, II: a "complicated" (and more realistic) track has a branch, B (a secondary particle such as a delta ray), and owing to multiple scattering deviates strongly from a quaslinear shape, particularly toward the end of its trajectory. A change in the density results in a *different* number of transfer points inside its volume, i.e., an erroneous result. This is termed the wall effect because of the difference in density between the surrounding medium (wall) and the measuring volume (gas). Its magnitude is in direct relation to the relative occurrence of such "complicating" features as secondary particles and large-angle deviations.

case, or from a different part of the same track, in the other. The increased geometrical site of the measuring object distorts the energy spectrum toward higher energies. These types of distortions are called wall effects because they are normally caused by the variation in density between the detector wall (solid) and the sensitive volume (gas).

There are two important points to be noted. First, the main track and its branch, as an example, need not be necessarily simultaneous inside the cavity. No detector available to date, though, has a resolving time short enough to discriminate between such events. The total energy imparted will be detected, falsely, as resulting from one single event.

Second, when the atomic composition of the solid material and gas in the cavity are the same, the distribution of energy depositions is the same, irrespective of density (Fano's theorem). Because of the different geometrical size of the cavity (i.e., difference in density), what is changed in fact is the *time* distribution of the energy depositions: for a larger cavity (of lower density) an increased fraction of otherwise similar energy depositions will be associated with the same event, and therefore will be statistically correlated in time (in practice, coincident).

It is clear that the magnitude of the wall effects depends directly on the radial structure of energy depositions around the track (e.g., δ rays, by-products of nuclear interactions, etc.) and/or the deviations from a

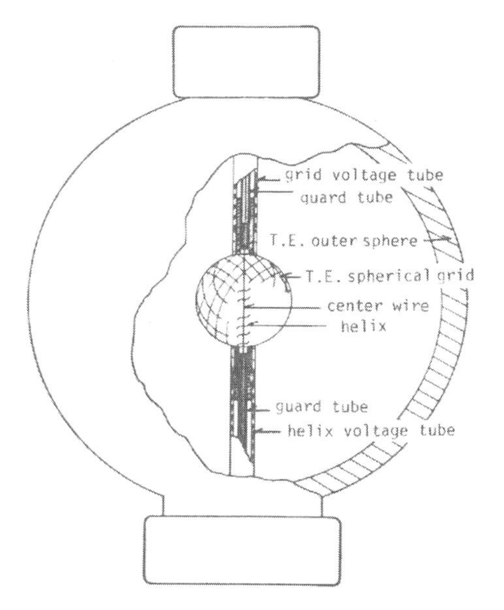

Figure 8. Schematic representation of a spherical wall-less counter.

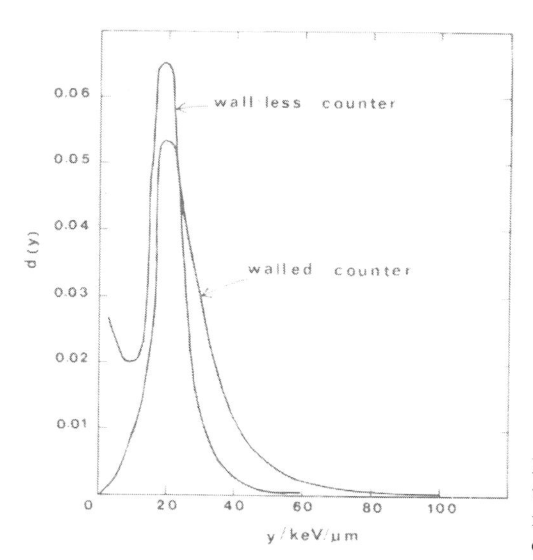

Figure 9. Comparisons of dose distributions for 3.9-GeV nitrogen ions, measured with walled or wall-less counters (redrawn from Ref. 34).

straight line of the tracks because of scattering. Simplified theoretical estimations of these effects are available[32] together with the necessary corrections. The experimental approach, however, has been to build counters for which the sensitive volume and the surrounding material have, in addition to the same atomic composition, the same density. Such devices are called wall-less counters.[33]

An example of a spherical wall-less counter is shown in Fig. 8. The sensitive volume of the counter is defined by a spherical fine grid of high transparency (>90%) made from TE material. The whole counter is enclosed in a larger cavity uniformly filled with TE gas at the desired pressure. The cavity dimensions are usually dictated by the condition of charged particle equilibrium. The longest range of any secondary particles (primarily electrons) should be smaller than the distance between the counter and the solid walls of the cavity. It is also apparent that the use of such a counter, particularly in phantoms, might distort significantly the radiation field to be measured.

The wall effects can be directly measured by using walled and wall-less detectors. An example is shown in Fig. 9.

4. MICRODOSIMETRIC DISTRIBUTIONS

4.1. General Considerations

A microdosimetric spectrum describes the distribution of energy imparted to a small volume of matter in a radiation field. Strictly speaking, such a distribution is entirely defined if the fluence and energy spectra of the charged particles which traverse (even incompletely) the volume are known. The distribution of charged particles in the reference volume called "slowing-down" spectrum, is related to the original source of particles through a number of parameters: the initial energy of the beam, the amount and composition of absorber placed between the reference volume and the source, collimators, beam contaminants, the phase-space of the beam, and the position of the volume within the irradiated field. A complete specification of all these parameters is not always essential or even necessary. In the simplest cases (uncharged particles), the energy of the source particles might define quite uniquely the shape of the spectra if the detector is situated along, or close to, the central axis of the beam. Generally, at least two parameters are used: the beam energy *at the source* and the amount of absorber. The effect of the other parameters mentioned previously depends usually on the experimental facility used to produce the particles and is not, therefore, characteristic of the basic microdosimetric aspects of that radiation field.

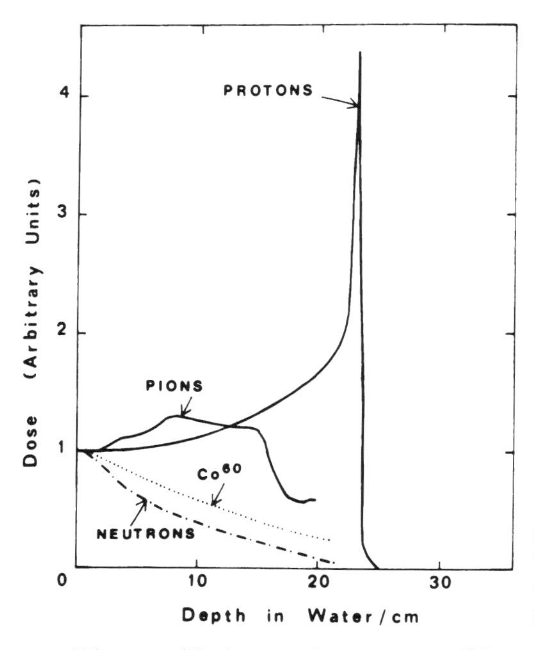

Figure 10. Typical depth–dose distributions for Co^{60} photons, pions, neutrons, and protons (redrawn in part from Ref. 35).

The specific interactions responsible for energy deposition processes and their dependence on the primary beam energy are discussed in the next subsections for each individual radiation. In the following a number of general, qualitative features are examined. Figure 10 shows depth–dose distributions for γ rays, pions, neutrons, and protons. In their passage through a medium, uncharged particles (photons and neutrons) deposit energy in a two-step sequence. First, an "initial' spectrum of charged particles is produced via primary interactions. The actual distribution of charged particles at a given point will include, in addition to the initial spectrum, contributions from primary particles originating outside that point; these particles reach the point of interest after slowing down as they traverse the medium. This second distribution is the "slowing down" spectrum referred to before. The exponential shape of the depth–dose curve for photons or neutrons (Fig. 10) reflects, then, the process of *attenuation* of the beam in its interaction with the medium. The decrease in the average energy of the beam with depth is only a secondary effect, and within a good approximation one can state that no change in the quality of the beam (from microdosimetric viewpoint) with depth is expected. This fact is confirmed by the actual measurements.[36]*

* Variations occur, though, if one moves in the beam penumbra. They are due essentially to changes in the relative amount of charged vs. uncharged particles as one moves, laterally, from the main axis of the beam. Such variations are important mainly in therapy situations.

It is therefore appropriate to characterize the microdosimetry of a photon or neutron field by a spectrum obtained in one position only and as a function of the primary beam energy (see, however, footnote below).

In contrast to uncharged particles, beams of charged particles interact with the medium in a quasicontinuous manner, primarily through collision with atomic electrons resulting in ionizations and excitations. A depth–dose curve for heavy charged particles (see Fig. 10) reflects in general the variation in the rate of energy loss as a function of particle energy (or equivalently, depth): after an initial, relatively flat region (called plateau), the rate of energy loss increases sharply as the particles reach the end of their range, producing a maximum known as the Bragg peak. The shape of the depth–dose curve is determined by the effects of both atomic collisions and nuclear interactions; the attenuation effect of this latter process is normally offset by the increased contribution from energy transfers to the atomic electrons with decreasing particle energy. It is clear that here, unlike in the case of photons or neutrons, the change in energy of the particle (and the associated changes in the slowing down spectra), and not an attenuation in number, dominates the variation of absorbed dose as a function of depth. Consequently, microdosimetric distributions are expected to be significantly altered as the amount of absorber is varied. Therefore, such distributions will be examined as a function of both energy *and* position along the Bragg curve.

Microdosimetric distributions for neutrons and photons have been measured for a wide spectrum of energies, in direct relation to their use in radiotheapy, radiobiology, and, because of their penetrating nature, for radiation protection purposes. Similar measurements for charged particles have been carried out primarily for beams utilized in clinical applications where the primary energy is selected in conformity with the characteristics of the depth–dose distribution (sufficient penetration, favorable peak to plateau ratio, etc.).

The object of the present section is to familiarize the reader with the salient characteristics of the microdosimetric distributions for different types of radiations. An extensive microdosimetric literature is available covering almost any ionizing particle of practical use. Only typical spectra are shown here with emphasis on the microdosimetric *quality* of different fields and the underlying physical processes.

4.2. Representations of Microdosimetric Spectra

A typical microdosimetric spectrum is presented in Fig. 11. This spectrum shows the energy depositions of negative pions, and was obtained with a spherical proportional counter filled with tissue equivalent gas at a pressure which simulates a 2-μm diameter, unit density

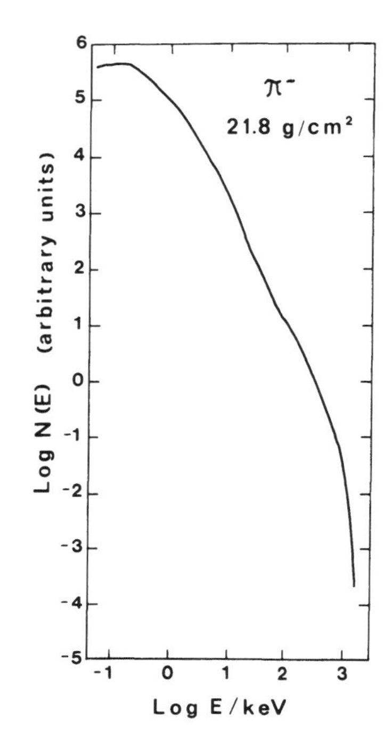

Figure 11. Typical microdosimetric distribution of the number of events, $N(E)$, vs. the energy, E, deposited in a spherical counter simulating a 2-μm diameter site of unity density.[37] This spectrum was obtained with 78-MeV negative pions after traversing a 21.8-g/cm^2 thickness of water. Because of the extreme range of values for both E and $N(E)$ a double logarithmic representation is used.

material. In the representation of Fig. 11 the abscissa indicates the energy, E, deposited in the counter and the ordinate is the number of events, $N(E)$, depositing a given amount of energy.

A remarkable feature to be observed in Fig. 11 is the extreme range of values for both the energy deposited (4.5 decades) and the number of events (9.5 decades). This situation is not atypical for other radiations as well. In order to enhance the clarity of the spectrum display, a logarithmic energy scale is frequently used. Converting then the experimental spectrum from a linear to a logarithmic scale requires a number of modifications. Consider first the frequency distribution $f(E)$ obtained by normalizing the spectrum $N(E)$ to 1:

$$f(E)\,dE = N(E)\,dE \Big/ \int_0^\infty N(E)\,dE \tag{59}$$

In a linear representation, $f(E)$ vs. E, the area under the curve $f(E)$ delineated by two values of E is equal to the fractional number of events which occurred in that energy interval. This property can be preserved in a logarithmic representation if $Ef(E)$ is plotted versus $\log E$. Indeed

$$\int_{E_1}^{E_2} f(E)\,dE = \int_{E_1}^{E_2} Ef(E)\,d\ln E = k\int_{E_1}^{E_2} Ef(E)\,d\log_{10} E \tag{60}$$

where k converts the natural logarithm to a logarithm of base 10. Similarly, if a linear representation is used for the *dose* distribution $Ef(E)$ vs. E, the area under the curve delineated by two values of E is proportional to the fraction of the total *dose* delivered in that energy interval. In a logarithmic representation one then plots $E^2f(E)$ versus $\log E$ [see Eq. (60)]. It is important to realize clearly the difference between the logarithmic display of the *frequency* distribution, $Ef(E)$ vs. $\log E$, and the *dose* distribution, $Ef(E)$ vs. E, which corresponds to a linear energy scale. The following distributions are then pairwise equivalent:

$$\begin{aligned} f(E) \text{ vs. } E \quad &\text{or} \quad Ef(E) \text{ vs. } \log E \\ Ef(E) \text{ vs. } E \quad &\text{or} \quad E^2f(E) \text{ vs. } \log E \end{aligned} \tag{61}$$

Figures 12 and 13 represent the frequency and dose distributions, respectively, corresponding to the spectrum of Fig. 11. Here the energy scale has been converted to lineal energy. The average values of the

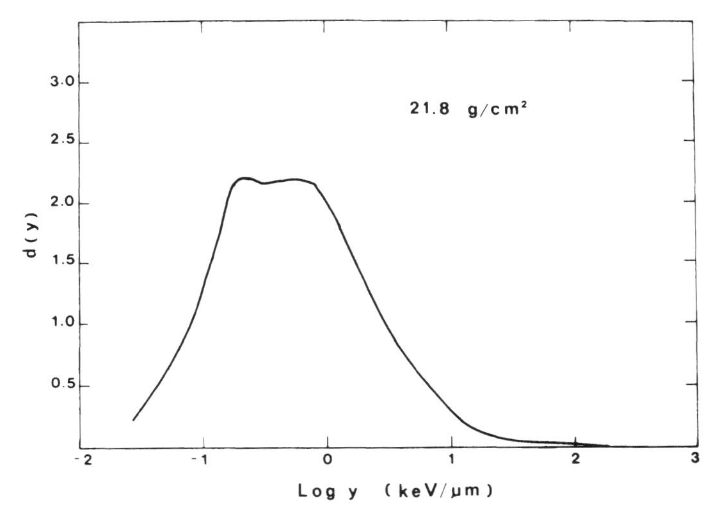

Figure 12. The same spectrum as in Fig. 11 in a $d(y)$ vs. $\log y$ representation. This is a *frequency* distribution (and not a dose distribution) because of the logarithmic scale in y [see Eqs. (61) and (38)].

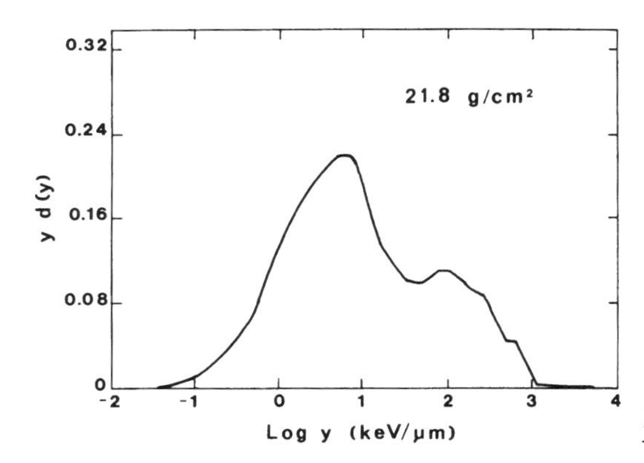

Figure 13. The same spectrum as in Fig. 11 in a dose representation, $y\,d(y)$ vs. $\log y$.

lineal energy can be calculated from

$$\bar{y}_F = \sum_i y_i^2 f(y_i) \Big/ \sum_i y_i f(y_i)$$

$$\bar{y}_D = \sum_i y_i^3 f(y_i) \Big/ \sum_i y_i^2 f(y_i) \tag{62}$$

where the summations are over points equally spaced on a *logarithmic y* scale.

4.3. Photons

The X- and γ-ray photons deposit energy in a medium by initial formation of secondary charged particles (mostly electrons) which in turn produce ionizations and excitations. There are three main processes[38] through which photons interact with a medium: photoelectric effect, Compton scattering, and pair production. At very high energies nuclear interactions occur with relatively low probability.

In a photoelectric interaction the photon energy is completely transferred to an atomic electron: part of this energy is expended in removing the electron from the atom (binding energy); the rest provides the kinetic energy of the emitted electron (photoelectron). The photoelectric effect is accompanied by the production of secondary X rays resulting from the filling of the vacancy left by the photoelectron with outer elecrons. Occasionally these secondary photons (also called fluorescent radiation) are absorbed by an outer electron, resulting in its emission. The ejected electron is called an Auger electron. The photoelectric effect represents an interaction of the photon with the atom as a whole; the cross section for this effect increases, therefore, with the atomic number, Z, of the medium (roughly as $Z^4 - Z^5$) and decreases rapidly with increasing photon energy ($\sim E^{-3}$).

For photon energies much larger than the binding energies of the atomic electrons an elastic scattering occurs between the photon and the electron. This process, termed the Compton effect, can be completely described by the kinematics resulting from the energy and momentum conservation laws. Thus, the maximum value for the energy of the Compton electron results when the photon is scattered in the backward direction and vice versa. Since in Compton processes the electrons behave essentially as if they were free, the fraction of Compton scattered photons depends only on the total number of electrons per unit mass in the medium. The cross section for Compton scattering is thus independent of Z, and approximately proportional to $1/E$, where E is the photon energy.

Pair production occurs when a photon materializes into an electron and a positron in the vicinity of an atomic nucleus. The minimum photon energy necessary for this process is 1.02 MeV (the sum of the rest masses of the electron and positron). The cross section for pair production increases both with the atomic number ($\sim Z^2$) and the energy ($\sim E$).

For tissuelike media the photoelectric effect dominates the interactions of photons for energies up to about 50 keV. For higher energies Compton scattering becomes more prominent, with pair production becoming comparable at energies of the order of 10 MeV or more. For energies of interest in radiobiology, pair production can usually be neglected.

Several microdosimetric distributions for photons are shown in Fig. 14. These spectra were measured with a wall-less spherical proportional detector at a gas pressure which simulates a 1-μm diameter cavity. The photon primary energies range from 11.9 to 1250 keV (this latter is the average energy of ^{60}Co gamma rays). The distributions—here and in the rest of this section—are dose distributions of y represented in a $y^2 f(y)$ vs. $\log y$ plot.

The photon microdosimetric spectra span about three decades in lineal energy, from 0.01 to about 10 keV/μm. This is typical for the so-called low-LET radiations. The widths of the microdosimetric spectra in Fig. 14 are, to a large extent, a reflection of the energy spread of initial electron energies and the corresponding slowing down distributions for each photon energy. Soft photons (10–60 keV) interact predominantly via photoelectric effect producing monoenergetic electrons of essentially equal energy. In contrast, higher energy gamma rays (e.g., ^{60}Co) result in a wide spectrum of Compton electrons.

A second important factor affecting the spread in lineal energy is the combined effect of path length and energy straggling. For smaller energy depositions, i.e., shorter path length and/or higher electron energy, the energy straggling spread is larger. Energy straggling effects can be clearly

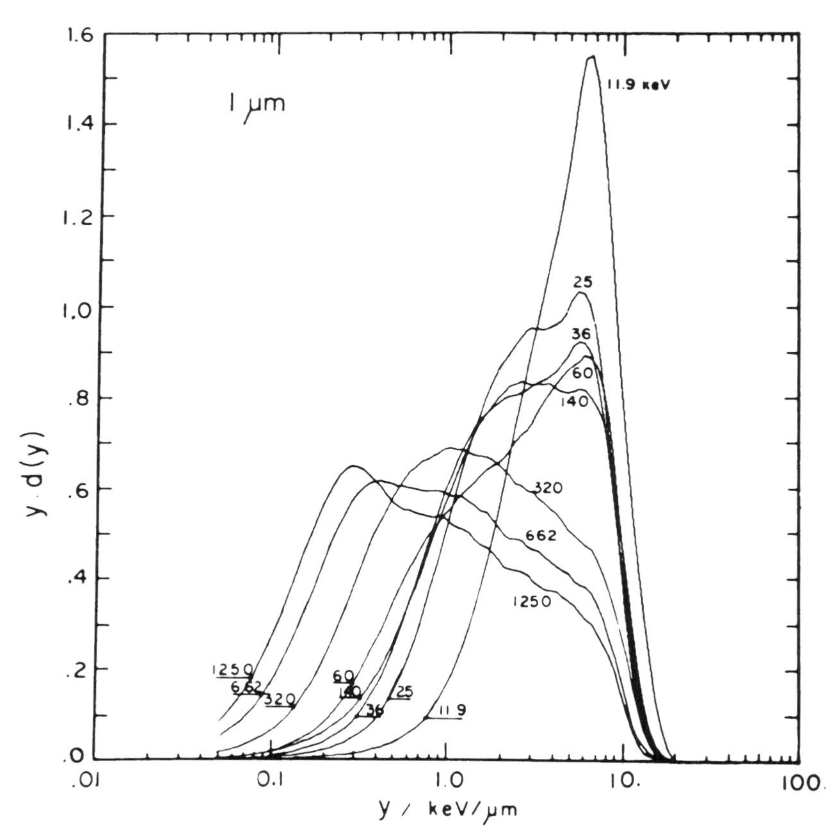

Figure 14. Microdosimetric distributions for photons of various energies (redrawn from Ref. 39).

seen at high y values. Indeed, inspection of an energy-range table shows that for a given path length, the average amount of energy deposited reaches a maximum when the particle range roughly equals the path length (i.e., all the energy is deposited across the track segment). For the distributions of Fig. 14, one expects a maximum lineal energy of about 8.6 keV/μm. Because of energy straggling the spectra extend in fact well above this value. Two more factors, of experimental nature, influence the shape of the measured spectra: the fluctuations in the number of primary ion pairs[40] (Fano factor) and the statistics of the electron multiplication process. This latter factor is usually negligibly small.

Finally, the position of the peaks in the spectra can be understood in relation to the energy spectrum of the secondary particles. For instance, the average of Compton electrons from ^{60}Co is 580 keV, and this corresponds to the peak at about 0.3 keV/μm in the y distribution.

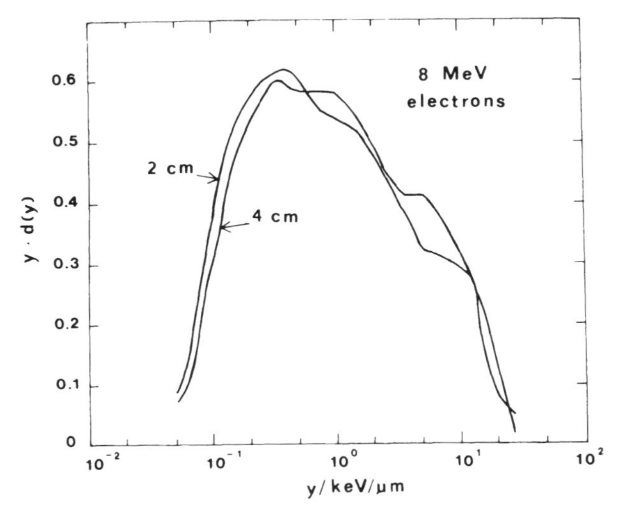

Figure 15. Microdosimetric distributions for 8-MeV electrons at 2 and 4 cm depth in water (redrawn from Ref. 41).

The spectra shown in Fig. 14 are typical for the microdosimetry of photons. A fortiori, they are representative for electron beams as well. Examples of microdosimetric distributions for 8 and 38 MeV electrons are shown in Figs. 15 and 16 as a function of the amount of absorber. Since at these energies the electrons are relativistic over most of their path length, no significant change is observed with depth in the phantom.

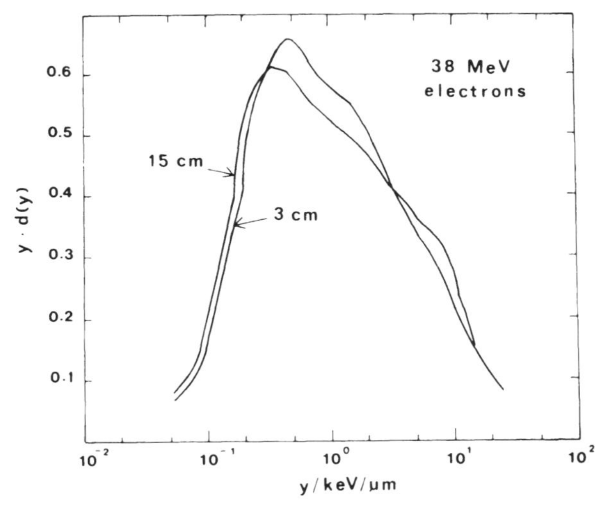

Figure 16. Microdosimetric distributions for 38-MeV electrons at 3 and 15 cm depth in water (redrawn from Ref. 41).

4.4. Neutrons

As in the case of photons, neutrons deposit energy through the agency of secondary charged particles, but these have high LET. The interaction of neutrons with a medium depends strongly on their energy (see Ref. 42). Conventionally, neutrons are classified in four categories according to their energy: (a) thermal neutrons, with energies of the order of eV; (b) intermediate neutrons, with energies between a few eV and 10 keV; (c) fast neutrons, with energy between 10 keV and 10 MeV; and (d) relativistic neutrons, having energies larger than 10 MeV. The last two classes of neutrons are of particular importance in radiobiology and radiotherapy.

With the exception of thermal energies, the interactions of neutrons are dominated by the process of elastic scattering. Such a collision occurs between the neutron and the nucleus as a whole with energy and momentum conserved (billiard-ball collision). Simple kinematics shows that the neutron transfers maximum energy to nuclei of comparable mass, i.e., hydrogen nuclei (protons). Typically, an average of 1/2 of the neutron energy is transferred to such recoil protons. Elastic collisions of this type are particularly significant in tissuelike media where the hydrogen content is very high.

Inelastic collisions begin to contribute for neutron energies larger than about 0.5 MeV. Here the target nucleus is left in an excited state which subsequently decays by emitting photons. For light nuclei the first excited states are a few MeV above the ground state (^{12}C, 4.4 MeV; ^{14}N, 2.31 MeV; ^{16}O, 6.1 MeV), while for heavier nuclei energies of the order of 100 keV are more typical. Inelastic scattering is therefore more important in high-Z nuclei.

For increasingly higher energies (>5 MeV) nonelastic collisions (i.e., emission of particles other than the incident neutron) occur. Examples of such nuclear reactions are ^{12}C (n, α) ^{9}Be, ^{16}O (n, p) ^{16}N, or ^{16}O (n, α) ^{13}C. These processes are almost always accompanied by the emission of nuclear gamma rays. The resulting charged particles (heavy recoils), together with the proton recoils, contribute a significant fraction of the energy deposition in the medium and are of particular importance in microdosimetry.

A special class of inelastic processes, termed capture reactions, occur at thermal energies. In a capture process the neutron is absorbed in the nucleus and the excess energy is given off either by gamma radiation (radiation capture) or by the emission of charged particles. Finally, at the other end of the neutron energy spectrum (100 MeV or more) spallation becomes significant. In such an interaction the nucleus is fragmented into light charged fragments which deposit energy locally. Several neutrons or photons might also be emitted.

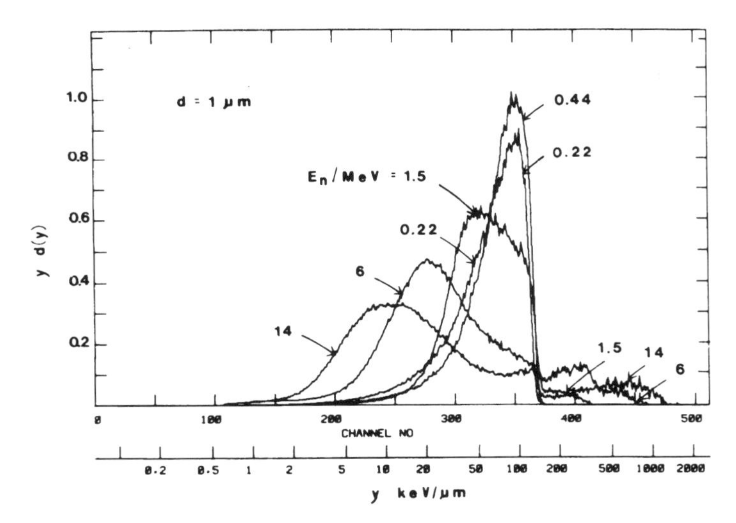

Figure 17. Microdosimetric distributions for neutrons of various energies.[4]

Microdosimetric distributions for neutrons can be fairly well understood in terms of the basic physical processes described above. Figure 17 shows a number of microdosimetric spectra for neutron energies ranging from 0.22 to 14 MeV. The spectra were obtained[4] with a wall-less counter simulating a site size of 1 μm diameter. The distributions are dominated by the proton recoil peak. This peak is shifted toward lower lineal energies as the neutron energy is increased, in direct relation to the fact that the initial spectrum of protons (and the corresponding slowing down spectrum) covers an increasingly larger range of energies. The right-hand edge of the proton peak is similar for all neutron energies and corresponds to protons of maximum stopping power ($y \sim 150$ keV/μm). Heavy recoils are clearly apparent for neutron energies above 1 MeV. Their contribution, at y values larger than 100 keV/μm, becomes more prominent as the neutron energy increases. For 14-MeV neutrons, distinct peaks belonging to alpha particles and heavier recoils can be observed.

All the neutron fields are contaminated by gamma radiation with lineal energies below 10 keV/μm (see previous section). The contribution to the *dose* is usually small compared with that due to the protons or heavier recoils (see Fig. 17).

4.5. Heavy Charged Particles

As noted previously the main mechanism by which fast, charged particles lose energy when moving through matter is the collision with atomic electrons. Nuclear interactions are a second important process in this respect. Although relatively infrequent, compared with atomic collisions, these processes may contribute significantly to the dose because of the fact that they result, in part, in low-energy (high-LET) secondary particles. In a relatively simplistic way the nuclear interactions of heavy ions can be classified in four main groups according to their mechanism[43]: (a) projectile fragmentation, (b) target fragmentation, (c) central collisions, and (d) multiple Coulomb scattering.

The process of projectile fragmentation, as its name indicates, represents an interaction where the incoming heavy ion is fragmented into several lighter fragments. Experimental evidence shows that projectile fragments are emitted mostly forward with velocities roughly equal to that of the original particle.[44,45] The cross sections for these processes are essentially independent of the projectile energy for energies above 300 MeV/amu (this is usually the region of interest). Projectile fragments effectively modify the composition of a fast heavy-ion beam with a lower-LET component. Since such "contaminants" travel parallel with the main beam, their contribution will increase as a function of the material thickness traversed.

For lighter particles (e.g., protons) projectile fragmentation is either insignificant or impossible. Instead target fragmentation dominates. Here, the nuclei in the medium (targets) disintegrate into low-energy fragments emitted mostly isotropically.[46] In contrast to projectile fragmentation, this process results in particles of higher-LET compared with the primary beam. Central collision interactions have been described by the so-called "fireball" model:[47] the projectile and the target "melt" into a compound system which evaporates through the emission of light fragments. Characteristic for the last two processes described is the fact that they tend to broaden the dose distribution around the beam.

The above processes are accompanied by the production of a large number of indirectly ionizing particles (neutrons and photons). It is important also to note that the heavy ions and their secondaries produce copious amounts of delta rays, a significant fraction of which might be energetic enough to deposit energy in a small volume independently of the originating particle.

Figures 18b and 18c show examples of microdosimetric distributions for protons. The first spectrum corresponds to the "plateau" position of a 160-MeV proton beam. The second spectrum was obtained at the distal position of the Bragg peak in a range-modulated beam (see Fig. 18a). The distributions in Figs. 18b and 18c are remarkably similar to photon

→

Figure 18. Microdosimetric distributions for 160-MeV protons in the "plateau" (b) and the distal (c) position along the depth–dose curve (a) (redrawn from Ref. 48).

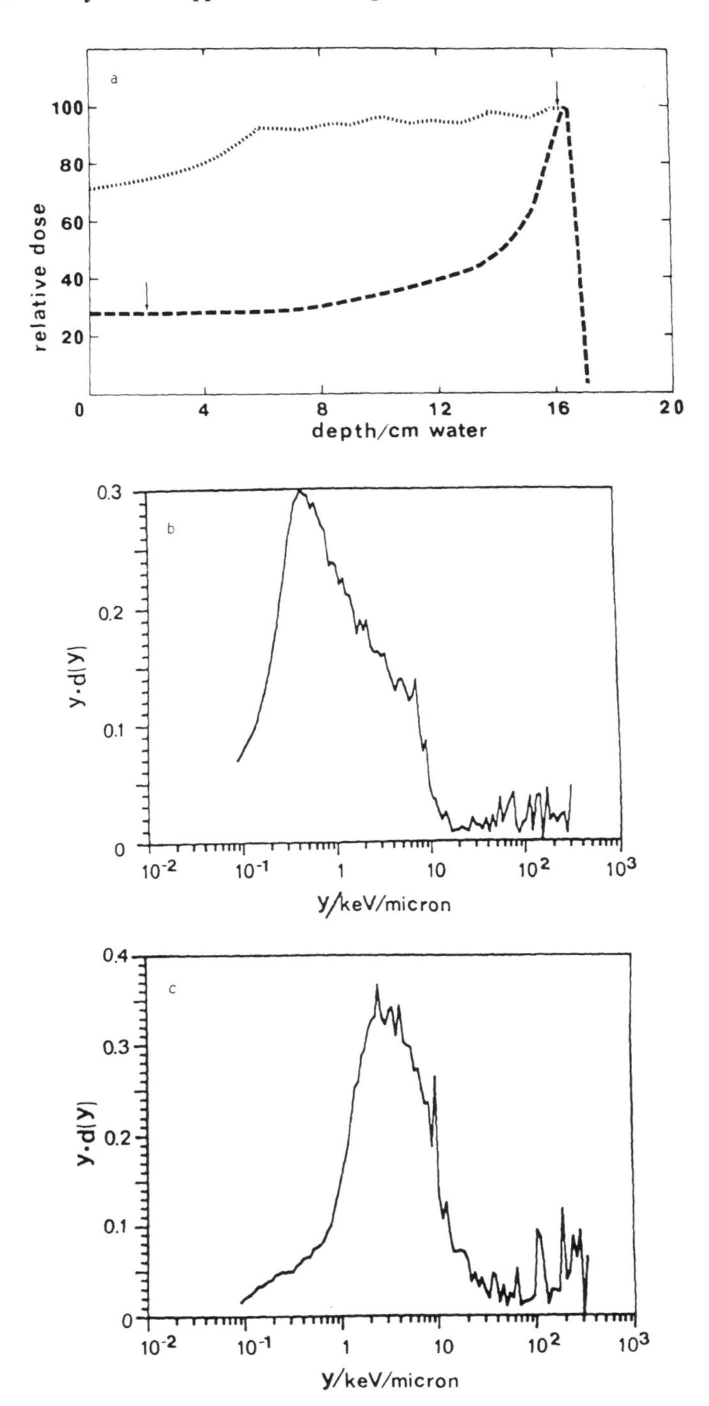
a
100
80
60
40
20
relative dose
0
4
8
12
16
20
depth/cm water
b
0.3
0.2
0.1
0
y·d(y)
10^{-2}
10^{-1}
1
10
10^2
10^3
y/keV/micron
c
0.4
0.3
0.2
0.1
0
y·d(y)
10^{-2}
10^{-1}
1
10
10^2
10^3
y/keV/micron

spectra. They consist essentially of three distinct regions: (a) a low-y section (around 1 keV/μm) contributed by protons, (b) the region between 1 and 10 keV/μm attributed to scattered protons and/or delta rays, and (c) a small component at higher lineal energies due to spallation products and very low energy protons. From the microdosimetric viewpoint high-energy protons are therefore true low-LET radiation.

For heavier particles the microdosimetric distributions show, in addition to the passing primary ions, contributions from projectile fragments (at y values lower than the main peak) and target fragments (at y values larger than the main peak). The contribution from these two

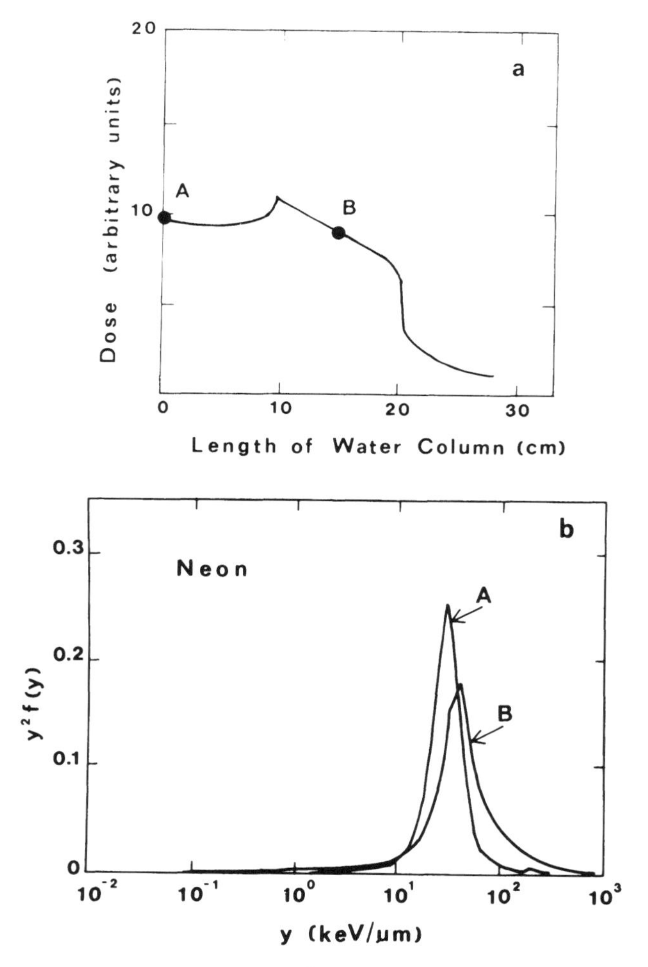

Figure 19. Microdosimetric distributions[5] for 557-MeV/amu Ne^{10} ions (b) at two positions along an extended Bragg curve (a).

components increases with the atomic number of the projectile and as a function of the absorber thickness in front of the detector (i.e., position along the Bragg curve). The spectra in Fig. 19 correspond to measurements at two positions along an extended Bragg curve for 557 MeV/amu ^{10}Ne ions.

4.6. Negative Pions

Negative pions (π^-) are charged particles with a rest mass of 139.6 MeV, intermediate between the electron (0.511 MeV) and the proton (938.2 MeV). The general characteristics of energy deposition by heavy charged particles apply to negative pions as well.[35] A unique feature of negative pions is that when they reach the end of their range (energies of the order of eV) they are captured into the outer orbits of the atoms of the medium, forming pionic atoms. The pion then cascades down to the lower atomic orbit, where, because of its large mass, it spends a large fraction of the time inside the nucleus. This leads to the process of pion capture: the pion rest mass (140 MeV) is absorbed in the nucleus causing its fragmentation into short-range, heavily ionizing particles, neutrons, and gamma rays. The whole process is sometimes called star formation because of the pattern of nuclear fragment tracks observed in nuclear emulsions.

Star formation results in enhanced dose and high-LET content at the Bragg peak of a pion depth dose distribution; it is precisely this feature which makes negative pions attractive for radiotherapy.[35]

The microdosimetric studies performed to date have been carried out for π^- beams utilized in clinical trials. Figure 20 shows microdosimetric spectra obtained at different positions along the Bragg curve of 78-MeV negative pions. The broad peak in the spectrum obtained at the plateau represents particles passing directly through the chamber (the stopping power of 78 MeV π^- in tissue is 0.3 keV/μm). y values between 1 and 10 keV/μm are contributed to by scattered protons and secondary electrons. A small fraction of the dose, which is a result of in-flight interactions, is delivered above 10 keV/μm.

As the pions slow down, their stopping power increases and the main peak is shifted to higher y values (see Fig. 20). The distribution in the Bragg peak region shows clearly the effects of star products (protons and alpha particles) of much higher lineal energy. The relative contribution of these products increases further at depths beyond the peak as most of the passing pions are stopped. The effect of the star products is most clearly seen by comparing spectra obtained with π^- and π^+ at a given depth. Such a comparison is shown in Fig. 21 at 18.75 cm depth in the Bragg curve. Finally, it should be noted that pion beams are contaminated with

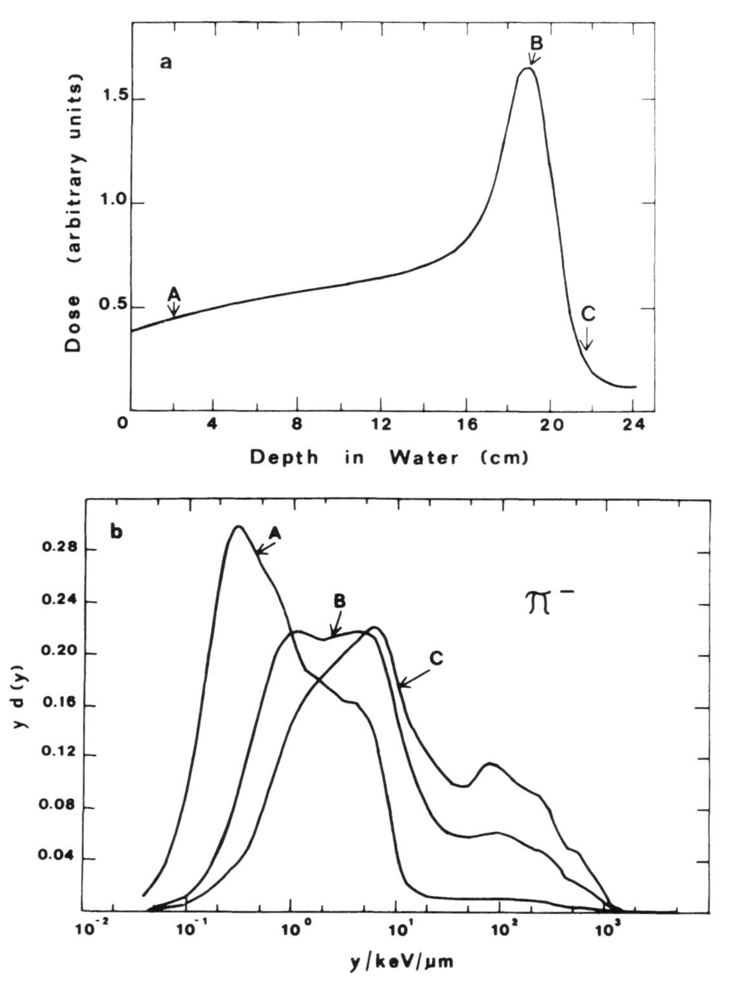

Figure 20. Microdosimetric distributions for 78-MeV negative pions[3] (b) at different positions along the Bragg curve (a).

electrons and muons of equal momentum, and therefore, longer range. These particles contribute a low-y component to the microdosimetric distributions for all the positions along the pion depth–dose curve.

PART II. THE BIOPHYSICAL IMPORT OF MICRODOSIMETRY

The second part of this chapter deals with the significance of the concepts of microdosimetry to theories of the biological action of ionizing

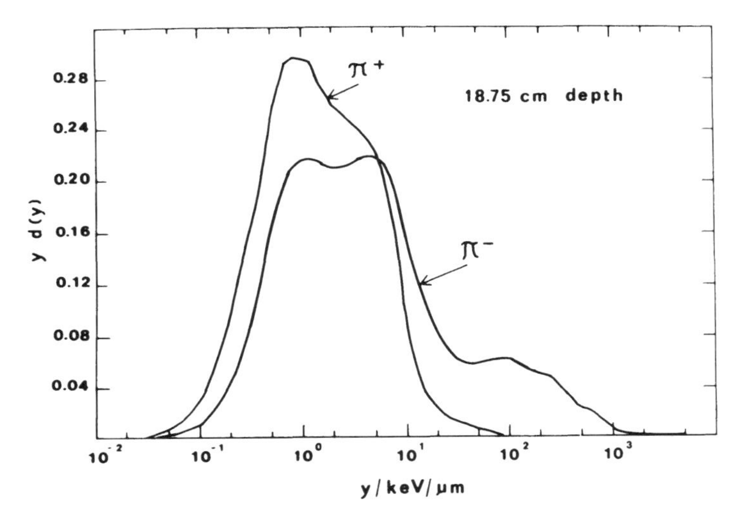

Figure 21. Comparison between π + and π − microdosimetric spectra at 18.75 cm depth in water (see also Fig. 20a).

radiation and especially with the consequences of the statistical correlation between the energy transfers in individual events.

In order to achieve rigor it has been necessary to employ mathematical formalisms. However, while somewhat complex, the relations presented do not embody particularly difficult mathematical concepts. An attempt has also been made to present the material in such a way that detailed study of equations is not essential.

5. MICRODOSIMETRY AND HIT–TARGET THEORIES

5.1. Formal Hit–Target Theories

The hit and target theories were the first systematic and sometimes successful attempt to explain dose–effect relationships based on the statistical nature of energy depositions in biological objects. Conceptually, these theories envisage the biological object (e.g., a cell) as containing a number of volumes termed *targets* in which a prescribed number of elementary energy deposition events (*hits*) has to occur in order to produce the effect under observation. In principle a mathematical expression describing a dose–effect relationship can be obtained if assumptions are made concerning (a) the numerical values of the effect parameters, i.e., number of targets and number of effective hits per target necessary to produce elementary injuries, and the probability of obtaining the effect, given a number of such injuries; (b) the probability for a quantal energy deposition (hit) to occur in a target; and (c) for a given dose of radiation, the number of hits in the biological structure of interest. Concerning point (c), all hit and target theories have in common the hypothesis that *absorption events (hits) are randomly distributed and independent of each other.* Consequently, the distribution of hits over a given volume is treated using Poisson's law.

With the above assumptions *formal* equations describing hit–target mechanisms can be developed and no further specifications are necessary as to what hits and targets actually are. However, whatever physical description for the hit is employed it should conform with the basic postulate of randomness. Also, all the hit definitions currently in use imply that the average number of hits in the cells is proportional to the absorbed dose.

Consider first a population of N cells each having *one target.* Let v be the cellular volume and v_t the target volume. Let further H be the *total* number of hits delivered to the N cells. N is related to the dose delivered, D, by the simple relation

$$D = \frac{HW}{\rho v N} \tag{63}$$

where W is the average energy transferred in a hit and ρ is the density of the cellular material. Since hits are assumed statistically random and independent, each hit has a probability of $1/N$ to occur in a given cell, and a probability $p = v_t/(vN)$ to occur in a target volume. The probability of h effective hits in a cell (i.e., h hits in v_t) after a total of H hits is simply

$$P(h; H) = p^h(1-p)^{H-h}C_H^h \tag{64}$$

with

$$C_H^h = \frac{H!}{h!\,(H-h)!} \tag{65}$$

This is the well-known binomial law of probabilities. Let now $s(h)$ be the probability of survival for a cell which received h effective hits. Obviously

$$s(0) = 1 \tag{66}$$

The survival probability of a cell in a population which received a total of H hits is

$$S(H) = \sum_{h=0}^{H} P(h; H)s(h) \tag{67}$$

Further development requires assumption concerning $s(h)$. If, for instance, a single effective hit is enough to inactivate a cell, then

$$s(h) = \begin{cases} 1, & h = 0 \\ 0, & h \neq 0 \end{cases} \tag{68}$$

and, from Eqs. (64) and (67)

$$S(H) = P(0; H) = (1 - p)^H \tag{69}$$

This is usually written as

$$S(H) = e^{-cH} \tag{70}$$

with

$$c = -\ln(1 - p) \tag{71}$$

If H is replaced with D [see Eq. (63)] one obtains

$$S(D) = e^{-c_1 D} \tag{72}$$

with an obvious relation between c_1 and c. The equation (72) is the so-called single-target single-hit scheme. In a graphical representation, $\log S(D)$ vs. D, this equation is a straight line. The expression (72) can be further written

$$S(D) = \exp\left[D \frac{\rho v}{W} N \ln\left(1 - \frac{v_t}{vN}\right)\right] = \exp\left(-\rho \frac{v_t}{W} D\right) \tag{73}$$

since the total number of cells exposed, N, can be made arbitrarily large.

The assumption of Eq. (68) is a particular case of a more general mechanism in which hits contribute *independently* to produce the effect. Then if a cell received $h_1 + h_2$ hits, the probability of survival can be factorized as

$$s(h_1 + h_2) = s(h_1)s(h_2) \tag{74}$$

The solution of Eq. (74) is

$$s(h) = e^{-kh} \tag{75}$$

where k is a constant. When k is infinitely large, and with the condition of Eq. (66), one obtains the expression (68). It can be easily shown that with $s(h)$ as defined by Eq. (75) one obtains again an exponential dose–effect relationship. It is important to remember therefore that in the framework of hit–target theories, pure exponential survival curves reflect independent action of the hits and not necessarily single-hit single-target mechanism. As pointed out by Lea[16] (although for different reasons), postulating such a simple mechanism requires additional tests related to dose-rate effects or variation in response when the radiation quality is changed.

A second commonly used scheme is the so-called multitarget single-hit formalism. Here it is assumed that a cell (volume v) has t targets (of volume v_t each) each of which has to be hit at least once in order to produce the effect. In analogy with the single-target single-hit treatment, the probability for one single target to escape a hit, after a dose, D, is [see Eq. (72)]

$$S_1(D) = e^{-c_1 D} \tag{76}$$

The probability of hitting a target is then $1 - S_1(D)$ and the probability of hitting all t targets (and therefore inactivating the cell) is simply $[1 - S_1(D)]^t$. The fraction of surviving cells is

$$S(D) = 1 - (1 - e^{-c_1 D})^t \tag{77}$$

the well-known multitarget single-hit expression.

Expressions of the type of Eq. (72) or (77) are frequently used in radiobiological studies to describe survival data. Usually, constants such as c_1 or t are left as free parameters to be obtained from the fitting procedure; as such, the use of hit–target formalisms is merely a convenient way to classify experimental results by using a small number of parameters. Even for classification purposes these equations may not be too useful since, as has been repeatedly pointed out,[49] sets of data can be equally well described by significantly different combinations of parameters.

In a number of studies, attempts have been made to actually infer from a particular hit–target scheme information about the interaction mechanisms or the sensitive structures in the cell (see, for instance, the classical text of Lea[16]). Common to all these efforts is the basic problem of identifying the hit event with some physical energy deposition process, thus making the connection between the formal and the realistic (physical) hit–target theories. Most frequently the hit has been associated with an *ionization event,** and this interpretation will be employed in the

* Ionization events are obviously *correlated* quantities (e.g., delta rays associated with primary charged particles). To partially avoid this difficulty it was proposed[16] to identify a hit with *primary* ionizations only. This prescription becomes questionable, though, when the range of the secondary particles is comparable with the size of the target.

following. Other interpretations, such as identifying a hit with the passage of a charged particle through the target, can be reduced in the final analysis to the previous assumption: the ionizations in a particle track are considered frequent enough to justify the idea that each particle traversal results in at least one ionization (hit) in the cell.

In the next section the formal hit–target models discussed here will be reexamined using the "rigid"[49] definition of a hit event and microdosimetric concepts.

5.2. Microdosimetric Considerations

Before examining the consequences of identifying the hit with a concrete physical event (ionizations in the present case) it is useful to establish a more rigorous terminology for the sensitive target. The following picture is asumed:[19] a cell contains a number of subvolumes, termed *loci,* defined such that whenever a hit occurs therein this hit becomes *effective* (i.e., of consequence for the end point observed). The definition of locus is obviously related to the effect observed. The geometrical volume occupied by loci is termed the *sensitive matrix.* In the language of hit–target theories the sensitive matrix is identical with the total volume of the target(s).

In general the sensitive matrix might have a quite complex, nonconvex geometrical shape. The smallest convex volume containing the sensitive matrix is called *gross sensitive volume.* To date, this volume is identified with the cell nucleus.[50]

As mentioned in the previous section, in classical hit–target theories, hits are assumed to be randomly distributed in space. Consequently the only relevant parameter, as far as the targets are concerned, is their volume [see, for instance, Eq. (73)]. If the hypothesis of hit randomness is invalid (as in fact *is* the case with the present definition of a hit), one should clearly consider the details of the distribution of loci over the gross sensitive volume. The term *site model* refers to the situation when loci are randomly dispersed over the gross sensitive volume (taken as a sphere, for simplicity).

Consider first a somewhat different approach to the situation where one deals with single-target cells. The main idea in this approach is to separate out the distribution of hits over the gross sensitive volume (GSV) from the probability that a hit (already in GSV) becomes effective (i.e., occurs in the target). The probability of h' hits in the GSV is

$$p_1(h'; H) = \left(\frac{1}{N}\right)^{h'} \left(1 - \frac{1}{N}\right)^{H-h'} C_H^{h'} \tag{78}$$

where, as a reminder, N is the total number of cells considered and H is the total number of hits delivered in the GSVs of these cells. The probability that, out of h' hits in the GSV, h will be effective is

$$p_2(h; h') = \left(\frac{v_t}{v}\right)^h \left(1 - \frac{v_t}{v}\right)^{h'-h} C_{h'}^h \quad (79)$$

with v_t and v being the volumes corresponding to the sensitive matrix and GSV, respectively. Finally, the probability of having h effective hits, *irrespective* of h', is

$$p(h; H) = \sum_{h'=h}^{H} p_1(h'; H)p_2(h; h') \quad (80)$$

and it can be easily verified, using Eqs. (78)–(79), that the result, Eq. (61), is obtained, as expected. One can clearly see that in the formal hit–target approach a binomial law is assumed for the distribution of hits over the GSV, i.e., for $p_1(h'; H)$. With the definition hit = ionization event one should replace $p_1(h'; H)$ with the *microdosimetric* distribution $f(h'; H)$ describing the probability of h' ionizations in the GSV.

With this, Eq. (80) becomes

$$p(h; H) = \sum_{h'=h}^{H} f(h'; H)p_2(h; h') \quad (81)$$

where the site model (see above) was assumed. Consider again the single-hit mechanism. Then

$$S(H) = \sum_{h'=0}^{H} f(h'; H)\left(1 - \frac{v_t}{v}\right)^{h'} \quad (82)$$

or

$$S(H) = \sum_{h'=0}^{H} e^{-c_2' h'} f(h'; H) \quad (83)$$

with

$$c_2' = -\ln\left(1 - \frac{v_t}{v}\right) \quad (84)$$

In a more familiar microdosimetric notation the Eq. (83) becomes

$$S(D) = \int_0^\infty e^{-c_2 z} f(z; D)\, dz \quad (85)$$

where [see Eq. (63)]

$$z = \frac{h'W}{\rho v} \tag{86}$$

$$c_2 = c_2' \frac{\rho v}{W} \tag{87}$$

and $f(z; D)$ is any smooth, well-behaved function which satisfies

$$f(h'; H) = \int_{(h'-1/2)W/\rho v}^{(h'+1/2)W/\rho v} f(z; D)\, dz \tag{88}$$

A closed form for Eq. (85) can be obtained using Eq. (15):

$$f(z; D) = \sum_{\nu=0}^{\infty} e^{-n} \frac{n^{\nu}}{\nu!} f_{\nu}(z)$$

and the fact that

$$\int_0^{\infty} e^{-c_2 z} f_{\nu}(z)\, dz = \left[\int_0^{\infty} e^{-c_2 z} f_1(z)\, dz \right]^{\nu} \tag{88'}$$

This expression is a direct consequence of the convolution theorem for Laplace transforms.[12] Consider, as an example, the single event distribution, $f_1(z)$, defined for any real value of z and with the understanding that $f_1(z) = 0$ whenever $z \leqslant 0$. By definition [see Eq. (16)]

$$f_2(z) = \int_{-\infty}^{+\infty} f_1(z') f_1(z - z')\, dz'$$

Then

$$\begin{aligned}
\int_{-\infty}^{+\infty} e^{-c_2 z} f_2(z)\, dz &= \int_{-\infty}^{+\infty} f_1(z') \left[\int_{-\infty}^{+\infty} f_1(z - z') e^{-c_2 z}\, dz \right] dz' \\
&= \int_{-\infty}^{+\infty} f_1(z') e^{-c_2 z'} \left[\int_{-\infty}^{+\infty} f_1(z - z') e^{-c_2 (z - z')}\, d(z - z') \right] dz' \\
&= \left[\int_{-\infty}^{+\infty} f_1(z') e^{-c_2 z'}\, dz' \right]^2
\end{aligned}$$

Since in general

$$f_\nu(z) = \int_{-\infty}^{+\infty} f_{\nu-1}(z')f_1(z - z')\,dz'$$

one obtains by simple induction the result, Eq. (88′). Now

$$S(D) = \sum_{\nu=0}^{\infty} e^{-n}\frac{n^\nu}{\nu!}\int_0^\infty e^{-c_2 z}f_\nu(z)\,dz$$

$$= \sum_{\nu=0}^{\infty} \frac{e^{-n}}{\nu!}\left[n\int_9^\infty e^{-c_2 z}f_1(z)\,dz\right]^\nu$$

$$S(D) = \exp\left[-n\int_0^\infty (1 - e^{-c_2 z})f_1(z)\,dz\right] \tag{89}$$

where $n = D/\bar{z}_{1F}$ is the average number of events in the GSV. Under the assumption of the site model Eq. (89) is the expression describing the survival probability for systems in which single ionizations (hits) are sufficient to produce the effect. This expression, similarly to Eq. (72), represents an exponential survival–dose curve.

A further simplification of Eq. (89) occurs if one considers irradiations with mono-LET charged particles (the so called track-segment irradiation mode). The single-event distribution, $f_1(z)$, corresponding to a spherical site of diameter a is

$$f_1(z) = 2z/z_0^2, \qquad z \in [0, z_0] \tag{90}$$

where z_0 is related to the LET of the particle, L, through

$$z_0 = \frac{6}{\pi\rho a^2}L \tag{91}$$

The specific energy deposited by a particle traversing a spherical cavity along a chord of length x is

$$z = \frac{xL}{\rho v} = \frac{xL}{\rho\frac{4}{3}\pi(a/2)^3} \tag{91'}$$

The distribution in z, i.e., $f_1(z)$, is then identical—up to a normalization constant—to the distribution in chord lengths, x. This latter distribution is[1]

$$f(x) = 2x/a^2, \qquad x \in [0, a]$$

With C a normalization factor one has

$$f_1(z) = Cz, \qquad z \in [0, z_0]$$

where z_0—the maximum specific energy deposited—corresponds to a particle traversing the sphere along the diameter. From Eq. (91′) with $x = a$ one obtains Eq. (91). The constant C is then obtained from

$$1 = \int_0^{z_0} f_1(z)\,dz = C\int_0^{z_0} z\,dz = Cz_0^2/2$$

The final result is the expression (90). The equation (89) becomes

$$S(D) = e^{-\Phi\sigma} \tag{92}$$

where

$$\Phi = \rho\frac{D}{L} \tag{93}$$

is the number of particles traversing a unit area element (i.e., the particle fluence), and

$$\sigma = \pi\left(\frac{a}{2}\right)^2\left\{1 - \frac{2}{(z_0c_2)^2}[1 - e^{-c_2z_0}(1 + z_0c_2)]\right\} \tag{94}$$

can be interpreted as an effective interaction cross section. The equations (92)–(94) are similar to results obtained with the associated volume method of Lea[16] [compare, for instance, the cross section σ and the overlapping factor F of Lea—Eq. (111-4) in Ref. 16].

Measured cross sections,[51,52] obtained by fitting Eq. (92) to survival data (for exponential curves) or obtained from the initial slope (for sigmoid survival curves) are shown in Fig. 22 together with predictions based on Eqs. (92)–(94) with a and c_2 adjusted for the "best" fit. It is apparent from Fig. 22 that the single-target single-hit mechanism is unable to describe correctly these experimental data and can therefore be rejected.

A similar treatment can be applied to the multitarget single-hit formalism. The result will be quoted here without demonstration. If t targets must be inactivated (by at least one hit) in order to produce the effect, the fraction of surviving cells is given by

$$S(D) = 1 - \sum_{i=0}^{t} C_t^i(-1)^i \exp\left\{-\frac{D}{\bar{z}_{1F}}\int_0^\infty [1 - e^{-c_iz}]f_1(z)\,dz\right\}$$

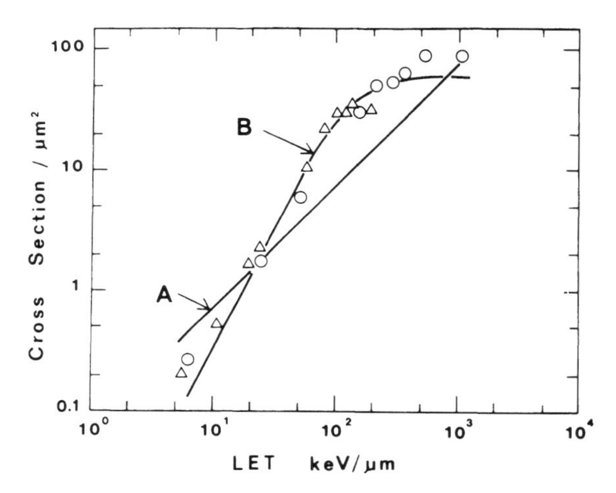

Figure 22. Experimental cross sections for the inactivation of mammalian cells *in vitro* with heavy charged particles as a function of LET (△, Ref. 51; ○, Ref. 52). Curves A and B are least-squares best fits obtained with Eqs. (94) and (101), respectively.

where

$$c_i = -\frac{\rho v}{W} \ln\left(1 - i\frac{v_t}{v}\right)$$

This expression is sufficiently complex (and flexible) to describe a variety of experimental data when the available parameters are adjusted correspondingly. Therefore, such a fit becomes a very weak test for the validity of a specific mechanism, let alone for deriving information on the targets. The usefulness of hit–target theories, with hits identified with ionizations, is therefore severely limited.

A different approach is examined in the next section.

5.3. Microdosimetry and Cellular Survival

In a more pragmatic approach, and without any reference to the notions of hit and target, one can investigate whether, given the assumption of the site model, a relation can be found between the specific energy in the GSV of a cell, z, and the probability of this cell to be inactivated. Assuming, as a working hypothesis, that such a relation indeed exists, let $s(z)$ be the probability of survival of a cell which received a specific energy z. The fraction of surviving cells after a dose D is

$$S(D) = \int_0^\infty s(z)f(z; D)\, dz \tag{95}$$

The basic question now is to find a function $s(z)$ which conforms with observations on survival. The problem can be greatly simplified if, for the moment, only the initial part of the survival curve is considered, that is, the low-dose range for which

$$n = D/\bar{z}_{1F} \ll 1 \tag{96}$$

Under these conditions [see Eq. (17)]

$$f(z; D) = e^{-n}[\delta(z) + nf_1(z)]$$

and

$$S(D) = e^{-n}\left[1 + n\int_0^\infty s(z)f_1(z)\,dz\right] \tag{97}$$

In Eq. (97) the first term corresponds to the fraction of cells in which no event occurred in the GSV and which survive with probability 1. The second term, corresponding to cells with one event only, describes the effectiveness of each event to actually inactivate a cell. It must be emphasized that, according to Eq. (97), the surviving fraction at very low doses depends linearly on dose: doubling the dose simply doubles the number of cells experiencing events without changing the distribution in specific energy over the individual cells. This remains true even if the inactivation process has a nonzero threshold in z since, because of the statistical nature of energy deposition, there will always be a finite fraction of cells in which the specific energy deposited exceeds the threshold.

With the condition of Eq. (96), one can write

$$S(D) = \exp\left\{-n\int_0^\infty [1 - s(z)]f_1(z)\,dz\right\} \tag{98}$$

This expression can be now compared directly with experimental data and inferences about $s(z)$ can be made. In fact, a particular form of $s(z)$,

$$s(z) = \exp(-cz)$$

has been examined in the previous section. Consider again irradiations with mono-LET charged particles. One has

$$S(D) = \exp[-\Phi(D)\sigma]$$

where the fluence, $\Phi(D)$, is given by Eq. (93) and the cross section σ is

$$\sigma = \pi\left(\frac{a}{2}\right)^2 \int_0^\infty [1 - s(z)] f_1(z)\,dz \tag{99}$$

A good fit to the experimental data of Fig. 22 is obtained using

$$s(z) = \exp(-kz^2) \tag{100}$$

where k is a "saturation" parameter equal to the value of $1/z^2$ for which the survival probability of a cell is 37%. From Eqs. (99) and (100) one obtains

$$\sigma = \pi\left(\frac{a}{2}\right)^2 \left[1 - \frac{1}{kz_0^2}(1 - e^{-kz^2})\right] \tag{101}$$

where z_0 is given by Eq. (91). The fit of this expression to the experimental data, with a and k as free parameters, is shown in Fig. 22.

If the function $s(z)$ of Eq. (100) is now postulated to represent the probability of survival as a function of z one can write, for any dose,

$$S(D) = \int_0^\infty e^{-kz^2} f(z; D)\,dz \tag{102}$$

For high-LET radiations and moderate doses ($n < 1$) one has

$$S(D) = \exp\left[-\frac{D}{\bar{z}_{1F}} \int_0^\infty (1 - e^{-kz^2}) f_1(z)\,dz\right] \tag{103}$$

For low-LET radiation and doses such that $kz^2 \ll 1$ one can write

$$e^{-kz^2} = 1 - kz^2$$

and

$$S(D) = \exp\left[-k \int_0^\infty z^2 f(z; D)\,dz\right]$$

or, using Eq. (50),

$$S(D) = e^{-k(z_{1D}D + D^2)} \tag{104}$$

The biophysical interpretation of the expressions, Eqs. (100)–(102) constitutes the basis for the theory of dual radiation action.[19,53]

6. THEORY OF DUAL RADIATION ACTION (TDRA)

6.1. The Basic Assumptions

The Eq. (102) can be interpreted in terms of a hit–target theory. If the number of hits is proportional to z^2,

$$h = k_1 z^2$$

one has

$$S(D) = \sum_h \exp\left(- \frac{1}{k_1 z_0^2} h \right) p(h; D) \tag{105}$$

In Eq. (105) $p(h; D)$ is the probability of producing h hits in the GSV of the cell. The probability for a cell to survive h hits [i.e., $\exp(-h/k_1 z_0^2)$] defines the hits as acting *independently* to produce the effect. The main point to be remarked, in fact the basic thesis of the theory of dual radiation action, is that the definition of a "hit" corresponds to a *quadratic* (rather than linear) dependence on the number of ionizations. In the TDRA terminology such a hit is called a *lesion,* while the elementary damage proportional to z (or the number of ionizations) is called a *sublesion.* Sublesions correspond to the classical definition of hits.

The following is the basic set of assumptions in the TDRA:[19,53] (a) The action of radiation on living objects results in elementary injuries termed sublesions. The number of sublesions is proportional to the energy concentration, z, in the gross sensitive volume of the object. (b) Sublesions interact pairwise to produce entities termed lesions (i.e., proportional to z^2 in the site model), which in turn are responsible for the effect observed.

With the above two assumptions a general mathematical theory has been developed describing the radiation effects on cellular systems.[19] In this general formulation the site model assumption is no longer used. However, a great deal of experimental data can actually be successfully explained if this simplifying hypothesis is used. A description of the mathematical formalism and a number of applications are the subject of the remaining sections.

6.2. The Formalism of the TDRA

In this section the average number of lesions per cell induced by a dose, D, of radiation is calculated. Consider N irradiated cells and let n be the average number of *events* occurring in the sensitive matrix of the cell. This domain will be assumed spherical for simplicity; the final results, though, are valid for any geometry of the sensitive material.

Energy absorption events are, by definition, statistically independent. The probability of a cell experiencing ν events is then given by Poisson's law:

$$p(\nu; n) = e^{-n} n^{\nu}/\nu! \tag{106}$$

Each event consists of a number of elementary energy absorptions (ionizations), E_i, and at each transfer point, $\vec{x}_i$, sublesions are produced at a rate proportional to the energy deposited:

$$\text{number of sublesions} = cE_i \tag{107}$$

Let $g(x)$ be the probability that two sublesions that are produced at initial separation x produce a lesion. Then the number of lesions, ε, in cell k is given by

$$\varepsilon(k) = \tfrac{1}{2}c^2 \sum_{(i,j)_k} E_i E_j g(x_{ij}), \qquad x_{ij} = |\vec{x}_i - \vec{x}_j| \tag{108}$$

where the summation extends over all the transfer points $\vec{x}_i$ in the matrix of the cell, k. The factor 1/2 was inserted because each product E_iE_j appears twice in summations. Equation (108) is the basic relation for the so-called "distance model," which is more general than the "site model".[53]

No events (and therefore no lesions) occur in a $p(0; n) = \exp(-n)$ fraction of the cells. $Np(1; n)$ of the cells have only one event each, and the *total* number of lesions in these cells is

$$\varepsilon_1 = \tfrac{1}{2} \sum_{k=1}^{Np(1;n)} c^2 \left[\sum_{(i,j)_k} E_i E_j g(x_{ij}) \right] \tag{109}$$

Lesions resulting from one event are called *intratrack lesions*; for a given cell (k) E_i and E_j in Eq. (109) are statistically correlated.

Consider next the $Np(2; n)$ cells which experience two events only. An expression similar to Eq. (109) can be written. Further, for each such cell the sum over cross products E_iE_j can be conveniently split into two terms: one referring to transfer points corresponding to the *same* event (intratrack lesions) and one corresponding to transfer points from *different* events (so called *intertrack* lesions). The total number of lesions of the first kind is

$$\varepsilon_2' = \tfrac{1}{2}c^2 \sum_{k=1}^{2Np(2;n)} \left[\sum_{(i,j)_k} E_i E_j g(x_{ij}) \right] \tag{110}$$

The square brackets refer to one event only and since there are two events per cell the summation in Eq. (110) is from 1 to 2 $Np(2; n)$. With the notation

$$\varepsilon'(k) = \tfrac{1}{2}c^2 \sum_{(i,j)_k} E_i E_j g(x_{ij}) \tag{111}$$

representing the number of lesions from one event, the total number of *intratrack* lesions in the N cells is

$$\varepsilon' = \sum_{i=1}^{\infty} \sum_{k=1}^{iNp(i;n)} \varepsilon'(k) = \sum_{k=1}^{nN} \varepsilon'(k) \tag{112}$$

and the average number *per cell* is

$$\overline{\varepsilon'} = \frac{1}{N} \quad \varepsilon' = n \frac{1}{Nn} \sum_{k=1}^{Nn} \quad \varepsilon'(k) = n\overline{\varepsilon'(k)} \tag{113}$$

$\overline{\varepsilon'(k)}$ can be obtained in exact analogy with the calculation of $\bar{z}_{1D}$ in Section 2.3. The result is

$$\overline{\varepsilon'} = \tfrac{1}{2}c^2 mD \int_0^{\infty} \frac{1}{4\pi x^2} t(x)\phi(x)g(x)\, dx \tag{114}$$

where $t(x)$ is the proximity function of energy transfers, Eq. (43), and $\phi(x)/V_a$, Eq. (51), is the distribution of distances between two random points in a spherical volume (corresponding in the present calculation to the shape of the GSV).

Consider now intertrack lesions. Let $\varepsilon''(k)$ be the number of such lesions resulting from the interaction of sublesions from *two* events

$$\varepsilon''(k) = c^2 \sum_{(i,j)_k} E_i E_j' g(x_{ij}) \tag{115}$$

where the superscript (E_j') emphasizes that E_i and E_j are from different events. In a cell with ν events there will be $\nu(\nu - 1)/2$ terms of the type of $\varepsilon''(k)$. The total number of intertrack lesions in N cells is

$$\varepsilon'' = \sum_{i=1}^{\infty} \sum_{k=1}^{i(i-1)Np(i;n)/2} \varepsilon''(k) = \sum_{k=1}^{Nn^2/2} \varepsilon''(k) \tag{116}$$

and the average number per cell is

$$\overline{\varepsilon''} = \frac{1}{N} \quad \varepsilon'' = \frac{n^2}{2} \frac{1}{Nn^2/2} \sum_{k=1}^{Nn^2/2} \quad \varepsilon''(k) = \frac{n^2}{2} \overline{\varepsilon''(k)} \tag{117}$$

Since E_i, E_j' and x_{ij} are *independent* random variables, from Eq. (115) one has

$$\overline{\varepsilon''(k)} = c^2\overline{\left(\sum_i E_i\right)}\,\overline{\left(\sum_j E_j'\right)}\,\overline{g(x_{ij})}$$

$$= c^2\overline{\left(\sum_i E_i\right)}^2\,\overline{g(x_{ij})}$$

$$\overline{\varepsilon''(k)} = c^2 m^2 \bar{z}_{1F}^2 \frac{1}{V_a}\int_0^\infty \phi(x)g(x)\,dx \qquad (118)$$

The equation (117) becomes

$$\overline{\varepsilon''} = \frac{1}{2}\frac{c^2m^2}{V_a}D^2\int_0^\infty \phi(x)g(x)\,dx \qquad (119)$$

The total average number of lesions per cell is

$$\bar{\varepsilon} = \overline{\varepsilon'} + \overline{\varepsilon''}$$

$$\bar{\varepsilon} = \tfrac{1}{2}c^2 m\left[D\int_0^\infty \frac{t(x)\phi(x)g(x)}{4\pi x^2}\,dx + \frac{mD^2}{V_a}\int_0^\infty \phi(x)g(x)\,dx\right] \qquad (120)$$

or

$$\bar{\varepsilon} = k(\xi D + D^2) \qquad (121)$$

with

$$K = \frac{c^2m^2}{2V_a}\int_0^\infty \phi(x)g(x)\,dx \qquad (122)$$

and

$$\xi = \int_0^\infty \frac{t(x)\phi(x)g(x)}{4\pi x^2}\,dx \Big/ \frac{m}{V_a}\int_0^\infty \phi(x)g(x)\,dx \qquad (123)$$

It has been demonstrated[19] that Eq. (121) is valid for any geometry of the sensitive volume if $\phi(x)$, Eq. (51), is the corresponding distribution of distances x.

Equation (121) is one of the fundamental results of the TDRA. It states that the average number of lesions per cell depends in a *linear–quadratic fashion on the absorbed dose.* The linear term represents the contribution of intertrack interactions. The functions $\phi(x)$ and $g(x)$ characterize the irradiated object; and the function $t(x)$ describes the radiation field. With the assumption $g(x)$ = const. for $x \leqslant d$ and zero elsewhere, Eq. (123) becomes

$$\xi = \frac{1}{m} \int_0^d \frac{t(x)\phi(x)}{4\pi x^2} dx = \bar{z}_{1D} \tag{124}$$

[see Eqs. (50 and 51)]. Equation (121) then reads

$$\bar{\varepsilon} = k(\bar{z}_{1D} D + D^2) \tag{125}$$

i.e., the distance model is reduced to the site model.

A number of observations should be made. In spite of its simple form the basic expression, Eq. (121), involves two functions $\phi(x)$ and $g(x)$ (actually only their product) on which no information was available until recently. Experiments with molecular ions[54] have yielded preliminary information on the product $\phi(x)g(x)$ which indicates that it is a highly skewed function of x. The constant k of Eq. (122) is considered as a parameter to be inferred from further experimental observations.

Equations (121)–(125) predict the average yield of lesions per cell. Generally speaking, this number cannot be correlated with the observed effect. In the case of survival, for instance [see Eq. (105)],

$$S(D) = \sum_{\varepsilon=0}^{\infty} e^{-\text{const}\cdot\varepsilon} p(\varepsilon; D) \tag{126}$$

and only for situations where the number of lesions per cell is sufficiently small one can write

$$S(D) \simeq e^{-\text{const}\cdot\bar{\varepsilon}} \tag{127}$$

Erroneous results are obtained if the two expressions, Eqs. (126) and (127), are used as equivalent without careful analysis. A direct calculation based on Eq. (126) might be quite complex, but under particular irradiation conditions simplified expressions can be obtained.

7. APPLICATIONS OF THE TDRA

7.1. RBE–Dose Relations

The relative biological effectiveness (RBE) of a radiation, 2, relative to a reference radiation, 1, is defined as the ratio of doses D_1/D_2 of the two radiations, respectively, necessary to produce the same biological effect:

$$\text{RBE} = D_1/D_2 \tag{128}$$

The definition of the RBE is chosen such that it increases as the radiation 2 becomes more effective—that is, a smaller dose, D_2, is necessary to produce the same effect.

The doses D_1 and D_2 which produce the same effect are related through Eq. (126):

$$\sum_{\varepsilon} e^{-k_1\varepsilon} p_1(\varepsilon; D_1) = \sum_{\varepsilon} e^{-k_2\varepsilon} p_2(\varepsilon; D_2) \tag{129}$$

where $p_\alpha(\varepsilon; D)$ is the probability of producing ε lesions in a cell after delivering a dose, D, of the radiation α, and k_1, k_2 are constants. An RBE–dose relation can be obtained therefore only if the distributions $p(\varepsilon; D)$ are known, and to date, no such calculations are available. As mentioned in the previous section, a simplified situation arises when the approximation

$$e^{-k\varepsilon} = 1 - k\varepsilon \tag{130}$$

can be made. In this case Eq. (129) becomes

$$\bar{\varepsilon}_1(D_1) = \bar{\varepsilon}_2(D_2) \tag{131}$$

and the expressions Eq. (121) or (125) can be applied. While no *a priori* statement as to the validity of the approximation, Eq. (130), can be made, formal relations based on this can be developed and, by comparison with experimental results, a decision might be reached, *a posteriori,* as to the correctness of such a simplified approach. This procedure will be followed in this section.

Since the inception of the TDRA the difference between the general expression, Eq. (129), and the approximate relation, Eq. (131), has often not been fully appreciated. It might therefore be useful to further illustrate the above statements with a simple hypothetical example. Consider a population of 1000 cells in which following irradiation with a dose, D_1, of radiation 1, or D_2 (radiation 2) 1000 lesions are produced. Let the distributions $p(\varepsilon; D)$ be such that for the first radiation each cell has one lesion while for the second radiation all 1000 lesions are in one cell. The *average* number of lesions *per cell* is the same in the two situations ($\bar{\varepsilon}_1 = \bar{\varepsilon}_2 = 1$). Yet, if for instance one lesion is sufficient to inactivate a cell, the surviving fraction is 0 for radiation 1 and 0.999 for radiation 2.

Consider now the assumption that the survival probability of a cell with ε lesions is proportional to ε [such as in Eq. (130)]. Then the number of surviving cells N_1 and N_2 corresponding to the two radiations, respectively, is equal:

$$N_1 = 1000(1 - k_1 \cdot 1), \qquad N_2 = 999 + (1 - k_2 \cdot 1000)$$

when $k_1 = k_2$. Indeed, the survival probability in this case depends on the *average* number of lesions per cell only.

Equation (131) can be written, using Eq. (121),

$$k_1(\xi_1 D_1 + D_1^2) = k_2(\xi_2 D_2 + D_2^2) \tag{132}$$

By simple manipulation one obtains

$$\text{RBE} = \frac{2(k_2/k_1)(\xi_2 + D_2)}{\xi_1 + [\xi_1 + 4(k_2/k_1)(\xi_2 D_2 + D_2^2)]^{1/2}} \tag{133}$$

which is the desired relation under the assumption of Eq. (130). A further simplified expression is obtained if ξ_2 and D_2 are such that ξ_1 can be neglected in Eq. (133). Then

$$\text{RBE} = \left[\frac{k_2}{k_1}(1 + \xi_2/D_2)\right]^{1/2} \tag{134}$$

For small doses ($d_2 \gg \xi_2$) the Eq. (134) corresponds to a straight line of slope -0.5 in a logarithmic representation $\log \text{RBE}$ vs. $\log D_2$. This basic prediction of the present formalism has been confirmed in extensive comparisons with experimental results obtained with neutrons of various energies.[53] Several examples are shown in Fig. 23. Retrospectively it is apparent therefore that, at least for the region of low doses the approximation of Eqs. (130) and (131) is justified.

A detailed description of the application of the TDRA to RBE–dose relations can be found in Ref. 53.

7.2. Temporal Aspects

The conceptual and mathematical derivations developed up to this point have been made with the tacit assumption that energy depositions and subsequent biological effects occur in an infinitely short time interval. The present formalism will be now modified in order to include the effects of sublesion recovery, finite interaction time between sublesions as

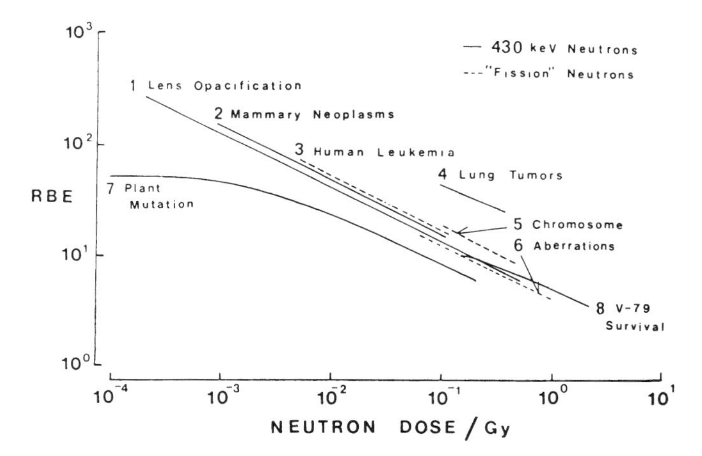

Figure 23. RBE vs. neutron dose for fission neutrons and 430-keV neutrons. Data are from (1) Bateman *et al.*[55]; (2) Shellabarger *et al.*[56]; (3) Rossi[57]; (4) Ullrich *et al.*[58]; (5) Awa[59]; (6) Biola *et al.*[60]; (7) Sparrow *et al.*[61]; (8) Hall *et al.*[62]

well as the temporal distribution of the dose delivery (i.e., irradiation time, fractionation, etc.).

The quantity under consideration here is the average number of lesions per cell after a time t following the beginning of the irradiation, i.e., $\bar{\varepsilon}(t)$. Let $\varepsilon_s(t)$ be the number of sublesions in a cell at time t. In the time interval between energy absorption events (which are considered instantaneous) the number of available sublesions changes due to the processes of sublesion repair (recovery) and lesion formation. If the rates of these two processes are taken proportional to $\varepsilon_s(t)$ and $\varepsilon_s^2(t)$, respectively, the following expression can be written:

$$\frac{d\varepsilon_s(t)}{dt} = -a(t)\varepsilon_s(t) - b(t)\varepsilon_s^2(t) \tag{135}$$

The number of lesions produced at the time t is then

$$\varepsilon(t) = \int_0^t b(t)\varepsilon_s^2(t)\,dt \tag{136}$$

Equations (135) and (136) are quite general formulations describing the evolution in time of the number of sublesions and lesions in the irradiated object. To actually solve these equations might be a quite formidable task: one needs the functions $a(t)$ and $b(t)$, that is, the variation in time of the recovery and interaction rates. Furthermore, initial conditions such as the number of sublesions produced following each event and the time intervals between events are necessary. Finally, the number of lesions, $\varepsilon(t)$, has to be averaged out over all possible combinations of such initial conditions according to the microdosimetric qualities of the radiation and the overall temporal scheme of irradiation.

In practice simplifying assumptions are used and predictive results are compared with experiments. It is generally assumed to date that the repair rate $a(t)$ does not change with time. This rate is measured in split-dose experiments and is typically of the order of $0.1\ \text{min}^{-1}$. No similar information is available concerning the lesion formation rate, $b(t)$.

One possible assumption, to be examined in the following, is that $b(t)$ is constant and much smaller than the repair rate a. This is in line with the assumption that only a small fraction of the number of sublesions available actually produce lesions. The treatment will also be restricted to the range of doses for which no significant depletion in the number of available sublesions occurs because of lesion formation. The main reason for this restriction is that the interaction and repair of sublesions produced by each event (both intra- and inter-track) can be treated as independent of each other.

Consider first intratrack processes. Let $\varepsilon_s(0)$ be the number of sublesions produced by one event, with time zero coincident with the occurrence of the event. For this event (again, considered independently of other events) one has

$$\frac{d\varepsilon_s}{dt} = -a\varepsilon_s - b\varepsilon_s^2 \tag{137}$$

and the solution is

$$\varepsilon_s(t) = \varepsilon_s(0)\frac{e^{-at}}{1 + (b/a)\varepsilon_s(0)(1 - e^{-at})} \tag{138}$$

From Eq. (136) the number of lesions is

$$\varepsilon(t) = \varepsilon_s(0)\left[1 - \frac{e^{-at}}{1 + (b/a)\varepsilon_s(0)(1 - e^{-at})}\right] - \frac{a}{b}\ln\left[1 + \frac{b}{a}\,\varepsilon_s(0)(1 - e^{-at})\right] \tag{139}$$

After a time long enough compared with the characteristic repair time

$(1/a)$ Eq. (139) reduces to

$$\varepsilon = \varepsilon_s(0) - \frac{a}{b}\ln\left[1 + \frac{b}{a}\varepsilon_s(0)\right]$$

and since $b/a \ll 1$:

$$\varepsilon = \frac{1}{2}\frac{b}{a}\varepsilon_s^2(0) \tag{140}$$

The result, Eq. (140), indicates that under the present assumptions (a) the number of intratrack lesions does not depend on the dose distribution in time, and (b) it is a function of the relative contribution of repair and interaction processes.

Consider further intertrack lesions resulting from the interaction of sublesions produced by two events. If for the moment the two events are taken as simultaneous let $\varepsilon_{s1}(0)$ and $\varepsilon_{s2}(0)$ be the initial number of sublesions from each event, respectively. The equations describing the variation in time of $\varepsilon_{s1}(t)$ and $\varepsilon_{s2}(t)$ are

$$\frac{d\varepsilon_{s1}}{dt} = -a\varepsilon_{s1} - b\varepsilon_{s1}\varepsilon_{s2} \tag{141}$$

$$\frac{d\varepsilon_{s2}}{dt} = -a\varepsilon_{s2} - b\varepsilon_{s1}\varepsilon_{s2} \tag{142}$$

As can be easily verified, the solution of this system of differential equations is

$$\varepsilon_{s1}(t) = \frac{\varepsilon_{s1}(0)[\varepsilon_{s1}(0) - \varepsilon_{s2}(0)]e^{-at}}{\varepsilon_{s1}(0) - \varepsilon_{s2}(0)\exp\{-(b/a)[\varepsilon_{s1}(0) - \varepsilon_{s2}(0)](1 - e^{-at})\}} \tag{143}$$

$$\varepsilon_{s2}(t) = \frac{\varepsilon_{s2}(0)[\varepsilon_{s1}(0) - \varepsilon_{s2}(0)]e^{-at}}{\varepsilon_{s1}(0)\exp\{(b/a)[\varepsilon_{s1}(0) - \varepsilon_{s2}(0)](1 - e^{-at})\} - \varepsilon_{s2}(0)} \tag{144}$$

Again, from Eq. (136), the number of lesions can be calculated and the result is

$$\varepsilon(t) = \varepsilon_{s1}(0) + \varepsilon_{s2}(0) + \frac{a}{b}\ln\frac{\varepsilon_{s1}(0) - \varepsilon_{s2}(0)}{\varepsilon_{s1}(0)\exp[(b/a)\varepsilon_{s1}(0)(1 - e^{-at})] - \varepsilon_{s2}(0)\exp[(b/a)\varepsilon_{s2}(0)(1 - e^{-at})]}$$

$$- e^{-at} \frac{\varepsilon_{s1}^2(0) \exp[(b/a)\varepsilon_{s1}(0)(1 - e^{-at})] - \varepsilon_{s2}^2(0) \exp[(b/a)\varepsilon_{s2}(0)(1 - e^{-at})]}{\varepsilon_{s1}(0) \exp[(b/a)\varepsilon_{s1}(0)(1 - e^{-at})] - \varepsilon_{s2}(0) \exp[(b/a)\varepsilon_{s2}(0)(1 - e^{-at})]} \tag{145}$$

This equation is, as expected, symmetrical in $\varepsilon_{s1}(0)$ and $\varepsilon_{s2}(0)$. For $t \gg 1/a$ and $b/a \ll 1$, the Eq. (145) reduces to the simple form

$$\varepsilon(t) = \frac{1}{2}\frac{b}{a}\, \varepsilon_{s1}(0)\varepsilon_{s2}(0) \tag{146}$$

If a finite time, t_{12}, passed between the first and second event, $\varepsilon_{s1}(0)$ should be replaced with $\varepsilon_{s1}(0)e^{-at_{12}}$ since some of the original sublesions, $\varepsilon_{s1}(0)$, repaired at a constant rate during this time [see Eq. (137) with $b = 0$]. Equation (146) becomes

$$\varepsilon(t) = \frac{1}{2}\frac{b}{a}\, \varepsilon_{s1}(0)\varepsilon_{s2}(0)e^{-at_{12}} \tag{147}$$

(times zero signify now "initial number" for each event).

For *any* two events the number of intertrack lesions is reduced by the quantity $\exp(-at_{12})$. Since the time interval between events is *independent* of the number of sublesions involved, one can write an expression for the average number of lesions following an irradiation with dose, D:

$$\bar{\varepsilon} = k\frac{1}{2}\frac{b}{a}(\xi D + \overline{e^{-at_{12}}} D^2) \tag{148}$$

where

$$\overline{e^{-at_{12}}} = \int_0^\infty e^{-at_{12}} h(t_{12})\, dt_{12} \tag{149}$$

and where $h(t_{12})\, dt_{12}$ is the probability to have a time interval between t_{12} and $t_{12} + dt_{12}$. The function $h(t_{12})$ can be easily calculated for a variety of temporal schemes of irradiation.[53,63] As an example, if the dose is delivered at a constant dose rate during the time, T, one has[53]

$$h(t_{12}) = \begin{cases} (2/T^2)(T - t_{12}), & 0 \leqslant t_{12} \leqslant T \\ 0, & t_{12} > T \end{cases} \tag{150}$$

and from Eq. (146):

$$\overline{e^{-at_{12}}} = \frac{2}{aT}\left[1 - \frac{1}{aT}(1 - e^{-aT})\right] \tag{151}$$

It is interesting to remark that Eqs. (148)–(151), in spite of the rather restrictive assumptions used, describe correctly the qualitative features of experiments in which time effects are explored. Thus, for instance, according to these equations the average number of lesions becomes increasingly dominated by the linear component in dose as the irradiation time is increased (i.e., for lower dose rates). Also, since dose rate affects only the quadratic term in dose, one expects smaller variations with the dose rate for high-LET radiations (where the linear term dominates).

Different assumptions might lead to similar results. For instance, one can define a sublesion interaction rate, $b(t)$, as a rapidly decreasing function of time compared with the average time between consecutive events (for "normal" dose rates) and triggered on by each incoming event. In contrast with the previous formalism the lesion formation process will then be consumated very quickly following each event. Nevertheless, an expression similar to Eq. (148) is obtained.

No position is taken here regarding the validity of any of the above hypotheses. They were considered because the mathematical consequences furnish an illustration of possible contributions of the TDRA to the analysis of time-dependent phenomena in radiobiology.

7.3. Sequential Irradiations with Different Radiations

A number of experimental studies[64–70] have been performed in recent years concerning the effects of sequential exposures to different radiations. The motivation behind these studies was twofold: (1) an investigation of the sublethal damage production and interaction for *different* radiations, including time-dependent repair effects; and (2) examining the possibility of using a combined modality procedure in radiotherapeutical applications.

The TDRA formalism presented in the previous sections can easily be extended to this particular experimental situation. Consider a population of cells exposed to a dose, D_1, of radiation 1 and D_2 of radiation 2. Let $f_1(z_1; D_1)$ and $f_2(z_2; D_2)$ be the distributions in specific energy corresponding to the two radiations, respectively. Obviously, z_1 and z_2 are independent stochastic variables. In a particular cell which received a specific energy z_1 from the first radiation and z_2 from the second one, a total of $(c_1 \cdot z_1 + c_2 \cdot z_2)$ sublesions are created. If, for simplicity, the site model assumption is used, the total number of lesions produced is

$$\varepsilon(z_1, z_2) = A(c_1 z_1 + c_2 z_2)^2 \tag{152}$$

where A is the probability of interaction between two sublesions. The average number of lesions per cell is then

$$\bar{\varepsilon}(D_1, D_2) = A \int\int (c_1 z_1 + c_2 z_2)^2 f_1(z_1; D_1) f_2(z_2; D_2)\, dz_1\, dz_2 \quad (153)$$

With the notation

$$\alpha = Ac^2 \bar{z}_{1D}, \qquad \beta = Ac^2 \quad (154)$$

Eq. (153) becomes

$$\bar{\varepsilon}(D_1, D_2) = \alpha_1 D_1 + \beta_1 D_1^2 + \alpha_2 D_2 + \beta_2 D_2^2 + 2(\beta_1 \beta_2)^{1/2} D_1 D_2 \quad (155)$$

The interpretation of this result is fairly simple: the average number of lesions resulting from exposure to two radiations is equal to the total average number of lesions produced by each radiation *plus* an additional contribution representing the lesions formed by the interaction of sublesions from the two radiations (interradiation lesions). An enhanced effect with respect to the individual yields of lesions (i.e., synergism) is therefore obtained by combining different types of radiation.

In complete analogy with the treatment of the previous section, Eq. (155) can be modified to include time-dependent effects, such as sublethal damage repair. As an example one can write

$$\begin{aligned}\bar{\varepsilon}(D_1, D_2) = {} & \alpha_1 D_1 + \overline{e^{-at_1}} \beta_1 D_1^2 + \alpha_2 D_2 + \overline{e^{-at_2}} \beta_2 D_2^2 \\ & + 2\overline{e^{-at_{12}}} (\beta_1 \beta_2)^{1/2} D_1 D_2 \end{aligned} \quad (156)$$

where the recovery function $\exp(-at)$ is averaged over the distribution of time intervals during the delivery of the first dose or the second dose, or between the two doses, respectively. As was shown in a recent publication,[71] if T_1 and T_2 are the respective time intervals in which the two doses are delivered (at constant dose rate), and if T is the time between irradiations, one has

$$\overline{e^{-at_{12}}} = \frac{1}{a^2 T_1 T_2} e^{-aT}(1 - e^{-aT_1})(1 - e^{-aT_2}) \quad (157)$$

The average of $\exp(-at_i)$ for $i = 1, 2$ can be calculated with Eq. (148).

An example of the application of Eq. (155) is shown in Fig. 24.

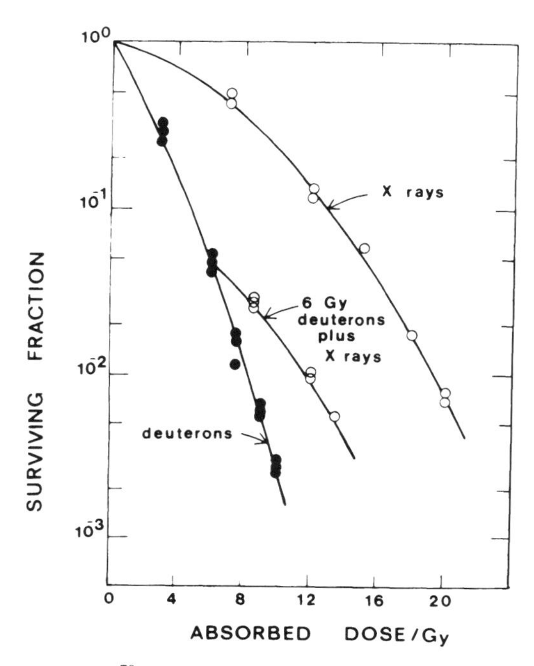

Figure 24. Experimental[70] and calculated survival fractions for sequential exposures of Chinese hamster V-79 cells with deuterons (50 keV/μm) and/or 250-kVp X rays. The theoretical curves were obtained by fitting the expression $S(D) = \exp[-\bar{\varepsilon}(D)]$ [see Eqs. (151)–(152)] to the deuteron and X-ray data and using the (α, β) values to *predict* the shape of the combined irradiation curve [cf. Eq. (155)].

7.4. Intercellular Effects

The utility of microdosimetry to radiobiology is largely restricted to effects on individual cells. It is also generally assumed that the response of these cells is *autonomous,* i.e., that it depends only on energy transfers in the cell and is independent of energy deposited elsewhere (i.e., other cells, medium, etc.). It is only under these conditions that most of the radiobiological considerations given above apply, and this is particularly so for those that are quite firm. These include *absence of threshold* for substantially all effects, and, at sufficiently low absorbed doses, an effect probability that is proportional to dose and independent of dose rate. In the case of autonomous response one can also specify an upper limit of effect probability in the low-dose domain. This is based on the limiting condition in which each event in a cell causes the effect. Then the fraction of the cells containing an event, $D/\bar{z}_F$, is the fraction affected.

Conversely, the fact that any of these conditions is not met is direct proof that the effect under consideration is not due to autonomous cell response. Examples are cellular survival of less than 37% when the average event frequency is 1,[72] and a nonlinear dependence of tumor production on dose when the event frequency is less than 0.01.[56]

There is reason to believe that despite some exceptions[72] the response of the cells of the germ plasm is essentially autonomous and it is therefore correct to assume that for doses of interest in radiation protection genetic effects are proportional to dose and independent of dose rate. On the other hand certain effects such as radiation cataract clearly are due to a multicellular response. The assertion that radiogenic cancer is due to an autonomous response of individual transformed cells cannot be categorically denied, but in addition to an obvious case[56] there are other considerations that make such a postulate doubtful.[73] It certainly should not be asserted that it is a necessary consequence of the TDRA that cancer incidence following exposure to low-LET radiation must exhibit a quadratic (i.e., linear and square) dependence on dose. However, this does not exclude the possibility that such a relation may be a reasonable fit, at least over limited dose ranges.

There is, however, one finding of the TDRA that appears to apply to complex biological effects, and this is the dependence of RBE on dose and specifically the relation RBE $\sim (D_N)^{-1/2}$ (see Section 7.1). A plausible explanation is that differences in radiation quality are of importance only in the early phases of biophysical action, i.e., the process of lesion formation, and that subsequent processes and especially intercellular ones proceed similarly regardless of radiation quality. The dependence of RBE on dose has been invoked as a useful relation in the derivation of risk estimates.[74]

7.5. Radiation Protection

The basic physical quantity of radiation protection is the dose equivalent, H, defined by

$$H = QND \tag{158}$$

where D is the absorbed dose and N a product of—largely unspecified—modifying factors, which is currently given the value of 1. Q, the quality factor, is intended to weight the dose by the relative biological effectiveness of the radiation under consideration. It is assumed that this is related to the LET (strictly L_∞, the stopping power in water) of the charged particles that impart the dose. Values of Q vs. L have been recommended by organizations concerned with radiation protection.[7,75] In all practical cases one deals with a range of values of LET and Eq.

(158) becomes

$$H = \int_0^\infty d(L)Q(L)\,dL \tag{159}$$

where $d(L)$ is the distribution of absorbed dose in L.

This scheme has been criticized on two counts.[76] One is that the values of $Q(L)$ are generally too low, a conclusion that is at least in part based on an analysis of data that relies on the TDRA. However, the other objection is more directly related to microdosimetric considerations.

The LET is defined as the energy lost by a particle per unit distance of travel. This is subject to statistical fluctuations and the LET is a (linear) average of such losses. Hence (neglecting effects of secondary importance) $L = \bar{y}_F$ in domains of small dimensions (e.g., of the order of 1 μm). However, the very existence of RBE (and the consequent assignment of a range of values to Q) is due to the fact that the biological response is not linear. Thus, according to the TDRA, $\bar{y}_D$ is a better index of biological effectiveness and it is always true that $L < \bar{y}_D$. Corresponding relations hold when distribution of the LET are involved, i.e., $\bar{L}_F = \bar{y}_F$ and $\bar{L}_D < \bar{y}_D$.

The LET is thus inherently a quantity that cannot be directly related to biological effect. Although a suitable choice of the distribution $Q(L)$ can nevertheless be made, the basic difficulty arises as the practical impossibility of measuring $d(L)$ in terms of energy deposition because of statistical fluctuations. Because these fluctuations appear equally in the measuring device and in irradiated tissue, y (which is the quantity directly measured) is a more practical parameter for Q. This approach requires a somewhat arbitrary choice of the diameter of the region for which $d(y)$ is to be determined. The value that has been recommended is 2 μm. The reasons for this and other aspects of the problem are discussed in Ref. 76.

ACKNOWLEDGMENTS. This investigation was supported by contract No. DE-AC02-78EV04733 from the Department of Energy and by grant No. CA 12536 to the Radiological Research Laboratory/Department of Radiology, and by grant No. CA 13696 to the Cancer Center/Institute of Cancer Research, awarded by the National Cancer Institute, DHHS.

REFERENCES

1. H. H. Rossi, Microscopic energy distribution in irradiated matter, in *Radiation Dosimetry,* Vol. 1, pp. 43–92, Academic, New York (1967).

2. R. Dvorak and W. Gross, Event distributions from monoenergetic photons, Annual Report on Research Project, pp. 68–77, Radiological Research Laboratory, Columbia University, New York (1974). (Available as COO-3243-3 from National Technical Information Service, Springfield, Virginia 22161.)
3. H. I. Amols, J. F. Dicello, and T. F. Lane, Microdosimetry of negative pions, in *Fifth Symposium on Microdosimetry,* Verbania Pallanza, Italy (J. Booz, H. G. Ebert, and B. G. R. Smith, eds.), pp. 911–928, Commission of the European Communities, Luxembourg (1976).
4. D. Srdoc, L. J. Goodman, S. A. Marino, R. E. Mills, M. Zaider, and H. H. Rossi, Microdosimetry of monoenergetic neutron radiation, in *Proceedings of the Seventh Symposium om Microdosimetry,* (J. Booz, H. G. Ebert, and B. G. R. Smith, eds.), pp. 765–774 Harwood, Oxford (1980).
5. M. Zaider, J. F. Dicello, D. Brenner, M. Takai, M. R. Raju, and J. Howard, Microdosimetry of range-modulated beams of heavy ions I. Determination of the yield of projectile fragments from microdosimetric spectra for Ne^{10} beams, *Radiat. Res.* **81,** 511–520 (1980).
6. A. M. Kellerer, Mikrodosimetrie, Bericht B-1 Gesellschaft für Strahlenforschung, Neuherberg-München (1968).
7. ICRU, Radiation Quantities and units, Report 33, International Commission on Radiation Units and Measurements, Washington, D.C. (1980).
8. ICRU, Radiation Quantities and units, Report 19, International Commission on Radiation Units and Measurements, Washington, D.C. (1971).
9. A. Cauchy, Memoire sur la rectification des courbes et la quadrature des surfaces courbes (1850), in *Oeuvres Completes,* Vol. 2, Gauthier Villard, Paris (1908).
10. L. Landau, On the energy loss of fast charged particles, *J. Phys. U.S.S.R.* **8,** 201–205 (1944).
11. P. V. Vavilov, Ionization losses of high-energy heavy particles, *J. Exptl. Theoret. Phys. (U.S.S.R.)* **32,** 920–923 (1957).
12. R. N. Bracewell, *The Fourier Transform and Its Applications,* McGraw-Hill, New York (1978).
13. A. M. Kellerer and D. Chmelevsky, Concepts of microdosimetry I. Quantities, *Radiat. Environ. Biophys.* **12,** 61–69 (1975).
14. A. M. Kellerer and D. Chmelevsky, Concepts of microdosimetry II. Probability distributions of the microdosimetric variables, *Radiat. Environ. Biophys.* **12,** 205–216 (1975).
15. A. M. Kellerer and D. Chmelevsky, Concepts of microdosimetry III. Mean values of the microdosimetric distributions, *Radiat. Environ. Biophys.* **12,** 321–335 (1975).
16. D. E. Lea, *Action of Radiation on Living Cells,* 2nd ed., Cambridge University Press, Cambridge (1962).
17. M. Zaider, Proximity function for low energy electrons in nitrogen—Application to Iodine-125, Annual Report on Research Project, Radiological Research Laboratory, Columbia University, New York (1979–1980). (Available as COO-4733-3 from National Technical Information Service, Springfield, Virginia 22161.)
18. M. G. Kendall and P. A. P. Moran, *Geometrical Probability,* Griffin, London (1963).
19. A. M. Kellerer and H. H. Rossi, A generalized formulation of dual radiation action, *Radiat. Res.* **75,** 471–488 (1978).
20. D. Chmelevsky, A. M. Kellerer, M. Terissol, and J. P. Patau, Proximity functions for electrons up to 10 keV, *Radiat. Res.* **84,** 219–238 (1980).
21. M. Zaider, K. Hanson, and G. N. Minerbo, Algorithms for determining the proximity distribution from variance measurement, Annual Report on Research Project, Radiological Research Laboratory, Columbia University, New York (1979–1980). (Available

as COO-4733-3 from National Technical Information Service, Springfield, Virginia 22161.)
22. L. G. Bengtsson and L. Lindborg, Comparison of pulse height analysis and variance measurements for the determination of dose mean specific energy, in Proceedings of the 4th Symposium on Microdosimetry, p. 823, EUR 5122 d-e-f (1974).
23. U. Fano, Note on the Bragg–Gray cavity principle for measuring energy dissipation, *Radiat. Res.* **1,** 237 (1954).
24. H. Bichsel, Charged-particle interactions, in *Radiation Dosimetry,* Vol. 1, pp. 157–224, Academic, New York (1967).
25. J. W. Boag, Ionization chambers, in *Radiation Dosimetry* (G. J. Hine and G. L. Brownell, eds.), Academic, New York (1956).
26. P. J. Campion, The operation of proportional counters at low pressures for microdosimetry, *Phys. Med. Biol.* **16,** 611 (1971).
27. P. J. Campion, Some comments on the operation of proportional counters, in Proceedings of the 3rd Symposium on Microdosimetry, Stresa, Italy, EUR 4810, (H. G. Ebert, ed.) (1972).
28. J. B. Smathers, V. A. Otte, A. R. Smith, P. R. Almond, F. J. Attix, J. J. Spokas, W. M. Quam, and L. F. Goodman, Composition of A-150 tissue-equivalent plastic, *Med. Phys.* **4,** 74 (1977).
29. D. Srdoc, Experimental technique of measurement of microscopic energy distribution in irradiated matter using Rossi-counters, *Rad. Res.* **43,** 302 (1970).
30. W. A. Glass and L. A. Braby, Gas pressure control for flow type proportional counters, Proceedings N. W. Lab. Ann. Rep. BNWL715, Part 2, p. 215 (1967).
31. B. Forsberg, M. Jensen, L. Lindborg, and G. Samuelson, Determination of the dose mean specific energy for conventional x-rays by variance measurements, Proceedings of the 6th Symposium on Microdosimetry, Brussels (Belgium), EUR 6064 DE-EN-FR (1978).
32. A. M. Kellerer, An assessment of wall effects in microdosimetric measurements, *Radiat. Res.* **47,** 377 (1971).
33. W. A. Glass and W. A. Gross, Wall-less detectors in microdosimetry, in *Radiation Dosimetry,* Suppl. 1, pp. 221–260, Academic, New York (1972).
34. R. C. Rodgers, J. F. Dicello, and W. Gross, The biophysical properties of 3.9-GeV nitrogen ions II. Microdosimetry, *Radiat. Res.* **54,** 12–23 (1973).
35. M. R. Raju and C. Richman, Negative pion radiotherapy: Physical and radiobiological aspects, *Curr. Top. Radiat. Res. Q.* **8,** 159–233 (1972).
36. G. Burger, E. Maier, and A. Morhart, Radiation quality and its relevancy in neutron radiotherapy, in: Proceedings of the Sixth Symposium on Microdosimetry (J. Booz and H. G. Ebert, eds.), EUR 6064 DE-EN-FR, Brussels, Belgium (1978).
37. J. F. Dicello, private communication.
38. B. Rossi, *High-Energy Particles,* Prentice-Hall, New York (1952).
39. P. Kliauga and R. Dvorak, Microdosimetric measurements of ionization by monoenergetic photons, *Radiat. Res.* **73,** 1–20 (1978).
40. U. Fano, Ionization yield of radiation, II. The fluctuations of the number of ions, *Phys. Rev.* **72,** 26–29 (1947).
41. U. L. R. Lindborg, Microdosimetry in high energy electron and ^{60}Co gamma ray beams for radiation therapy, Proceedings of the Fourth Symposium on Microdosimetry, Verbania Pallanta, Italy, 24–28, EUR 5122 d-e-f (1973).
42. J. A. Auxier, W. S. Snyder, and T. D. Jones, Neutron interaction and penetration in tissue, in *Radiation Dosimetry,* Vol. 1, pp. 275–315, Academic, New York (1967).
43. S. B. Curtis and W. Schimmerling, Nuclear physics of accelerated heavy ions, Report LBL-5610, pp. 36–48, University of California, Berkeley (1977).

44. A. Chatterjee, C. A. Tobias, and J. T. Lyman, Nuclear fragmentation in therapeutic and diagnostic studies with heavy ions, in *Spallation Nuclear Reactions and Their Applications* (Shen/Merker, eds.), p. 169, D. Reidel, Dordrecht, Holland (1976).
45. H. H. Heckman, H. J. Crawford, D. E. Greiner, P. J. Lindstrom, and L. W. Wilson, Central collisions produced by relativistic heavy ions in nuclear emulsions, *Phys. Rev. C* **17,** 1651–1664 (1978).
46. J. B. Cumming, P. E. Haustein, R. W. Stoenner, L. Mausner, and R. A. Naumann, Spallation of Cu by 3.9-GeV ^{14}Neons and 3.9-GeV Protons, *Phys. Rev. C* **10,** 739–755 (1974).
47. H. H. Heckman, D. E. Greiner, P. J. Lindstrom, and H. Shwe, Fragmentation of ^{4}He, ^{12}C, ^{14}N, and ^{16}O nuclei in nuclear emulsion at 2.1 GeV/nucleon, *Phys. Rev. C* **17,** 1735–1747 (1978).
48. P. J. Kliauga, R. D. Colvett, Y. M. P. Lam, and H. H. Rossi, The relative biological effectiveness of 160-Mev protons. I. Microdosimetry, *Int. J. Radiat. Oncol. Biol. Phys.* **4,** 1001–1008 (1978).
49. K. G. Zimmer, *Studies on Quantitative Biology,* Hafner, New York (1961).
50. M. M. Elkind and G. F. Whitmore, *The Radiobiology of Cultured Mammalian Cells,* Gordon and Breach, New York (1967).
51. G. W. Barendsen, Mechanism of action of different ionizing radiations on the proliferative capacity of mammalian cells, in *Theoretical and Experimental Biophysics* (A. Cole, ed.), Vol. 1, pp. 167–231, Dekker, New York (1967).
52. P. W. Todd, Reversible and irreversible effects of ionizing radiations on the reproductive integrity of mammalian cells cultured *in vitro,* thesis, University of California, Lawrence Radiation Laboratory UCRL 11614 (1964).
53. A. M. Kellerer and H. H. Rossi, The theory of dual radiation action, *Curr. Top. Radiat. Res. Q.* **8,** 85–158 (1972).
54. A. M. Kellerer, Y. P. Lam, and H. H. Rossi, Biophysical studies with spatially correlated ions. 4. Analysis of cell survival data for diatomic deuterium, *Radiat. Res.* **83,** 511–528 (1980).
55. J. L. Bateman, H. H. Rossi, A. M. Kellerer, C. V. Robinson, and V. P. Bond, Dose dependence of fast neutron RBE for lens opacification in mice, *Radiat. Res.* **51,** 381–390 (1972).
56. C. J. Shellabarger, A. M. Kellerer, H. H. Rossi, L. J. Goodman, R. D. Brown, R. E. Mills, A. R. Rao, J. P. Shanley, and V. P. Bond, Rat mammary carcinogenesis following neutron or x-irradiation, in *Biological Effects of Neutron Irradiation*; IAEA, Vienna (1974).
57. H. H. Rossi, The effects of small doses of ionizing radiation: Fundamental biophysical characteristics, *Radiat. Res.* **71,** 1–8 (1977).
58. R. L. Ullrich, M. C. Jernigan, and L. M. Adams, Induction of lung tumors in RFM mice after localized exposures to x-rays or neutrons, *Radiat. Res.* **80,** 464–473 (1979).
59. A. A. Awa, Chromosome aberrations in somatic cells, *J. Radiat. Res. Suppl.* **16,** 122–131 (1975).
60. M. T. Biola, R. Lego, G. Ducatez, G. Dacher, and M. Bourguignon, Formation de chromosomes dicentrique dans les lumphocytes humains soumis *in vitro* a un flux de rayonnement mixte (gamma, neutrons), in *Advances in Physical and Biological Radiation Detectors,* pp. 633–645, IAEA (1971).
61. A. H. Sparrow, A. G. Underbrink, and H. H. Rossi, Mutations induced in Tradescantia by small doses of x-rays and neutrons, Analysis of dose–response curves, *Science* **176,** 916–918 (1972).
62. E. J. Hall, H. H. Rossi, A. M. Kellerer, L. J. Goodman, and S. Marino, Radiobiological studies with monoenergetic neutrons, *Radiat. Res.* **54,** 431–443 (1973).

63. M. Zaider and J. F. Dicello, RBEOER: A FORTRAN program for the computation of RBEs, OERs, survival ratios, and the effects of fractionation using the theory of dual radiation action, Report LA-7196-MS, Los Alamos Scientific Laboratory (1978).
64. R. Railton, R. C. Lawson, and D. Porter, Interaction of x-ray and neutron effects on the proliferative capacity of Chinese hamster cells, *Int. J. Radiat. Biol.* **27,** 75–82 (1975).
65. R. E. Durand and P. L. Olive, Irradiation of multi-cell spheroids with fast neutrons versus x-rays: A qualitative difference in sub-lethal damage repair capacity or kinetics, *Int. J. Radiat. Biol.* **30,** 583–592 (1976).
66. F. Q. Ngo, A. Han, and M. M. Elkind, On the repair of sub-lethal damage in V79 Chinese hamster cells resulting from irradiation with fast neutrons or fast neutrons combined with x-rays, *Int. J. Radiat. Biol.* **32,** 507–511 (1977).
67. F. Q. H. Ngo, A. Han, H. Utsumi, and M. M. Elkind, Comparative radiobiology of fast neutrons: Relevance to radiotherapy and basic studies, *Int. J. Radiat. Oncol. Biol. Phys.* **3,** 187–193 (1977).
68. S. Hornsey, U. Andreozzi, and P. R. Warren, Sublethal damage in cells of the mouse gut after mixed treatment with x-rays and fast neutrons, *Br. J. Radiol.* **50,** 513–517 (1977).
69. F. Q. H. Ngo, E. A. Blakely, and C. A. Tobias, Do sublethal lesions and repair occur after high-LET radiation?, Report LBL-7454, Lawrence Berkeley Laboratory, University of California (1979).
70. R. Bird, M. Zaider, H. H. Rossi, and E. J. Hall, The sequential irradiation of mammalian cells with x-rays and charged particles of high LET, *Radiat. Res.* **93,** 444–452 (1983).
71. M. Zaider and H. H. Rossi, The synergistic effect of different radiations, *Radiat. Res.* **83,** 732–739 (1980).
72. J. L. Bateman, H. A. Johnson, V. P. Bond, and H. H. Rossi, The dependence of RBE on the energy of fast neutrons for spermatogonia depletion in mice, *Radiat. Res.* **35,** 86–101 (1968).
73. H. H. Rossi, Comment on the somatic effects section of the BEIR III report, *Radiat. Res.* **84,** 395–406 (1980).
74. A. M. Kellerer and H. H. Rossi, Biophysical aspects of radiation carcinogenesis, *CANCER,* 2nd edition (F. Becker, ed.), pp. 569–616, Plenum, New York (1982).
75. National Council on Radiation Protection and Measurements, Report 17, Permissible dose from external sources of ionizing radiation (1954) including Maximum permissible exposure to man, Addendum to National Bureau of Standards Handbook 59 (1958).
76. H. H. Rossi, A proposal for revision of quality factor, *Rad. Environ. Biophys.* **14,** 275–283 (1977).

Chapter 5

Ultraviolet Radiation Dosimetry and Measurement

Brian Diffey

1. THE ULTRAVIOLET SPECTRUM

Ultraviolet radiation (UVR) is part of the electromagnetic spectrum and lies between the visible and the X-ray regions. Different wavebands in the ultraviolet spectrum show enormous variations in causing biological damage, and for this reason the UV spectrum is divided into three spectral regions: UV-A, UV-B, and UV-C.

The notion of dividing the ultraviolet spectrum into different spectral regions was first put forward at the Copenhagen meeting of the Second International Congress on Light held during August 1932.[1] It was recommended that three spectral regions be defined as follows:

UV-A 400–315 nm

UV-B 315–280 nm

UV-C <280 nm

Various regulatory authorities including the International Commission on Illumination (CIE), the National Institute for Occupational Safety and Health (NIOSH) in the United States, and the Health and Safety Executive (HSE) and the National Radiological Protection Board (NRPB) in the United Kingdom, have by their use endorsed these regions, with the slight modification that the lower limit of the UV-C

Brian Diffey • Regional Medical Physics Department, Dryburn Hospital, Durham DH1 5TW, U.K.

region is taken to be 100 nm by the CIE, HSE, and NRPB, and 200 nm by the NIOSH. Radiation in the spectral region 100–200 nm (the vacuum UV) is readily attenuated in air with little opportunity to produce direct biological effects, and so the lower limit of the UV-C region for practical purposes is not critical.

The development of environmental photobiology has prompted some workers[2] to redefine the boundaries of the three spectral regions as follows:

UV-A	400–320 nm
UV-B	320–290 nm
UV-C	290–200 nm

The upper limit of the UV-B region has been extended to 320 nm, since most acute and chronic effects of sunlight exposure on biological systems are believed to occur at wavelengths less than 320 nm, while the division between the UV-B and UV-C regions has been set at 290 nm since this wavelength is the approximate lower limit of terrestrial radiation. Yet another publication[3] has defined the UV-B region as 280–320 nm. Although the divisions between the spectral regions are not necessarily rigid, it would seem sensible to adopt international recommendations, and so the convention in this chapter is as follows:

Spectral region	Also known as	Range of wavelengths (nm)
UV-A	Longwave or "black light"	400–315
UV-B	Middlewave or "erythemal"	315–280
UV-C	Shortwave or "germicidal"	280–100

2. RADIOMETRY

In clinical and photobiological UVR dosimetry it is customary to use the terminology of radiometry rather than that of photometry, since photometry is based on visible light measurements that simulate the human eye's photopic response curve, and strictly speaking, a source that emits only UVR at wavelengths less than 380 nm has a zero intensity in photometric terms.

TABLE I
Radiometric Terms and Units

Term	Unit	Symbol	Definition
Wavelength	nm	λ	
Radiant energy	J	Q	
Radiant flux	W	ϕ	dQ/dt
Radiant intensity	$\mathrm{W\ sr^{-1}}$	I	$d\phi/d\Omega$
Radiance	$\mathrm{W\ m^{-2}\ sr^{-1}}$	L	$d\phi/d\Omega\ dA \cos\theta$
Irradiance	$\mathrm{W\ m^{-2}}$	E	$d\phi/dA$
Radiant exposure	$\mathrm{J\ m^{-2}}$	H	$E \cdot t$

2.1. Radiometric Terms and Units

The common radiometric terminology is listed in Table I and defined below.[4,5] These radiometric quantities may also be expressed in terms of wavelength by adding the prefix "spectral."

Radiant Energy: energy propagated in the form of electromagnetic waves or particles. Only optical radiations of wavelength from about 1 nm to 1 mm are generally classified in this category.

Radiant Flux: the time rate of flow of radiant energy.

Radiant Intensity: quotient of the radiant flux leaving the source, propagated in an element of solid angle containing the given direction, by the element of solid angle.

Radiance: the radiance in a given direction, at a point on the surface of a source or a receptor, or at a point on the path of a beam is the quotient of the radiant flux leaving, arriving at, or passing through an element of surface at this point and propagated in directions defined by an elementary cone containing the given direction, by the product of the solid angle of the cone and the area of the orthogonal projection of the element of surface on a plane perpendicular to the given direction.

Irradiance: quotient of the radiant flux incident on an element of the surface containing the point by the area of that element.

Radiant Exposure: surface density of radiant energy received.

The time integral of the irradiance is strictly termed the "radiant exposure," but is sometimes expressed as "exposure dose," or even more loosely as "dose." The term "dose" in photobiology is analogous to the term "exposure" in radiobiology and not to "absorbed dose." As yet the problems of estimating the energy absorbed by the critical target in the skin (whatever that might be) remain unsolved.

2.2. Radiometric Calculations

2.2.1. The Inverse Square and Cosine Laws of Irradiation

The irradiance E on a surface of area dA subtending a solid angle $d\Omega$ at a point source is found by eliminating $d\phi$ between the definitions for a radiant intensity (I) and irradiance (E) given in Table I, such that

$$E = I\frac{d\Omega}{dA} \quad (1)$$

From geometrical considerations $d\Omega = dA'/d^2$, where d is the distance from the source to the surface and dA' is the projection of the area dA on the plane perpendicular to the direction of the source, given by

$$dA' = dA \cos\theta \quad (2)$$

where θ is the angle between the direction of the source and the perpendicular to the surface. Eliminating $d\Omega$ and dA from Eq. (1) leads to

$$E = I \cos\theta / d^2 \quad (3)$$

Equation (3) is the symbolic expression of the two fundamental laws of radiometry and may be formally stated as follows:

> The irradiance of an elementary surface due to a point source of radiation is proportional to the radiant intensity of the source in the direction of that surface, and to the cosine of the angle between this direction and the normal to the surface, and it is inversely proportional to the square of the distance between the surface and the source.

2.2.2. Surface Irradiation

When the surface cannot be regarded as small in comparison with its distance from the source, the irradiance varies over the surface. Walsh[6] has shown that for the simple case of a circular disk of radius a irradiated by a uniform point source of radiant intensity I placed at a distance d from the disk along the axis of the latter, the irradiance at any point P on the disc at a distance r from the center is

$$I \cos\theta/(r^2 + d^2) \quad (4)$$

where θ is the angle of incidence at P. Now since

$$\cos\theta = d/(r^2 + d^2)^{1/2} \tag{5}$$

expression (4) reduces to

$$I\cos^3\theta/d^2 \tag{6}$$

This is sometimes referred to as the "cosine-cubed" law. It follows that the average irradiance is

$$(1/\pi a^2)\int_0^a (I\cos^3\theta/d^2)2\pi r\,dr \tag{7}$$

which equals

$$(2I/a^2)[1 - d/(a^2 + d^2)^{1/2}] \tag{8}$$

Since fluorescent lamps are commonly used in medicine as sources of UVR, particularly for patient irradiation, another formula given by Walsh[6] for irradiation from cylindrical and flatstrip sources may prove useful. The irradiance at a point distance d from the center of a strip of breadth $2a$ and length $2l$, and radiance L, is given by

$$(2aL/d)[\tan^{-1}(l/d) + ld/(l^2 + d^2)] \tag{9}$$

2.2.3. Volume Irradiation

In many experiments in photochemistry and photobiology, the samples to be irradiated are often in suspension in a liquid. For example, a liquid suspension of single cells may absorb some of the radiation so that cells near the back of the cuvet (the vessel that holds the sample) receive less radiation than those near the front.[7,8] Even if the sample is stirred vigorously, the average irradiance per cell will be lower than the incident irradiance. For a parallel beam of monochromatic radiation larger than the cross-sectional area of the suspension perpendicular to the direction of propagation of the beam, Morowitz[9] has shown that the average irradiance per cell, $\bar{E}$, is related to the incident irradiance, E, by

$$\bar{E} = E[1 - \exp(-2.3A)]/2.3A \tag{10}$$

where A is the absorbance of the sample at the irradiating wavelength. The factor 2.3 approximates $\log_e 10$ since the absorbance A is related to the fraction of incident radiation transmitted, T, by

$$A = -\log_{10} T \tag{11}$$

Equation (10) has been modified by Johns[10] who considers the case of a converging beam of radiation on the cuvet. Jagger *et al.*[11] discuss the errors that can arise in the estimation of $\bar{E}$ by Eq. (10) if light scattering by suspensions of small cells is not taken into account.

Finally it should be remembered that the front surface of the entry face of a glass or quartz cuvet will normally reflect at least 4% of the incident radiation. The back surface of the entry face will reflect very little radiation since the refractive index of the solvent, e.g., water, is similar to that of glass or quartz. If measurements are made behind a liquid-filled cuvet, then reflection losses at two surfaces must be taken into account. If the cuvet is empty, there will be reflection losses at four surfaces.

3. BIOLOGICAL EFFECTS OF ULTRAVIOLET RADIATION IN HUMANS

The observable biological effects in man due to exposure from external sources of ultraviolet radiation are limited to the skin and to the eyes because of the low penetrating properties of UVR in human tissues. This section will outline the recognizable short-term and long-term effects of UV exposure in the skin and the eyes.

3.1. Effects of UVR on Normal Skin

The normal responses of the skin to ultraviolet radiation may be classified as either acute, e.g., erythema, melanin pigmentation, vitamin D production; or chronic, e.g., skin aging and skin cancer.[12]

3.1.1. Ultraviolet Erythema

Exposure to UVR, particularly from wavelengths less than 315 nm, can result in erythema. The redness of the skin which is characteristic of erythema is attributable to an increased blood content by dilation of the superficial blood vessels, mainly the subpapillary venules. Sunburn caused by exposure to solar radiation normally has a latent period of a few hours. Erythema induced by artificial sources is strongly dependent on the wavelength of radiation. At 300 nm an average threshold dose or minimal erythema dose (MED) in unacclimatized white skin is about

200 J m^{-2},[13] whereas for UV-A radiation the MED is about a thousand-fold higher.[14] Larger doses of UV-B may result in edema, pain, and blistering, although blistering seldom occurs with UV-C.[15]

3.1.2. Melanin Pigmentation

A socially desirable consequence of exposure to unfiltered sunlight is the delayed pigmentation of the skin known as "tanning," which becomes noticeable about two days after exposure and gradually increases for several days. Tanning is due not only to the formation of new melanin but also to the migration of the pigment already present in the basal cells to the more superficial layers of the skin.

3.1.3. Production of Vitamin D

The skin absorbs UV-B radiation in sunlight to convert sterol precursors in the skin, such as 7-dehydrocholesterol, to vitamin D_3.[16] Vitamin D_3 is further transformed by the liver and kidneys to biologically active metabolites such as 25-hydroxyvitamin D; these metabolites then act on the intestinal mucosa to facilitate calcium absorption, and on bone to facilitate calcium exchange.[17]

3.1.4. Aging of the Skin

Chronic exposure to sunlight can result in an appearance of the skin often referred to as premature aging or actinic damage. The clinical changes associated with skin aging include a dry, coarse, leathery appearance, laxity with wrinkling, and various pigmentary changes.[18]

3.1.5. Photocarcinogenesis

The idea that chronic exposure to sunlight can lead to skin cancer has been evident for about 90 years following the observations of Unna[19] and Dubreuilh.[20] Since then it has gradually been accepted that sunlight, or more precisely those wavelengths less than about 320 nm, plays an etiological role in the formation of skin cancer in man.[21] There appears to be little question now that the majority of squamous cell carcinomas are caused by chronic exposure to sunlight,[22] although such a clear relationship between the incidence of either basal cell carcinomas[23] or, more particularly, malignant melanoma[24] and sunlight exposure has not been demonstrated.

3.2. Effects of UVR on the Eye

3.2.1. Photokeratitis and Conjunctivitis

The acute effects of exposure to UV-C and UV-B radiation are primarily those of conjunctivitis and photokeratitis.[2]

Conjunctivitis is an inflammation of the membrane that lines the insides of the eyelids and covers the cornea, and may often be accompanied by an erythema of the skin around the eyelids. There is the sensation of "sand in the eyes" and also varying degrees of photophobia (aversion to light), lacrimation (tears), and blepharospasm (spasm of the eyelid muscles) may be present.

Photokeratitis is an inflammation of the cornea which can result in severe pain. Ordinary clinical photokeratitis is characterized by a period of latency that tends to vary inversely with the severity of UV exposure. The latent period may be as short as 30 min or as long as 24 h, but it is typically 6–12 h. The acute symptoms of visual incapacitation usually last from 6–24 h. Almost all discomfort usually disappears within two days and rarely does exposure result in permanent damage. Unlike the skin, the ocular system does not develop tolerance to repeated exposure to UVR. Many cases of photokeratitis have been reported following exposure to UVR produced by welding arcs and by the reflection of solar radiation from snow and sand. For this reason the condition is sometimes referred to as "welder's flash," "arc eye," or "snow blindness."

3.2.2. Cataracts

A cataract is a partial or complete loss of transparency of the lens or its capsule. Most of the available data on the production of cataracts have been obtained using the rabbit. The most effective wavelengths for producing lenticular opacities appear to lie in the range 295–315 nm.[25]

Chemical effects in the lens protein, tryptophan, have been shown to occur after UV-A exposure of human lenses.[26] The effects lead to the formation of chromatic photoproducts that bind to, and alter, the solubility of lens proteins, resulting in a yellowing of the lens material. While the basic biochemical mechanism remain to be found, it is nevertheless suggested that exposure to UV-A can enhance cataractogenesis in humans.

There is scant evidence that the incidence of cataracts is increased in temperate areas[27] and is more common in outdoor, rather than indoor workers.[28] Although circumstantial evidence does not prove a relationship between sunlight exposure and human cataracts, the evidence does suggest a causal relationship to be a likely possibility.

4. THE CONCEPT OF BIOLOGICALLY EFFECTIVE RADIATION

In radiobiology, the effectiveness of ionizing radiation to cause a particular biological effect is expressed by the relative biological effectiveness (RBE). The RBE for a given quality of radiation is normally defined as the absorbed dose of the 200 kVp X rays required to produce a specific biological end point divided by the absorbed dose of the radiation in question required to produce the same end point.[29] The values of RBE depend not only on the radiation compared but also on the particular biological effects considered. For many radiobiological actions, the RBE of electromagnetic radiation with photon energies from a few keV to several MeV is not appreciably different from unity.

Unfortunately, this is not the case in photobiology, where virtually all photobiological effects show a strong dependence on photon energy, or wavelength. For example, the radiant exposure of 320-nm radiation required to produce a given degree of erythema in human skin is about 100 times greater than the radiant exposure of 300-nm radiation needed to produce the same effect. In these situations where the interest lies in some particular action of the UVR, the effectiveness of the radiation is obtained by weighting the spectral irradiance (or radiant exposure) according to the appropriate function of wavelength and then integrating over all wavelengths for which the spectral content of the source is nonzero. The determination of this single quantity, the biologically effective irradiance (or radiant exposure), is often the goal of biological and clinical ultraviolet dosimetry. Since ultraviolet erythema probably represents the commonest biological action of interest in medical photobiology, this phenomenon will be used to illustrate the applicability and limitations of the concept of biologically effective radiation.

4.1. The Erythema Action Spectrum

The ability of UVR to produce an erythema response in human skin is highly dependent upon the wavelength of radiation and is expressed by the action spectrum. An action spectrum is a plot of the reciprocal of the dose required for a given effect against wavelength, and strictly applies only if the dose–response curves are similar at all wavelengths, which implies that the mechanism of action is the same at all wavelengths.

The effectiveness of ultraviolet radiation of different wavelengths in producing erythema in normal human skin has been a subject of study for about 60 years. Hausser and Vahle[30] first reported the erythema action spectrum which showed a major peak of activity at 297 nm, a minimum at about 280 nm, and a second but lesser peak at around 250 nm. The general form of this curve was confirmed by other investigators in the following ten years,[31–33] so much so that in 1934 Coblentz and Stair[34] proposed a "standard erythema curve," which was subsequently published by the CIE in 1935.

Experimental data from more recent estimates of the erythema action spectrum have all differed appreciably from the CIE standard, particularly in the spectral region 250–300 nm.[35–43] All of these investigators have shown that the effectiveness of radiation at 250 nm is in fact higher than that at about 300 nm, with a less apparent, or even no, minimum at 280 nm. At wavelengths greater than 300 nm these recent results concur with earlier estimates in that the erythemal effectiveness of ultraviolet radiation drops very rapidly, falling to an efficiency at 320 nm of about 1% of that at 300 nm. Figure 1 compares the CIE action spectrum with more recent estimates'

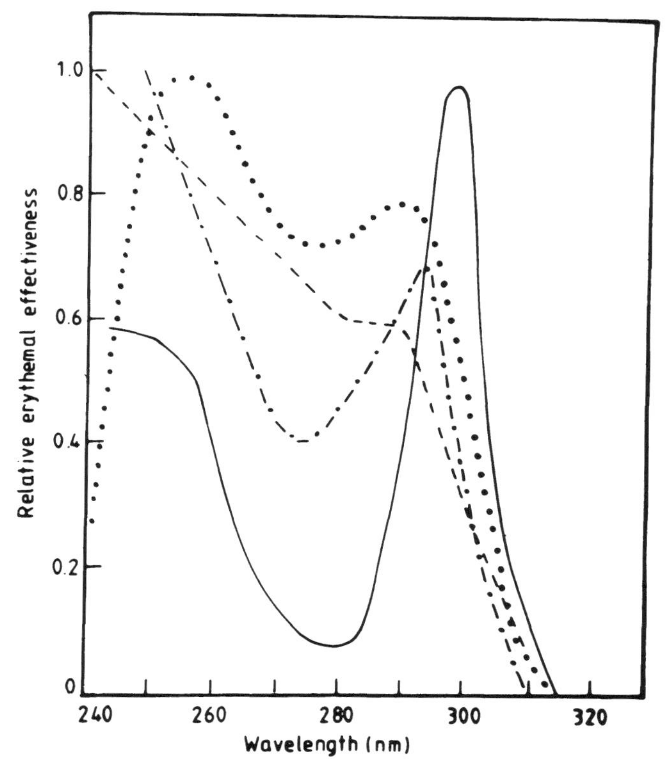

Figure 1. Comparison of the CIE standard erythemal curve with the erythema action spectra determined by other workers. ——, CIE curve[107]; . . . , Freeman *et al.*[37]; – – –, Everett *et al.*[36]; – · – ·, Cripps *et al.*[41] Each curve has been normalized to unity at its most effective wavelength.

Those erythema action spectra determined in the period 1964–1974 have been examined for interinvestigator consistency using one-way analysis of variance.[13]

The conclusion of this analysis indicated that the action spectra determined by the various investigators were not self-consistent, and seriously questioned the concept of a "standard erythema curve." Certainly the sensitivity of normal skin to ultraviolet radiation will vary with factors such as anatomical site,[39] time of observation,[38,39,41,43] size of irradiation field,[44] method of recording erythema,[38] and season.[45]

Also Goldsmith and Hellier[46] have reported that skin sensitivity is dependent upon the color of the skin, age, sex, thyroid function, pregnancy, menstruation, blood flow, humidity, and temperature. Technical factors during irradiation may further influence the results, for example, the bandpass of the irradiation monochromator used to determine the action spectrum,[47] or the presence of stray radiation.[38]

In particular, van der Leun[48] has examined the published erythema action spectra by confronting the data with the hypothesis that two distinct mechanisms contribute to the overall erythema following exposure to ultraviolet radiation. The two mechanisms are thought to be the production of a photochemical mediator in the epidermis which diffuses into the dermis causing vasodilation, and a direct action of the radiation on the vessel wall itself. Van der Leun argues that differences between action spectra can be accounted for to a greater or lesser degree by three factors:

1. choice of anatomical site for irradiation reflecting differences in epidermal thickness;
2. influence of time of observation on assessing erythema;
3. the criterion used to judge the erythema, i.e., whether a just perceptible reddening or an erythema with recognizable boundaries.

4.2. Photoaddition

Even though the concept of a standard action spectrum is questionable, not only for ultraviolet erythema, but also for many other photobiological processes such as skin carcinogenesis or inhibition of photosynthesis, generalized action spectra are commonly employed in photobiology to calculate some biologically effective irradiance, or dose.

The principle, known as photoaddition, is assumed whenever a photobiological action spectrum is multiplied by a spectral irradiance curve of some source of radiation to predict or explain an observed response. Ultraviolet erythema is the classic example of a photobiological response which is often assumed to obey photoaddition such that the predicted radiant exposure, $D_p(\mathrm{J\,m^{-2}})$, from a polychromatic source of optical radiation necessary to produce a given degree of erythema (e.g., minimal erythema) is calculated as

$$D_p = D(\lambda_o)\frac{\sum E(\lambda)\,\Delta\lambda}{\sum E(\lambda)\varepsilon(\lambda)\,\Delta\lambda} \quad \mathrm{J\,m^{-2}} \tag{12}$$

where $D(\lambda_0)$ is the radiant exposure ($\mathrm{J\,m^{-2}}$) of monochromatic radiation of wavelength λ_0 nm necessary to produce the desired response, $E(\lambda)$ is the spectral irradiance ($\mathrm{W\,m^{-2}\,nm^{-1}}$) at the subject's skin, $\Delta\lambda$ is the spectral bandwidth (nm) used for the summation, and $\varepsilon(\lambda)$ is the action spectrum for the desired erythemal response normalized to unity at a wavelength of λ_o nm. The summation in the numerator and denominator is over all wavelengths for which the source has a nonzero spectral emission.

The observed radiant exposure, $D_o(\mathrm{J\,m^{-2}})$, necessary to produce the erythemal response is simply

$$D_o = t\sum E(\lambda)\,\Delta\lambda \quad \mathrm{J\,m^{-2}} \tag{13}$$

where t is the exposure time in seconds. Agreement between D_p and D_o would support the notion of photoaddition.

The principle of photoaddition assumes there are no synergistic or protective interactions between wavelengths. This is unlikely to be true; Willis *et al.*[49] reported a photoaugmentation phenomenon associated with UV-A irradiation of skin followed immediately by UV-B radiation, whereas van der Leun and Stoop[50] showed that delayed erythema reactions to 250 nm and 300 nm radiation are lessened if the UV exposures are followed by five hours of exposure to full indoor daylight (UV-A and visible radiation, but no UV-B). This phenomenon is known as photorecovery or photoreactivation.

4.3. Dosage Units for Erythemally Effective Radiation

Because of the rapid variation of erythemal effectiveness with wavelength (see Fig. 1), the total irradiance, or irradiance in some limited spectral region, is of little value in estimating the exposure time to produce a given erythemal response. Instead some authors have chosen to speak of "weighted" or "effective" irradiance. In 1942 the Council on Physical Therapy of the American Medical Association introduced the "erythemal unit" (EU), which was defined as 10 μW cm^{-2} of monochromatic radiation of wavelength 296.7 nm (a characteristic line in the mercury vapor spectrum effective in producing erythema). The Illumination Engineering Society and the International Commission on Illumination had earlier suggested the term E-viton for that quantity of radiant flux which produced the same erythemal effect as 10 μW of radiant power at 296.7 nm. The E-viton was a biological unit in that it represented the integrated spectral radiant flux of a source weighted by the ability of the component wavelengths to produce an erythema. One EU or 1 E-viton per cm^2 was adopted as the unit of erythemal flux density and termed the "Finsen." It requires about 15,000 μW s cm^{-2} of radiation of wavelength 296.7 nm to produce a just perceptible erythema on untanned Caucasian skin; this is equivalent to 1500 E-viton s cm^{-2}, 1500 Finsen s, or 150 J m^{-2} using the SI notation. More recently, the concept of an erythemally effective UV-B irradiance, termed UV-B(EE), has been expressed mathematically as[51]

$$\text{UV-B(EE)} = \sum_{280}^{315} E(\lambda)\varepsilon(\lambda)\Delta\lambda \quad \text{W m}^{-2} \tag{14}$$

where $E(\lambda)$, $\varepsilon(\lambda)$, and $\Delta\lambda$ are as defined previously. Again this summation assumes that the actions of the separate spectral components are independent and combine in a simple additive manner. If the action spectrum is normalized to unity at a wavelength of λ_o nm (historically taken to be 297 nm), the erythemally effective UV-B irradiance given by Eq. (14) is equivalent to that irradiance of monochromatic λ_o nm radiation which would produce the same degree of erythema in a given time as the total irradiance given by

$$\sum E(\lambda)\Delta\lambda \quad \text{W m}^{-2} \tag{15}$$

where the summation is over all wavelengths emitted by the radiation source.

Analogous reasoning is used in the estimation of an effective hazardous irradiance in relation to occupational ultraviolet exposure standards (see Section 7).

Units such as the E-viton which are based on standardized weighting functions have long since fallen into disuse. New attempts to define similar biological units to represent UV-B(EE) will probably meet the same fate, principally because of the greater awareness of the variability in erythema action spectra with the consequent difficulty in achieving international agreement on a new "standard curve."

5. THE MEASUREMENT OF ULTRAVIOLET RADIATION

Techniques for the measurement of ultraviolet radiation may be divided into three classes: biological, chemical, and physical. In general, physical devices measure power, while chemical and biological systems measure energy.

5.1. Biological Dosimetry

5.1.1. Microorganisms as UV Dosimeters

The most dramatic effect of UVR on cells and viruses is inactivation, that is, loss of their ability to reproduce themselves. Inactivation is conveniently characterized by survival curves which represent the fraction of noninactivated individuals (survivors) as a function of radiant exposure, and forms the basis of many biological UVR dosimetry systems.[52] Virulent bacterial viruses (often called bacteriophages, or phages), such as phage T2 or T4 in the host bacterium *Escherichia coli,* are often used

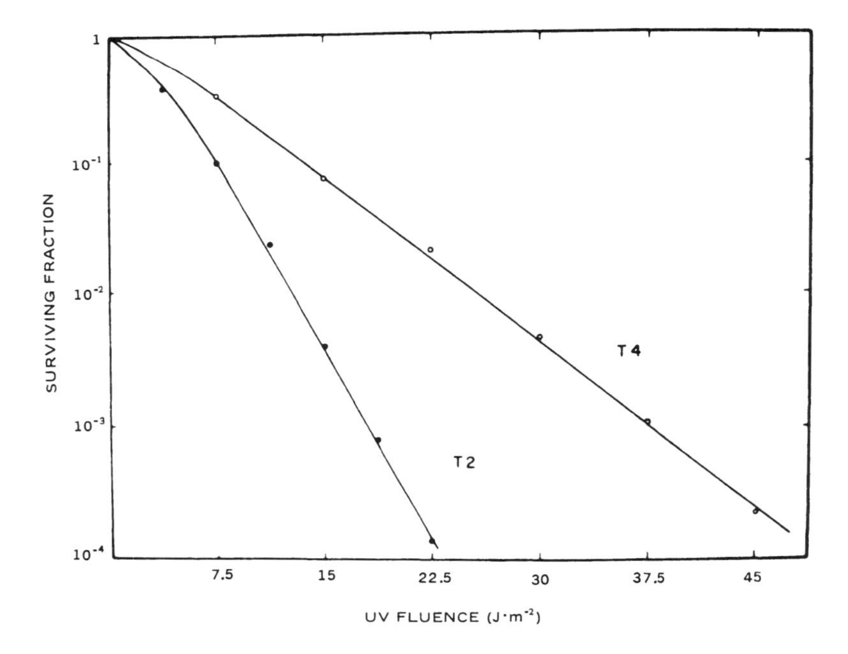

Figure 2. Survival curves of phages T2 and T4 as a function of UV radiant exposure (fluence) at 254 nm.[8]

as biological dosimeters. The slope of their survival curve is quite reproducible with only a slight shoulder at low radiant exposures (see Fig. 2). Once the quantitative response of a biological dosimetry system has been established, it can often be applied by other workers without the need for recalibration against physical or chemical methods. Biological dosimetry using microorganisms has a peculiar advantage not shared by any other measuring system, namely, that the biological dosimeter may often be mixed with the biological sample being studied and both irradiated simultaneously. This technique provides an accurate measure of the average radiant exposure incident on each cell in a suspension if due regard is paid to the considerations of volume irradiation discussed in Section 2.2.3.

5.1.2. Human Skin as a UV Dosimeter

Human skin is often used as a biological dosimeter in phototherapy, the unit of dose being a given degree of erythema. Before embarking upon a course of irradiation small areas of the patient's skin are exposed for varying times to the radiation from the lamp to be used for treatment, at the appropriate distance. The time that is sufficient to produce a just perceptible reddening is noted and so the exposure necessary for any other degree of erythema can be determined as discussed below.

The erythema reaction following exposure to UVR is classified into one of four categories, depending upon the severity of the reaction.

1. First degree erythema (E_1)—a just perceptible erythema, often referred to as the minimal erythema dose (MED), which lasts for about 24 h and leaves the skin apparently unchanged.
2. Second degree erythema (E_2)—a response resembling a mild sunburn which subsides after 3–4 d and may be followed by tanning.
3. Third degree erythema (E_3)—a severe reaction in which the erythema is accompanied by edema and tenderness. The reaction lasts for several days, pigmentation is more apparent, and exfoliation is marked; the skin often peels off in sheets or flakes.
4. Fourth degree erythema (E_4)—the initial changes are the same as in an E_3 reaction, but the edema and exudation are so severe that a blister is formed.

Although the dose response characteristic of erythema is a marked function of wavelength, as a rough guide, the relative exposure times for irradiation with a medium-pressure mercury arc lamp needed to produce an E_1 through to an E_4 are in the ratio 1:2.5:5:10.

5.2. Chemical Dosimetry

The measurement of optical radiation by the change produced in a chemical is called "actinometry" and may be accomplished in the gaseous, liquid, or solid phase. Like biological dosimeters, an actinometer integrates the radiant power received by it over the period of its exposure.

5.2.1. Gaseous Actinometer Systems

One of the most widely used and reliable gaseous actinometers for the spectral interval 250–320 nm is acetone photolysis.[53] At temperatures in excess of 125°C and pressures less than about 6600 N m^{-2} the quantum yield (ϕ) for the production of the photoproduct, carbon monoxide, is one. (The quantum yield is the number of molecules of reactant consumed or product formed per quantum of radiation absorbed.) If the acetone reactant and its photolysis products are cooled to liquid nitrogen temperature after the exposure, the amount of CO can be determined by standard chemical or physical methods.

The fraction of the incident radiation absorbed by the acetone is given by

$$1 - 10^{-\varepsilon c l}$$

and may be calculated from known values of the molar extinction coefficient ε ($cm^2\, mol^{-1}$), the concentration of acetone c (mol cm^{-3}), and the optical pathlength l (cm). The number of molecules of the photoproduct CO formed is η in the irradiation time t seconds. The flux, I, of the radiation beam just within the photolysis cell front window is then calculated as

$$I = \frac{\eta}{\phi t(1 - 10^{-ecl})} \quad \text{quanta s}^{-1} \tag{16}$$

Other molecules which undergo photolysis and have formed the basis of gaseous actinometers include diethyl ketone, hydrogen bromide, oxygen, nitrous oxide, and carbon dioxide.[53]

5.2.2. Liquid-Phase Chemical Actinometry

The potassium ferrioxalate system developed by Hatchard and Parker[54] is probably the most favored liquid-phase chemical actinometer for work in the ultraviolet and blue regions of the electromagnetic spectrum. When sulfuric acid solutions of $K_3Fe(C_2O_4)_3$ are irradiated in the spectral range 250–570 nm, a simultaneous reduction of iron to the ferrous state and oxidation of oxalate ion occur. The Fe^{2+} produced by photolysis of the ferrioxalate ion is converted to the phenanthroline complex and measured spectrophotometrically. The ferrioxalate actinometer can be used for measuring the energy emitted from optical radiation sources of almost any shape, as long as the spectral emission is in the appropriate range, but the method is too time-consuming for routine clinical UV dosimetry, where repeated measurements over relatively short time intervals are often necessary.

Another liquid chemical actinometer useful in the ultraviolet region is uranyl oxalate, which depends on the UO_2^{2+} sensitized decomposition of oxalic acid to CO, CO_2, and HCOOH.[55] The oxalate consumed is determined by titration with $KMnO_4$ or Ce^{4+} before and after exposure.

Other liquid actinometers suitable for the UV region include malachite green leucocyanide, monochloroacetic acid, and the complex chromium salt, $KCr(NH_3)_2(NCS)_4$. The quantum yields of all of the liquid actinometers mentioned above are shown as a function of wavelength in Fig. 3.[56]

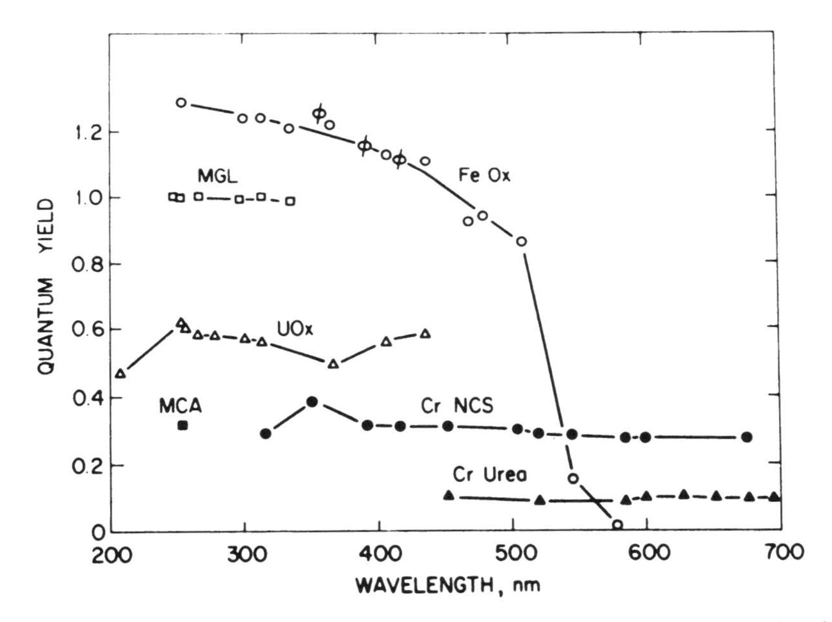

Figure 3. The quantum yield of several liquid actinometers at ca. 25°C as a function of wavelength.[56] FeOx, Ferrioxalate; UOx, uranyl oxalate; MGL, malachite green leucocyanide; MGA, monochloroacetic acid; CrNCS, $KCr(NH_3)_2(NCS)_4$; Cr Urea, $Cr(urea)_6Cl_3$.

Finally, Davies[57] has described the photochemistry of chemical actinometers based on halogenated hydrocarbons. One such compound and an indicator dye incorporated into a solid paraffin matrix has been used to give a qualitative measure of the anatomical variation of sunlight exposure on the surface of the head and neck.[58]

5.2.3. Diazo Systems

Diazo systems, which are based on diazonium compounds ($ArN_2^+Cl^-$), are one of the oldest photochemical nonsilver processes. The two fundamental properties of the diazo type process which make it suitable for use as a UV dosimeter are as follows[59]:

1. the ability to be decomposed by ultraviolet radiation:

$$ArN_2^+Cl^- + H_2O \xrightarrow{UV} ArOH + N_2 + HCl$$

2. the ability of the undecomposed diazonium compound to couple with a color former to produce a stable image:

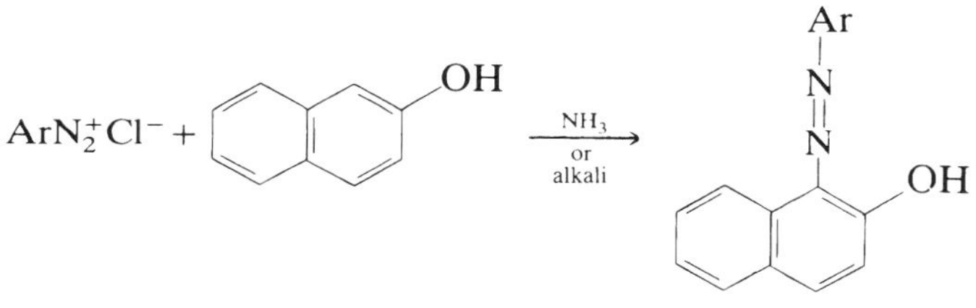

$$ArN_2^+Cl^- + \text{(OH)} \xrightarrow[\text{or alkali}]{NH_3} \text{Ar–N=N–(OH)}$$

Diazonium compounds are sensitive principally to the UV-A and blue regions of the spectrum,[60] as illustrated in Fig. 4. Their spectral sensitivity, together with the simplicity, economy, and convenience of the diazo system, have led to their use as film badge dosimeters for UV-A and blue radiation.

Jackson[61] has described a dosimeter based on the film, Diazochrome KBL, which is readily available and used in the photographic industry for making blue and white transparencies. When exposed to UVR and developed for 20 min in ammonia vapor, the clear film turns blue—dark blue if not exposed and clear if overexposed. The variation of film response with radiant exposure for a specified spectral quality is determined by expressing the percentage transmission of white light through the film as a function of radiant exposure.

A similar dosimeter system was described by Ali and Jacobson,[62] who used Kodak M diazo film, again processed in ammonia vapor. Since

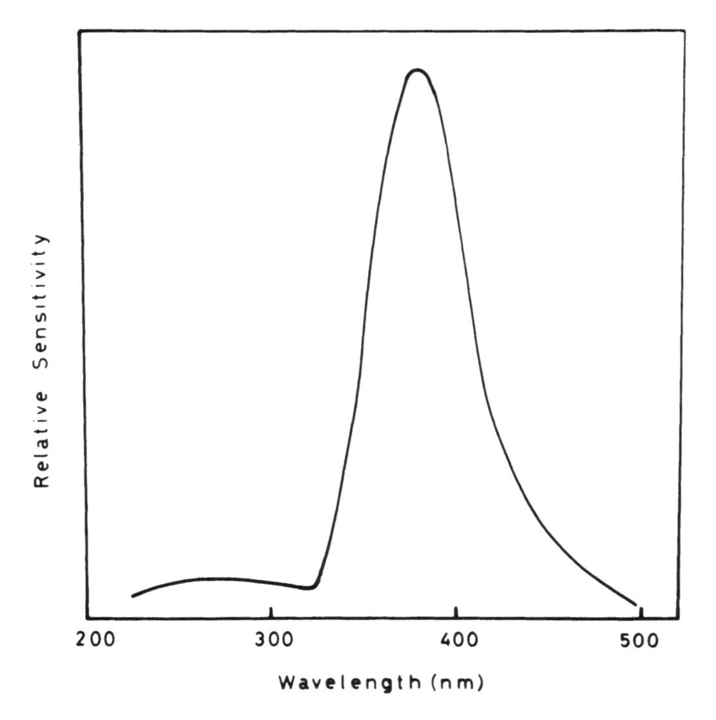

Figure 4. Approximate spectral sensitivity of the diazotype process.

these film badges were intended to be used out-of-doors for a period of about one week, it was necessary to reduce the sensitivity of the film by a factor of about 1000. This was achieved by incorporating Kodak Wratten neutral density gelatin filters in the film badge. An extensive evaluation of this and other diazo films for use as UV dosimeters is given by Card.[63]

5.2.4. Polysulfone Film

Perhaps the most commonly used material for studies of personal UV dosimetry has been the thermoplastic, polysulfone. Polysulfone, which has the structural unit shown in Fig. 5, was first suggested as a possible dosimeter for ultraviolet radiation by Davis *et al.*[64] Since then, the use of polysulfone film as a personal UV dosimeter has been well exploited for monitoring both environmental[65–69] and artificial[70–72] ultraviolet radiation. The basis of the method is that when polysulfone film is exposed to UVR at wavelengths less than 330 nm, the UV absorption of the film increases. The increase in absorbance measured at a wavelength of 330 nm is proportional to UV dose, and so the film readily lends itself

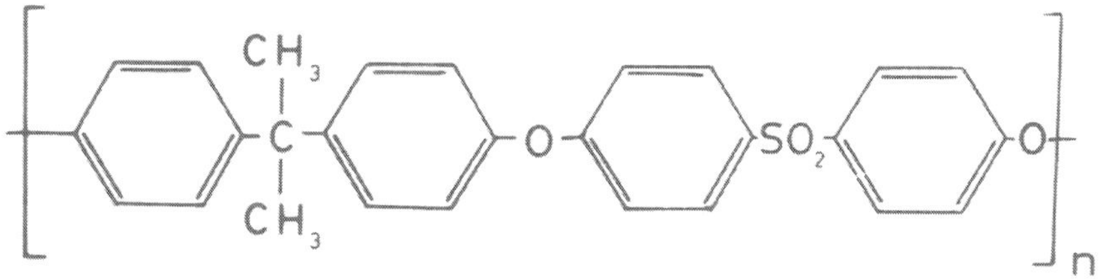

Figure 5. The structural unit of polysulfone.

to application as a UV dosimeter. In practice the film (40 μm thick) is mounted in cardboard photographic holders with a central aperture of 12 × 16 mm (see Fig. 6), and worn by subjects much as photographic film badges are worn as ionizing radiation dosimeters. The change in optical absorbance of the film at 330 nm is determined by noting the absorbance of the film badge before and after irradiation. This is normally accomplished using any standard UV spectrophotometer and takes about 5–10 s per badge, depending upon the dexterity of the operator. However, this task can be extremely tedious particularly when several hundred film badges have to be processed and so an automatic readout device for these dosimeters was developed by Pepper and Diffey.[73]

The principal limitation of the polysulfone film in this form has been

Figure 6. A polysulfone film badge. The badge measures approximately 3 cm on a side.

that its sensitivity extends to wavelengths up to 330 nm, whereas the biological effectiveness of UVR is largely confined to wavelengths less than about 315 nm. This small difference in spectral response can lead to significant errors in estimates of the biologically effective UV-B (see Section 4) when the film is used to monitor natural UVR, due to the rapid increase in spectral irradiance of the solar spectrum between 315 and 330 nm. Nevertheless it is possible to relate polysulfone response to an erythermally effective UV-B radiant exposure by carrying out correction calculations which involve estimates of the erythema action spectrum, the spectral sensitivity of polysulfone film, and the relative spectral power distribution of the incident radiation.[64] To circumvent this limitation a much thinner polysulfone film (1 μm thick) mounted on a

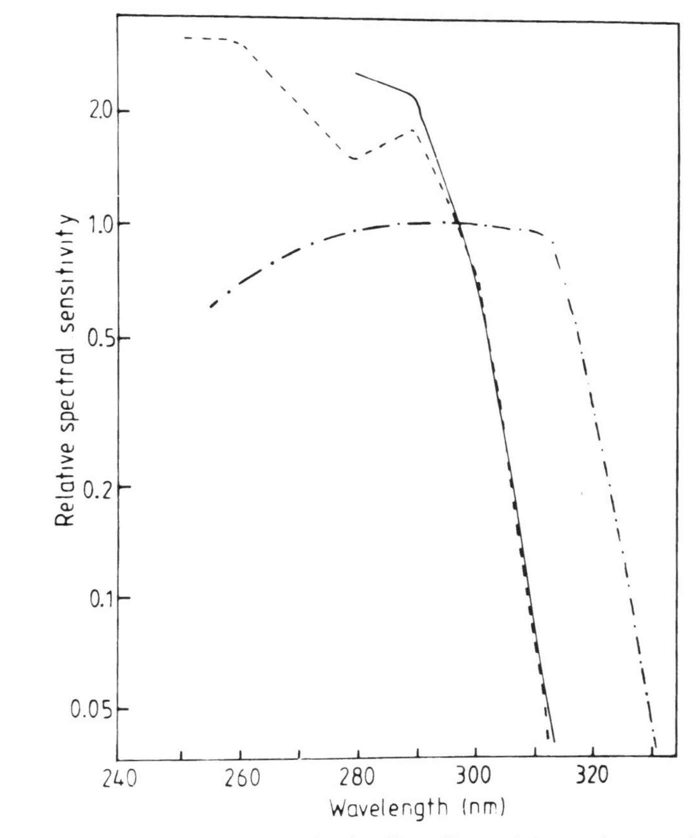

Figure 7. The spectral response of polysulfone film and the erythema action spectrum in normal human skin, normalized to unity at a wavelength of 297 nm. ——, 1-μm-thick polysulfone film[74]; –·–·, 40-μm-thick polysulfone film[64]; – – –, erythema action spectrum.[42]

cellophane substrate has recently been developed.[74] The spectral sensitivity of the thinner film is shown in Fig. 7 and is compared with the 40-μm-thick film, together with the erythema action spectrum in normal human skin. The agreement between the 1-μm-thick polysulfone film mounted on a cellophane substrate and the erythema action spectrum is good in the wavelength region between 295 and 315 nm, which is important for monitoring devices designed for natural ultraviolet radiation.

5.2.5. Plastic Films Incorporating Photosensitizing Drugs

It is sometimes useful to have available a dosimeter which is only sensitive to a limited spectral interval in the ultraviolet spectrum. To this end, several drugs which are known to have photosensitizing effects in humans have been incorporated as the chromophore in a polyvinyl chloride (PVC) film. PVC proves a suitable matrix for these compounds since it does not absorb significantly above 250 nm, and although the UV absorption spectrum of PVC can be affected by ultraviolet irradiation, the doses required are much larger than those the films are designed to measure. A summary of the PVC film dosimeters incorporating photoactive drugs which have been developed so far is given in Table II, and the relative spectral sensitivities of these films are compared in Fig. 8.

5.2.6. Photosensitive Papers

One of the drawbacks of the film dosimeters described above is that they require an expensive and bulky item of laboratory equipment (the spectrophotometer) to facilitate readout. An alternative approach is to

TABLE II
A Summary of Photosensitizing Drugs Used in PVC Film Dosimeters

Drug	Therapeutic action	Photosensitizing effects in humans	Dosimeter reference
Phenothiazine	Tranquillizer	Erythema, burning, itching	75
8-methoxy-psoralen	PUVA therapy for psoriasis (see Section 8)	Erythema, pigmentation ?carcinogenesis ?cataracts	76
Nalidixic acid	Antibacterial agent	Bullous eruption	77
Benoxaprofen	Analgesic	Immediate erythema, itching, burning, weals	78

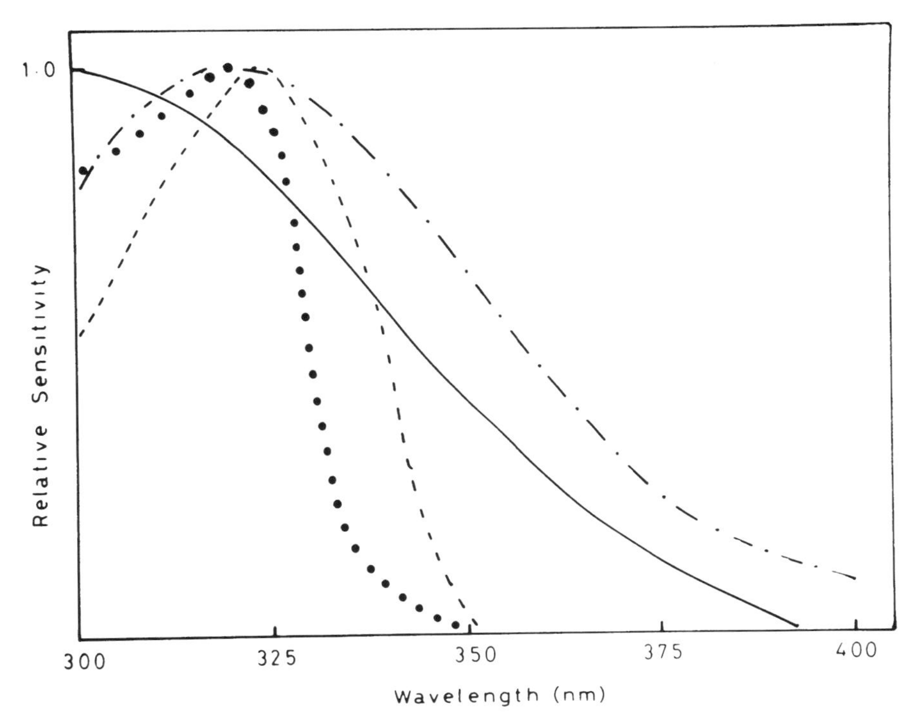

Figure 8. The relative spectral sensitivities of PVC films incorporating photosensitizing drugs. ——, 8-methoxypsoralen; –·–·, phenothiazine; – – –, nalidixic acid; · · ·, benoxaprofen. Each curve has been normalized to unity at its most effective wavelength.

use a system whereby the photochemical process initiates a color change so that visual comparisons with stable printed color standards enable the user to obtain a reasonably accurate and continuously readable integrated measure of his exposure to ultraviolet radiation. An example of a dosimeter based on this principle has been described by Zweig and Henderson.[79] This dosimeter is a polycarbonate film matrix incorporating 3′-[*p*-(dimethylamino)-phenyl] spiro-[fluorene-9,4′-oxazolidine]-2′,5′-dione, which converts to a red photoproduct of *p*-dimethylamino-*N*-fluoren-9-ylidine aniline following exposure to UVR of wavelengths less than 350 nm. The depth of red color developed depends solely on the radiant exposure.

5.3. Physical Ultraviolet Radiation Detectors

Ultraviolet radiation detectors consist of two basic physical types: thermal and photon. Thermal detectors respond to heat or power and have a uniform response over the spectral region for which the absorbance is near unity. The use of thermal detectors in clinical ultraviolet radiation dosimetry is not new: the thermopile and vacuum thermocouple were applied in this context 40 years ago.[80]

Photon detectors operate on the principle of the liberation of charge carriers by the absorption of a single quantum of radiation, and consequently, tend to have a nonlinear spectral response.

5.3.1. Thermopile

The most fundamental physical instrument for measuring radiant power is the thermopile, which, for many years, has provided the basis for calibrating all other types of UV measurement systems.[81] The principle of operation of the thermopile is based on the Seebeck or thermoelectric effect, whereby an EMF is generated when heat is applied to the junction of two dissimilar metals.

A schematic diagram of the Schwarz thermopile[82] is shown in Fig. 9. Radiation transmitted through the window is absorbed by the receiving element (usually gold or platinum), which is blackened for maximum absorption. Two cone-shaped semiconductors (one *p*-type and one *n*-type) are spot welded to its underside to form the "hot" junction. The "cold" junction is between the semiconductors and their platinum electrodes embedded in a metal block of high thermal capacity. The components are housed in a glass envelope which may be evacuated for increased sensitivity. A window is inset into one end of the envelope to allow radiation to fall onto the receiving element. The choice of window material is governed by the spectral region of interest; for the UV, fused

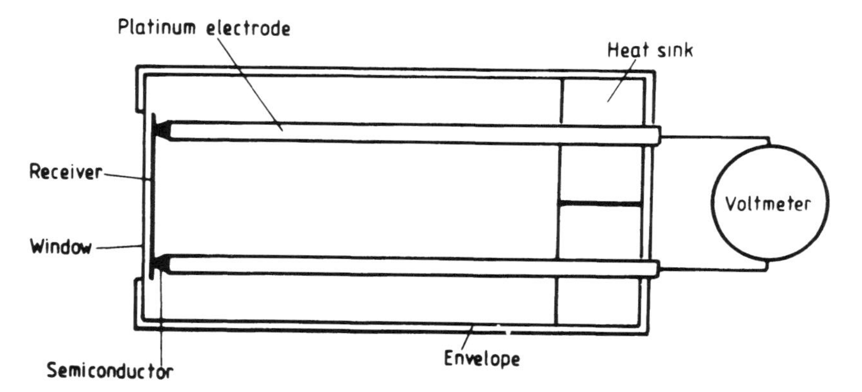

Figure 9. A schematic diagram of the Schwarz thermopile (courtesy of Rank-Hilger, Margate, England).

silica is most commonly used as this has a high, uniform transmission of radiation from 250–2200 nm.

5.3.1.1. Theory of Thermopile Operation. The thermopile is basically a heat engine which converts heat energy into electrical energy. The thermal efficiency η of a heat engine is defined as

$$\eta = \Delta\Theta/\Theta_h \tag{17}$$

where $\Delta\Theta$ is the temperature difference between the hot and cold junctions and Θ_h is the absolute temperature of the hot junction. In a typical thermopile $\Delta\Theta$ is usually <1 K and may be as low as 10^{-6} K, while Θ_h is of the order of room temperature. The generated EMF is related to $\Delta\Theta$ by

$$\text{EMF} = \Delta\Theta(P_p + P_n) \tag{18}$$

where P_p and P_n are the thermoelectric powers of the p-type and n-type semiconductors, respectively.

In order to optimize the efficiency, and hence the generated EMF, the heat losses must be kept low so that $\Delta\Theta$ is maximised.

Four types of heat loss occur in a thermopile: convection, reradiation, conduction, and the Peltier effect (current produced by thermopile acts to reduce $\Delta\Theta$). Convection losses depend upon the nature of the gas and the dimensions of the receiver and elements. These losses can be eliminated by evacuating the glass envelope which houses the components. Radiative losses from the hotter receiver and elements to the colder surroundings depend upon the temperature and emissivity of the surfaces involved. Radiative losses are unavoidable but can be minimized by making the back of the receiver highly reflecting. The most important loss is by conduction of heat through the elements from the hot to the cold junction. The losses due to the Peltier effect are negligible.

If the convection and Peltier losses are neglected, then $\Delta\Theta$ can be expressed as

$$\Delta\Theta = \varepsilon A/(L_r + L_c) \tag{19}$$

where ε is the energy absorbed in J m^{-2} s^{-1}, A is the area of the receiving element in m^2, and L_r and L_c are the heat losses in J K^{-1} s^{-1} due to radiative and conduction losses, respectively.

5.3.1.2. Calibration and Sensitivity of Thermopiles. Thermopiles can be calibrated at national standardizing laboratories. The UK National

Physical Laboratory calibrates thermopiles by comparing their response to the radiation from a tungsten filament lamp at a color temperature of 2850 K with a standard thermopile. The responsivity determined in this manner has a stated uncertainty of the order of 1% at an irradiance of around 10 W m^{-2}.

In medical photobiology the Hilger Schwarz FT17 vacuum thermopile has been widely used to measure irradiance. This device incorporates a circular receiving element of 2 mm in diameter together with a compensated element of similar dimensions which is screened from the direct radiation. The two elements are connected in opposition in order to minimize drift which can arise from variations in ambient temperature. The reciprocal sensitivity of the FT17 thermopile is usually in the range 10–20 W m^{-2} mV^{-1}.

The main drawback of the Hilger–Schwarz thermopile is its fragility. More recently, however, much more robust thermopiles have become available. For example, the range of thermopiles manufactured by the Dexter Research Center (Dexter, Michigan, U.S.A.) have their active elements hermetically sealed in a TO-5 package under a purged atmosphere of argon or nitrogen, and then heat treated to ensure long-term stability. This results in a thermopile which is said to be resistant to both mechanical and temperature shock. The Dexter model 2M thermopile, which has a sensitive area of $2 \times 2\ mm^2$, has a comparable sensitivity to that of the Rank Hilger FT17 thermopile.

5.3.1.3. Linearity of Thermopiles. Heat losses in the thermopile give rise to nonlinearity in response. The Hilger–Schwarz FT17 thermopile has a published linear response up to an irradiance of 300 W m^{-2}, but at higher irradiances the measured irradiance can be appreciably less than the true irradiance. On the other hand, the Dexter model 2M thermopile is stated to be linear in the range from 10^{-2} to 1000 W m^{-2}. The following experiment has been carried out by the author to evaluate the linearity of both of these types of thermopile.

Each thermopile was mounted in turn on an optical bench and irradiated with unfiltered radiation from a 100-W medium-pressure point-source mercury arc lamp operated by a stabilized power supply. The instrument reading (R) was noted for each of a series of distances (d) of the detector from the lamp housing. Readings were obtained over the range of values of d for which R varied between 30 and 1000 W m^{-2}. Deviations from an expected inverse square law response were examined by plotting $1/\sqrt{R}$ against d, rather than R vs. $1/d^2$, since the method takes into account the zero offset known to be present. An alternative linear transformation of $1/Rd^2$ against $1/d$ has been suggested[83] since it compensates for both zero offset and possible alignment errors. How-

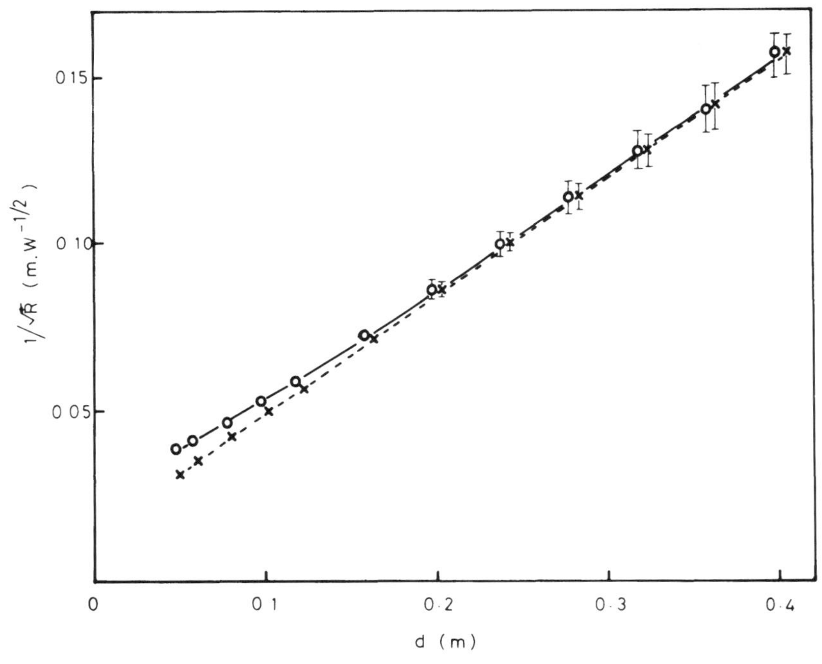

Figure 10. Reciprocal square root plot of the response of two thermopiles. O—O, Rank-Hilger model FT17; × --- × , Dexter Research Center model 2M.

ever, this method is only valid when the minimum value of *d* is considerably greater than the zero offset, which was not the case in these measurements.

The results of the linearity measurements, assuming an inverse square law response, are shown in Fig. 10. It may be seen that the results support the manufacturers' published data in that the Dexter model 2M thermopile remains linear up to at least 1000 W m^{-2}, whereas the Rank-Hilger model FT17 exhibits a nonlinear response at irradiances in excess of 300 W m^{-2}.

5.3.2. Pyroelectric Detector

Electrically calibrated pyroelectric radiometers (ECPR) are rapidly gaining acceptance as primary standards for the calibration of radiometric instruments, particularly in the United States. The sensor consists of a solid crystal that produces a change in current proportional to the rate of change of temperature of the crystal surface, which in turn is proportional to the rate of change of irradiance. The change in temperature alters the lattice spacings in the crystal which produces the change in electrical polarization. Since no current is produced in the steady state it is necessary to modulate the incident radiation with an optical chopper. In order to maximize the changes in absorbed radiant energy, and hence temperature, the surface of the crystal is coated with a material such as gold black, which exhibits the following desirable properties: low reflectance (typically 0.5% in the visible), flat spectral response (±2% in the range 250–1600 nm), and relatively high thermal conductance compared to organic blacks. The active element in a typical sensor consists of a polished slab of lithium tantalate mounted on a thin plastic diaphragm. A thin-film resistive layer on the front surface serves as both a detector electrode and an electrical heater. The gold black optically absorbing layer is deposited on top of this.

The principle of operation of an ECPR is as follows.[84] The incident radiation is chopped at a frequency of around 15 Hz with a 25% duty cycle. An electrical signal, 180° out of phase with the optical chopper, is generated and fed into the resistive heating element deposited onto the detector surface. This gives rise to a thermal signal in opposition to the chopped optical signal. The amplitude of the electrical signal is adjusted until a null is detected at the output of a synchronous rectifier. Under these conditions the optical power absorbed by the detector is nominally equal to the electrical drive power.

Absolute calibration of the ECPR—that is, determining the degree of equivalence between electrical and optical heating—is achieved by quantifying the following sources of variability and uncertainty: electrical calibration, reflectance and thermal resistance of the coating, spatial nonuniformity due to variations in detector thickness, heating in the leads to the heating element, and differences in the duty cycle for the chopped optical signal and generated electrical heating signal. Although these factors give rise to optical–electrical nonequivalence, their net effect is typically less than 1%.

A photograph of an ECPR suited for general radiometric use is shown in Fig. 11. This instrument is designed for use with three interchangeable pyroelectric sensors allowing measurement of irradiance in the range 10^{-4}–10^{6} W m^{-2}. The detectors have a flat spectral response with 99% of the incident radiation absorbed in the spectral range 400–3000 nm, dropping to 96% absorption in the ranges 250–400 and 3000–16 000 nm. The irradiance is stated to be accurate to ±1.5% of the reading.

5.3.3. Bolometer

A bolometer is essentially a Wheatstone bridge in which one arm of the bridge is made of a material with a high temperature coefficient of

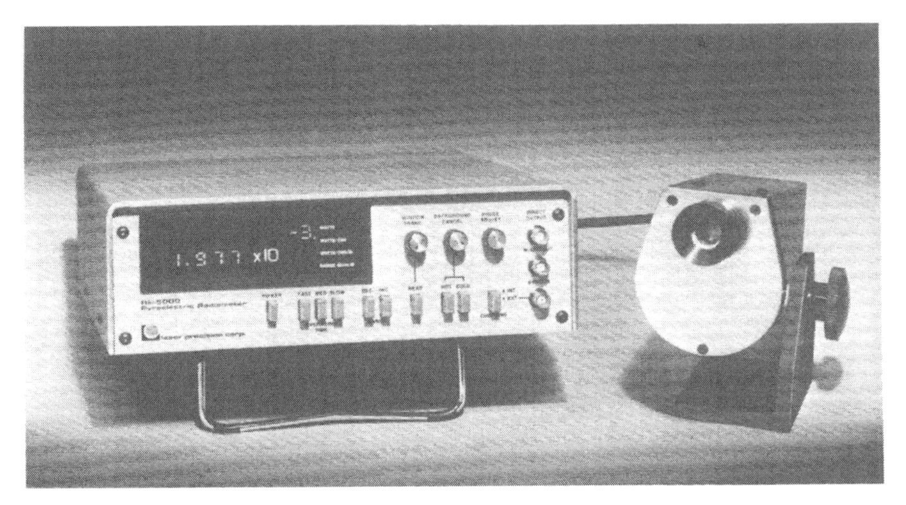

Figure 11. The Rk-5000 pyroelectric radiometer (courtesy of Laser Precision Corp., Irvine, California, U.S.A.).

resistance and exposed to the incident radiation. This causes a rise in temperature with a subsequent increase in resistance of the exposed arm of the bridge. By noting the voltage required to balance the bridge, the incident power can be calculated. A simple instrument based on this principle and intended for use in medical applications has been described by Byrne and Farmer.[85]

5.3.4. Phototubes

A phototube is basically a photoemissive cathode (photocathode) and a photoelectron collector (anode) housed in a vacuum or gas-filled envelope. A collection of various phototubes is shown in Fig. 12.

The principle of operation of a phototube is based upon the photoelectric effect, whereby photons incident on the photocathode liberate electrons which are accelerated towards, and collected by, the anode. At a given wavelength of incident radiation the number of electrons emitted from the photocathode (tube current) is directly proportional to the number of photons striking the photocathode, resulting in a high degree of linearity over several orders of magnitude of intensity of incident radiation.

The spectral response of a phototube is determined by the photocathode material and the window material. A useful combination for a UV-C/UV-B detector is a phototube incorporating a Cs–Te photocathode with a fused silica window, which results in a phototube with a spectral response in the range 160–320 nm (so-called "solar blind"

Figure 12. A collection of various phototubes (courtesy of Hamamatsu T.V. Co. Ltd., Japan).

device). The photocathode radiant sensitivity (in mA W^{-1}) of this and other common photocathode materials is shown in Fig. 13.

Vacuum phototubes are often the detectors of choice for portable ultraviolet radiometer systems. A vacuum phototube has been incorporated into an instrument (the IL730A UV actinic radiometer) which, with suitable filters and input optics, matches the wavelength response of the detector to the "hazard action spectrum" for occupational exposure to UVR (see Section 7).

5.3.5. Photomultiplier Tubes

A photomultiplier tube is essentially the same as a phototube, but with a built-in current amplifying section. The electron multiplier section consists of a chain of dynodes between the photocathode and anode. Electrons liberated at the photocathode are accelerated by a potential difference to the first dynode where they strike the dynode with sufficient kinetic energy to release additional electrons by the process of secondary emission. This process is repeated at each dynode such that the net effect is amplification of the primary photocurrent from the photocathode by a

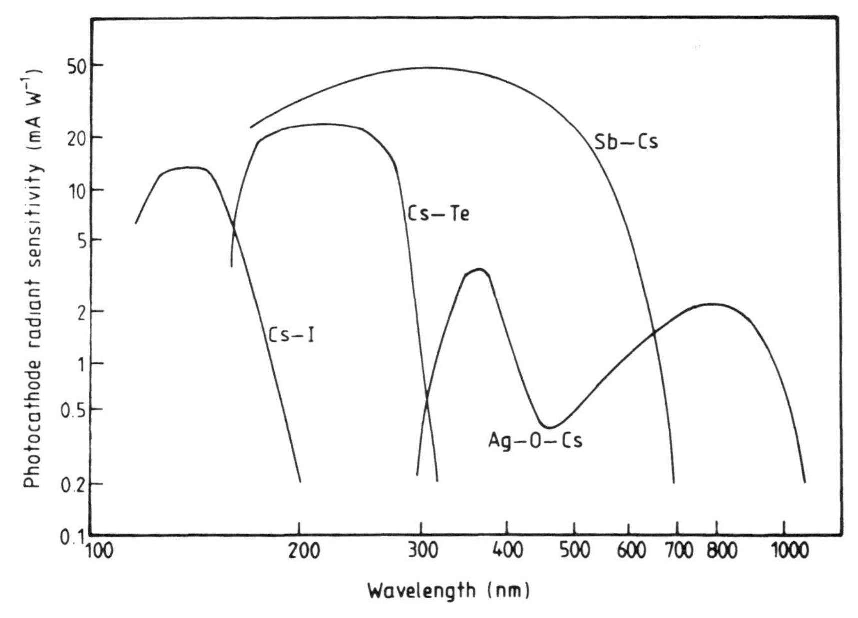

Figure 13. Some typical photocathode spectral response curves.

factor of 10^6 or more. The spectral responses of photomultiplier tubes are governed by the same constraints applicable to phototubes.

The high internal amplification of photomultiplier tubes results in high sensitivity and low noise, and these detectors are ideal for use in instruments where a monochromator is used to isolate a narrow wavelength region and where consequently a low irradiance is generally incident upon the detector, e.g., in a spectroradiometer.

The disadvantages of photomultiplier tubes are their fragility and their requirement for a stable high-voltage power supply. Both these characteristics are drawbacks for a portable instrument.

5.3.6. Solid State Photodiodes

In a solid state semiconductor with a reverse biased p–n junction (p material at a negative potential with respect to the n material), any free carriers which are generated are immediately swept by the electric field towards the region where there are majority carriers. If photons of sufficient energy are absorbed at or near the p–n junction, electrons will be excited from the valence band to the conduction band. The generation of excess electron–hole pairs across the junction gives rise to a current which can be measured in an external circuit. A p–n junction can be used therefore as a photoelectric detector, or photodiode. This mode of operation by applying a reverse bias across the junction is termed the "photoconductive mode."

Because of the internal potential rise in a *p–n* junction, photoexcited carriers are collected at the junction even in the absence of an external potential. If a resistor is placed across the terminals of the junction, a current passes through the resistor in proportion to the number of incident photons producing the potential difference across the *p–n* junction. This mode of operation of photodiodes is termed the "photovoltaic mode." This photovoltaic effect has widespread application in the familiar photographic exposure meter and in solar cell power supplies used on spacecraft, as well as in detectors for ultraviolet radiation.

Photodiodes may also be used in the photoconductive mode. In the photoconductive mode the device exhibits high sensitivity to incident radiation, but is prone to leakage currents resulting from changes in temperature. Temperature stability is worse than in the photovoltaic mode, but photosensitivity is higher.

Photodiodes can be constructed according to the following types[86]:

1. Planar Diffusion Type. A coating of SiO_2 is applied to the *p–n* junction surface which will result in a photodiode with a particularly low dark current.
2. Low Capacitance Planar Diffusion Type. This type of construction utilizes an *n*-type material of high purity and electrical resistance to enlarge the depletion layer, thereby decreasing the junction capacitance and lowering the response time to 1/10 of the normal value. By making the *p*-layer extra thin, increased sensitivity to the ultraviolet spectrum can be achieved.
3. *p-i-n* Type. This is an improved version of the lower capacitance planar diffusion type and incorporates a high-resistance intrinsic (*i*) layer between the *p* and *n* layers to improve the response time.
4. Inverted Layer Type. In this type of photodiode a SiO_2 coating and a thin *n*-type material are formed on a *p*-type substrate. Since SiO_2 has a high transmission for UVR, high sensitivity to this radiation can be obtained.
5. Schottky Type. A thin coating of gold is sputtered onto the *n*-type layer to form a Schottky effect *p–n* junction. The Schottky type photodiode exhibits high sensitivity to UVR, fast response time, and uniformity of response over its area.

The spectral response of solid state photodiodes depends upon the forbidden-energy gap of the semiconductor material. Materials which are commonly used as photodiodes include silicon (Si) and gallium arsenide

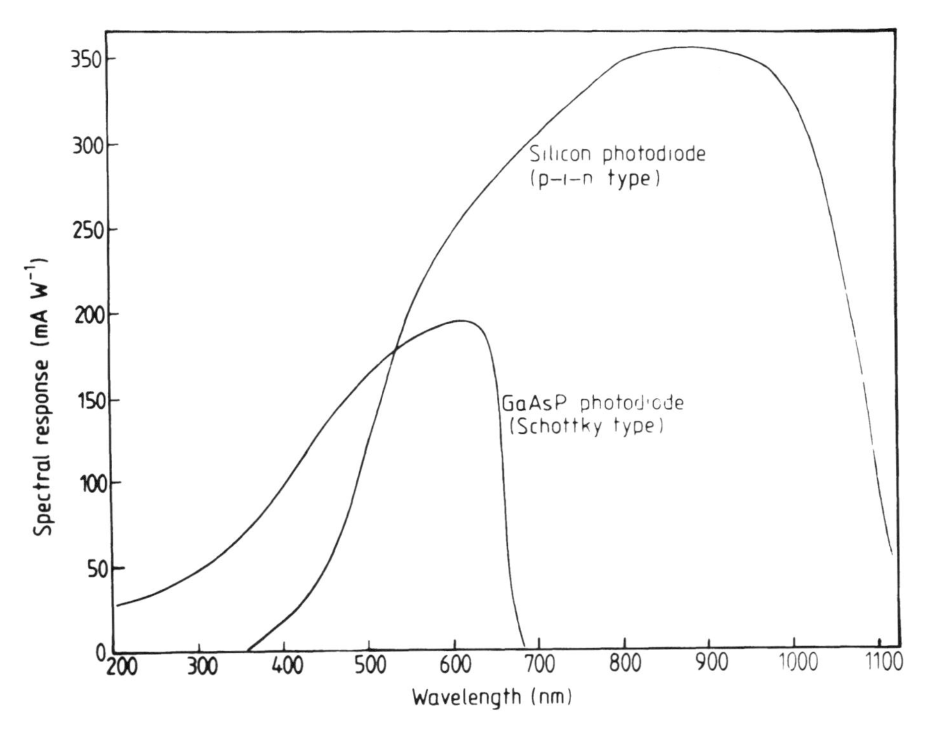

Figure 14. The spectral response of a typical *p-i-n*-type silicon photodiode and of a typical Schottky-type GaAsP photodiode.

phosphate (GaAsP). Silicon photodiodes have a peak spectral sensitivity in the near infrared and so their use for UV radiometry demands excellent rejection of longer wavelengths by filters or monochromators. This is not a problem with GaAsP photodiodes, which exhibit no infrared response. The spectral response of a *p-i-n*-type silicon photodiode and a Schottky-type GaAsP photodiode are compared in Fig. 14.

Figure 15 illustrates a GaAsP photodiode which when used in combination with an ultraviolet transmitting color glass filter (e.g., Schott UG1) results in a UV selective detector.[87] Unfiltered GaAsP photodiodes have been used as sensors in broad-band optical radiation monitors.[88]

Solid state photodiodes are linear over a wide range of incident intensities (as much as ten orders of magnitude), and have a response time which is usually better than 1 μS and an operating temperature range of around −20 to +80°C.

The low cost and small size of solid state photodiodes combined with their stability, robustness, and simple electronic circuitry requirements,

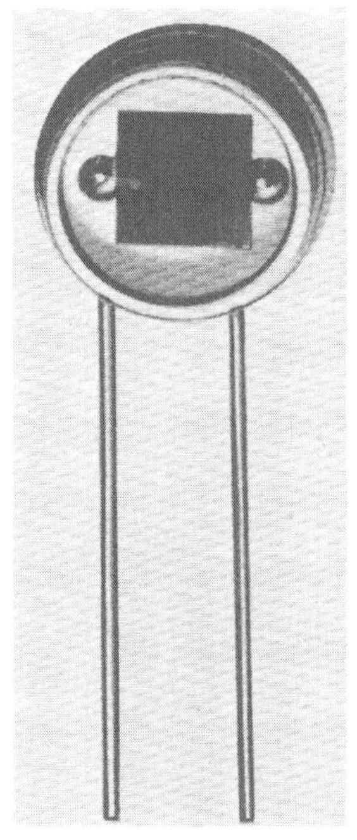

Figure 15. A GaAsP photodiode (type G1127) (courtesy of Hamamatsu T.V. Co. Ltd., Japan).

make them ideal detectors for portable instruments designed for use in the clinical environment.

5.3.7. Photoconductive Cells

Photoconductive cells are of two types: the photoconductive junction cell, which utilizes the properties of the *p–n* junction discussed in Section 5.3.6, and the bulk effect photoconductive cell (or light-sensitive resistor), which has no junction since the entire layer of crystals changes conductivity as a function of illumination. The properties of the latter type will be described in this section.

The most widely used photoconductive materials are cadmium sulfide (CdS) and cadmium selenide (CdSe). The photoconductive cells are normally made by sintering either CdS or CdSe powder into ceramiclike tablets of the required shape.[89,90]

The resistance of the cell decreases with increasing illumination in a nonlinear fashion. When plotted on log–log paper the resistance vs. illumination curve is approximately a straight line, the slope of the line depending upon the material and construction.[4] The spectral responses of photoconductive cells are not well suited for use in the ultraviolet region; the CdS cell, which has been used as a clinical UVR monitor,[91,92] has a

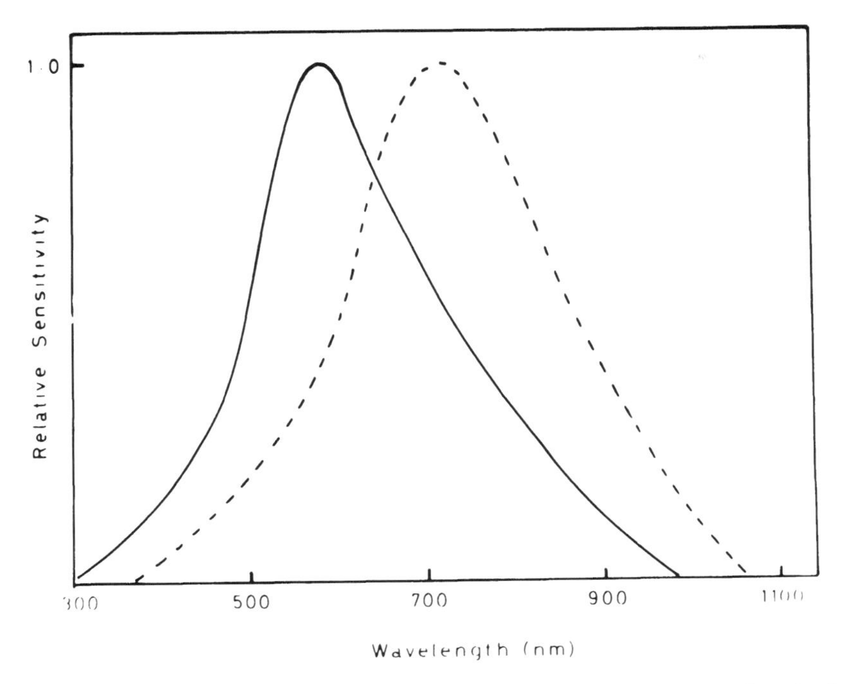

Figure 16. The typical spectral response of a CdS (——) and a CdSe (- - -) photoconductive cell.

peak spectral response around 500–600 nm, whereas the peak sensitivity of CdSe is shifted to even longer wavelengths (see Fig. 16).

5.3.8. Photovoltaic Cells

Photovoltaic cells are the only photoelectric (as opposed to thermal) detectors that do not require a power supply. Several semiconductor materials can be used to make photovoltaic cells, but the most widely used have been selenium and silicon.

Selenium photovoltaic cells were first produced and used in photographic exposure meters in Germany about 1930.[4] The power output was sufficient to drive a moving-coil meter when exposed to daylight. Selenium cells are made by painting or evaporating amorphous selenium onto a metal base, usually steel. A monomolecular layer of cadmium is applied to the surface to form the *p–n* junction, followed by the deposition of an extremely thin layer of a noble metal such as gold or platinum. The noble metal layer, which is thin enough to transmit 50% or more of the incident radiation, forms the cathode and the steel base forms the anode. A schematic diagram of the structure of a selenium

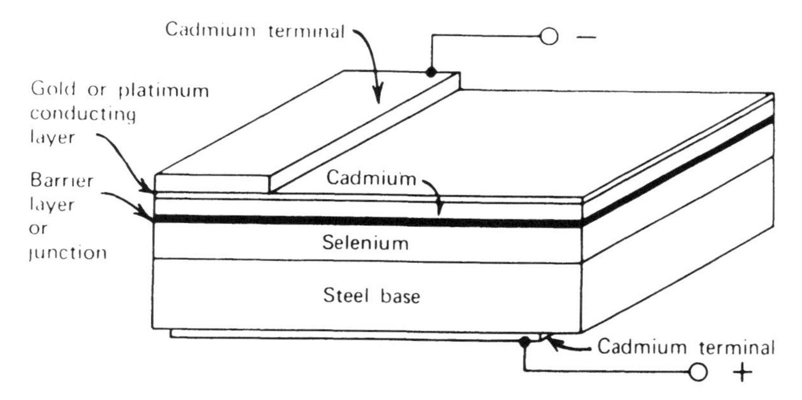

Figure 17. The schematic structure of a selenium photovoltaic cell.

photovoltaic cell is shown in Fig. 17. The spectral sensitivity of a selenium cell extends from about 250 to 800 nm. An instrument incorporating a selenium photovoltaic cell and designed to monitor UVR intensity in biological experiments has been described by Jagger.[93]

Silicon photovoltaic cells were first produced in the 1950s and have found wide application as solar cells since they are about ten times as efficient as selenium cells when exposed to sunlight, principally because of their appreciable response to infrared radiation (see Fig. 14). A silicon photovoltaic cell is the sensor in a commonly used commercial UV meter (the "Blak-Ray" instrument manufactured by Ultra-Violet Products Inc., San Gabriel, California, U.S.A.).

5.3.9. Fluorescent Detector

The principle of this type of UV detector is to employ a phosphor whose excitation spectrum matches or closely approximates the desired spectral response. A detector which operates on this principle and which has been extensively used to monitor global UV-B at several sites throughout the world for the past decade is the so-called Robertson–Berger Sunburn Meter.

About 1958, a proposal was made in Australia to record diurnal variations of erythemally effective natural UVR over a period of several years, and to correlate the resulting UV climatology at several sites with the incidence of skin cancer. The spectral sensitivity of the meter was specified to approximate the erythema action spectrum because of the belief that the carcinogenic action spectrum would be similar. The meter constructed for this study successfully fulfilled its purpose.[94,95] In the early 1970s the growing concern about the effect of atmospheric pollutants on the viability of the stratospheric ozone layer, with the subsequent requirement for more widespread environmental UVR monitoring, led to the design of a second generation meter by Berger.[96]

A schematic diagram of the Robertson–Berger meter is shown in Fig. 18. Solar radiation passes through an ultraviolet transmitting, visible absorbing color glass filter (prefilter) and excites a thin phosphor layer of magnesium tungstate bonded to another color glass filter (postfilter), which transmits the fluorescent radiation (380–620 nm) emitted by the phosphor but absorbs the UVR and small amount of red light transmitted by the prefilter. A vacuum phototube or selenium photovoltaic cell is used to detect the phosphor emission.

The spectral sensitivity of the meter is compared in Fig. 19 with the CIE erythema action spectrum (see Section 4.1), where it may be seen that the meter responds to longer wavelengths than are desirable. Nevertheless the meter is calibrated in "Sunburn Units" such that, for a reference solar spectrum of an overhead sun on a clear day and an ozone layer thickness of about 2.6 mm, the meter records 5 sunburn units in one

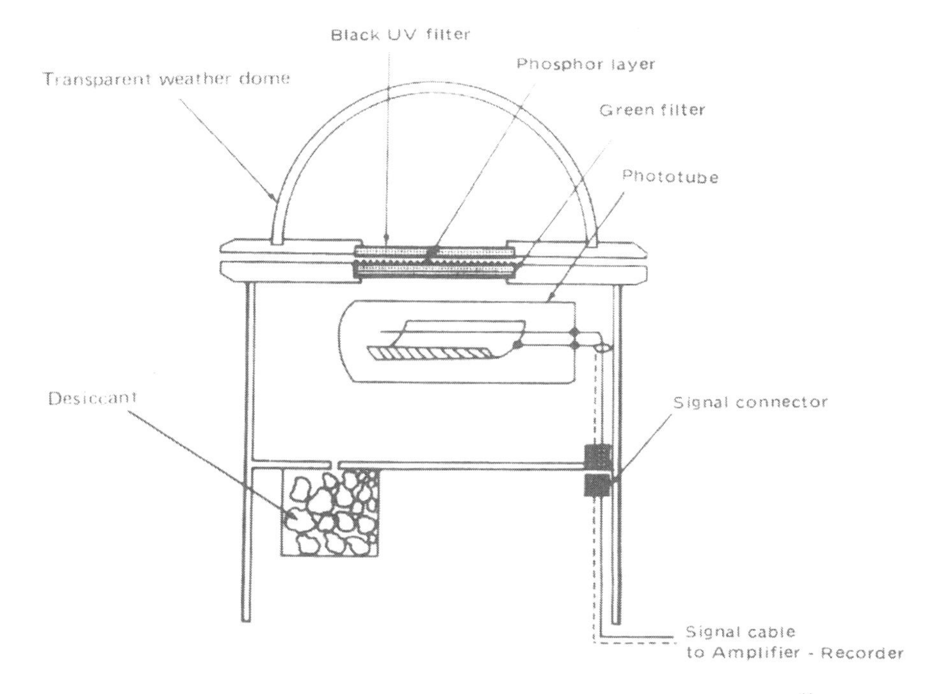

Figure 18. The schematic structure of the Robertson–Berger meter.[96]

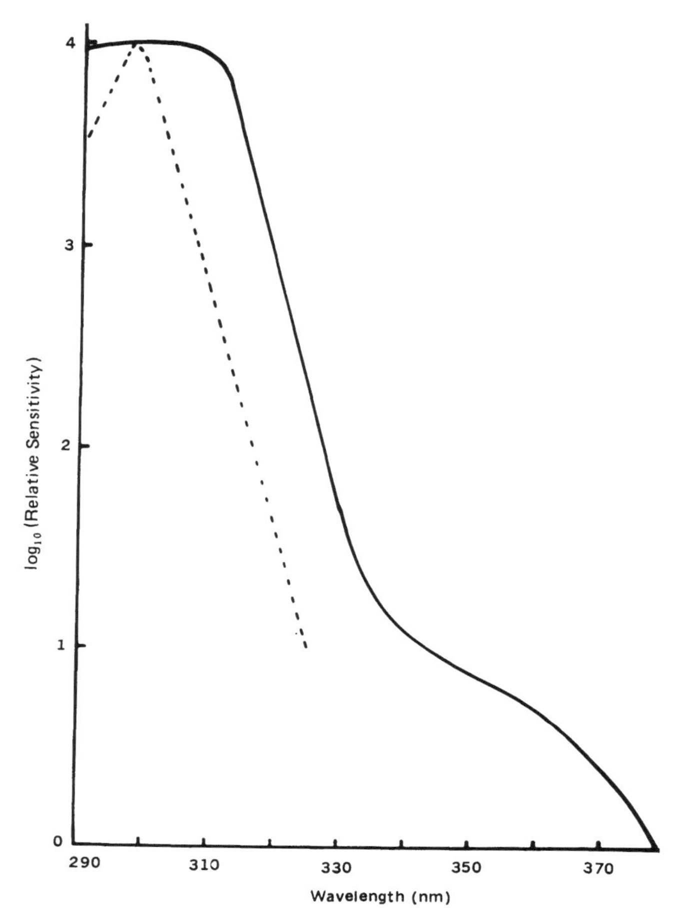

Figure 19. The spectral sensitivity of the Robertson–Berger meter (——) and the CIE erythema action spectrum (– – –).

hour since, from this reference spectrum, an untanned average Caucasian will develop a minimum erythema in about 12 min, i.e., 1/5 of an hour.[97] The corrections to the meter response which can be applied to account for differences between its spectral sensitivity and the erythema action spectrum are discussed by Berger.[96]

The fluorescent sunburn meter has been shown to have long-term stability, little maintenance requirements, low initial cost, long lifetime, and acceptable stability with changes in ambient temperature.[97]

5.3.10. Thermoluminescent Materials

Several thermoluminescent (TL) materials have been investigated as possible UV dosimeters. Many materials (e.g., LiF:Mg; $CaSO_4$:Tm; CaF_2: natural) require preirradiation with high doses of gamma radiation and partial annealing before showing sensitivity to UVR (so-called "transferred thermoluminescence"), whereas other materials (e.g., MgO; Al_2O_3:Si; CaF_2:Dy) have proved directly sensitive to UVR.

It is probably true to say that TL materials have yet to find an established role as dosimeters for ultraviolet radiation, but for a review of their potential see the article by Busuoli.[98]

5.4. Detector Parameters

There are several parameters that can be used to specify the performance of optical radiation detectors, although the three that are most often cited by manufacturers are probably the responsivity (R), the noise equivalent power (NEP), and the specific detectivity (D^*).

The responsivity (sometimes called radiant sensitivity) is defined as the ratio of the output voltage V_s (or current) to the radiant power ϕ incident upon the detector. However, both the electrical signal and the incident power are qualified in terms of the rms value of the fundamental component at the chopping frequency.[5] The responsivity is given by

$$R = V_s/\phi \quad \mathrm{V\,W^{-1}} \tag{20}$$

The noise equivalent power (NEP) is defined as the ratio of the rms noise V_n to the responsivity:

$$\mathrm{NEP} = V_n/R \quad \mathrm{W} \tag{21}$$

The magnitude of NEP depends upon the noise bandwidth of the measured electrical noise and can be thought of as the incident power that will produce an output voltage V_s (rms) equal to the electrical noise voltage V_n (rms).

The simple detectivity (D) is defined as the reciprocal of the NEP. However, a far more useful quantity is the specific or normalized detectivity D^*,[99] defined as

$$D^* = (A\Delta f)^{1/2}/\mathrm{NEP} \quad \mathrm{cm\,Hz^{1/2}\,W^{-1}} \tag{22}$$

where A is the detector area in cm^2 and Δf is the equivalent electrical bandwidth in hertz.

6. SPECTRORADIOMETRY

Spectroradiometry is concerned with the measurement of the spectrum of a source of optical radiation. In most cases, spectral measurements are not required as ends in themselves but for application to the calculation of biologically weighted radiometric quantities. These quantities are derived from the spectral data by a process of weighted integration or summation over the range of wavelengths present in the spectrum.

6.1. Components of a Spectroradiometer

The three basic requirements of a spectrometer system are (i) the input optics, designed to conduct the radiation from the source into (ii) the monochromator, which usually incorporates a diffraction grating as the wavelength dispersion element, and (iii) an optical radiation detector.

6.1.1. Input Optics

The spectral transmission characteristics of monochromators depend upon the angular distribution and polarization of the incident radiation as well as the position of the beam on the entrance slit. For measurement of spectral irradiance, particularly from extended sources such as linear arrays of fluorescent lamps or daylight, direct irradiation of the entrance slit should be avoided. There are two types of input optics available to ensure that the radiation from different source configurations is depolarized and follows the same optical path through the system: the integrating sphere or the diffuser.

The most accurate measurements of spectral irradiance make use of an integrating sphere (see Fig. 20), in which the radiation enters through a small aperture, with the entrance slit of the monochromator located at another aperture on the surface of the sphere. Because of multiple reflections within the sphere it is important that a diffuse coating with a high reflectance in the UV (e.g., MgO or Eastman 6080 $BaSO_4$ powder paint) be applied to the inside of the sphere to achieve good efficiency. Integrating spheres produce a cosine-weighted response, since the radiance of the source as measured through the entrance aperture varies as the cosine of the angle of incidence.

Alternatively, a plane diffuser of MgO, $BaSO_4$, or ground quartz irradiated normally and viewed at 45° may be used (see Fig. 20), as may a roughly ground quartz hemispherical diffuser placed at the entrance slit of the monochromator.

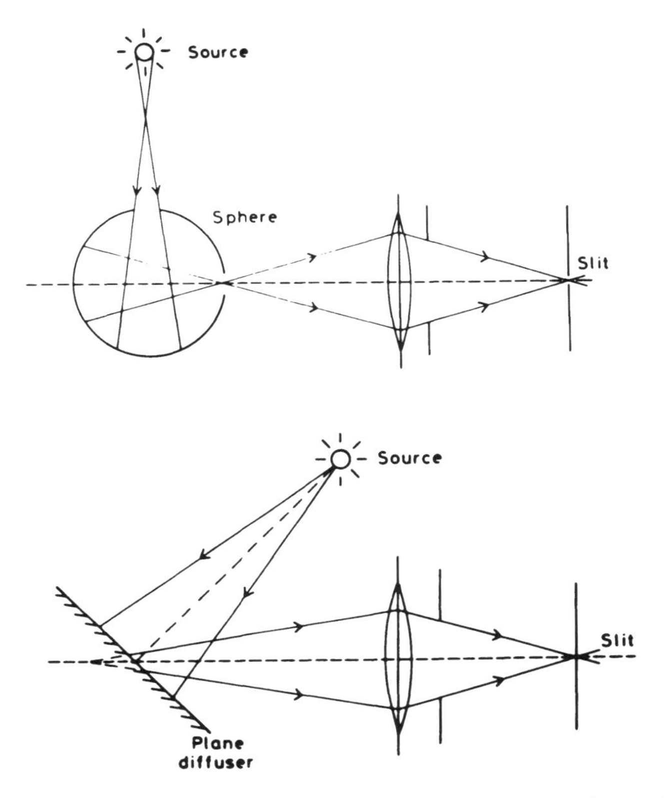

Figure 20. The input optics to a spectroradiometer using either an integrating sphere (upper figure) or plane reflecting diffuser (lower figure).

6.1.2. Monochromator

A blazed ruled diffraction grating is normally preferred to a prism as the dispersion element in the monochromator used in a spectroradiometer, mainly because of better stray radiation characteristics. Any radiation observed between the main diffracted orders in excess of that predicted by Fraunhofer diffraction is regarded as spectral impurity. In general there are four types of unwanted radiation, in addition to Fraunhofer diffraction, that can arise in the spectrum. These are known as ghosts, satellites, grass, and diffuse scatter[100] and arise from imperfections in the machinery used to make the gratings. The largest requirement for ruled diffraction gratings, in terms of numbers, is for small replicas with areas in the range 25×25–$50 \times 50\,\mathrm{mm}^2$. The most common type is a 1200-line mm^{-1} grating blazed for 300 or 500 nm. The tolerances are quite modest, for example, 50% of the theoretical resolution, and ghosts $<10^{-4}$ of the main order intensity.

In more recent years holographic, or interference, gratings have become available. The feature of the interference technique used in the manufacture of these gratings is that there are no short-term errors in groove position and so the gratings generate no ghosts or grass, which results in a stray radiation level of one to two orders of magnitude lower than a contemporary ruled grating.

High-performance spectroradiometers, used for determining low-UV spectral irradiances in the presence of high irradiances at longer wavelengths, demand extremely low stray radiation levels. Such systems may incorporate a double monochromator—that is, two single ruled grating monochromators in tandem, or laser-holographically produced concave diffraction gratings can be used in a single monochromator.

6.1.3. Detector

Photomultiplier tubes, incorporating a photocathode with an appropriate spectral response, are normally the detectors of choice in spectroradiometers. However, if radiation intensity is not a problem, solid state photodiodes may be used, since they require simpler and cheaper electronic circuitry. An example of a portable spectroradiometer system is shown in Fig. 21.

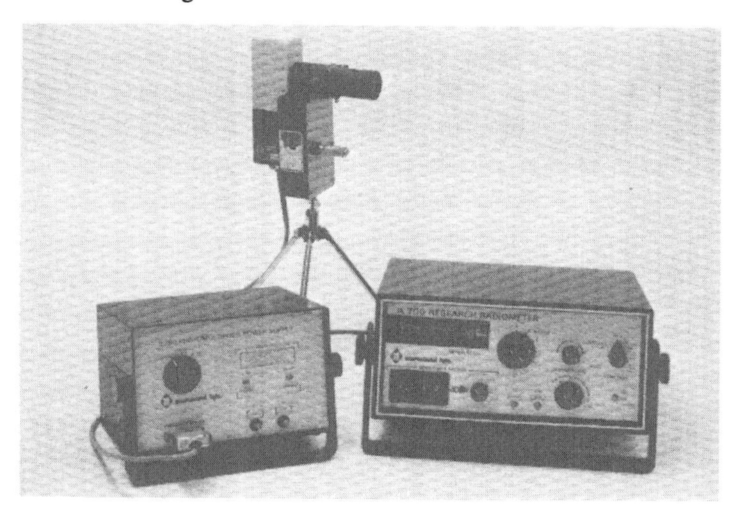

Figure 21. A photograph of a portable spectroradiometer showing the monochromator/detector head, the high-voltage power supply, and the readout unit (courtesy of International Light Inc., Newburyport, Massachusetts, U.S.A.).

6.2. Calibration and Correction

The output signal $S(\lambda_o)$ from a spectroradiometer set at wavelength λ_o is related to the spectral irradiance incident on the receiving aperture of the instrument by the equation

$$S(\lambda_o) = \int_0^\infty E_s(\lambda)R(\lambda)\sigma(\lambda - \lambda_o)\, d\lambda \tag{23}$$

where $E_s(\lambda)$ is the spectral irradiance at wavelength λ, $R(\lambda)$ is the spectral responsivity of the spectroradiometer at wavelength λ, and σ is the slit function which is principally dependent on the entrance and exit slit widths of the monochromator.

In many spectroradiometric measurements the monochromator slit widths are sufficiently narrow that the slit function can be approximated by a delta function. Equation (23) then simplifies to

$$S(\lambda_o) = E_s(\lambda_o)R(\lambda_o) \tag{24}$$

The spectral responsivity is determined by measuring the spectroradiometer response to a source of known spectral irradiance. The secondary standard source which is normally employed is an incandescent tungsten filament operating at a color temperature of about 3000 K. The spectral irradiance from the lamp is calibrated over the range 250–2500 nm by comparison with the radiance of a blackbody cavity radiator, operated at a known temperature and presumed to have the spectral distribution of power predicted by Planck's law.[101] Alternatively, deuterium lamps can be used as secondary standard sources for calibration in the ultraviolet. The primary standard for calibrating deuterium lamps is synchrotron radiation.[102]

The actual shape of the spectral responsivity function $R(\lambda)$ will depend upon many factors, such as the spectral efficiency of the diffraction grating and the photocathode material used in the detector. However, it is not unusual for $R(\lambda)$ to decrease by a factor of 10 from a wavelength of 350 nm to 250 nm, and so estimates of spectral irradiance in the UV, to even a moderate degree of accuracy, required prior knowledge of this function.

In high-precision spectroradiometry, for example, in solar spectroradiometry in the UV-B, it may be necessary to take into account the slit function, $\sigma(\lambda - \lambda_o)$. To a first approximation the slit function of a monochromator with equal entrance and exit slit widths is a triangle, or trapezoid when the slits are unequal.[103] In reality, the slit function has broad wings which slowly approach zero after an initial rapid decrease of

3–7 orders of magnitude, depending on whether the monochromator has one or two dispersive elements. These wings mean that the spectroradiometer can respond to radiant flux at wavelengths far removed from the wavelength setting λ_o of the instrument. If the slit function $\sigma(\lambda - \lambda_o)$ is known, it is possible to derive the spectral irradiance $E_s(\lambda_o)$ from the observed signal $S(\lambda_o)$ by deconvolution of Eq. (23).[104]

6.3. Spectral Sampling

In a conventional spectroradiometer the spectrum is obtained by noting the detector response at regular wavelength settings of the monochromator over the range of wavelengths emitted by the source. Such a method of discrete sampling can give acceptable results if due attention is paid to the relationship between the monochromator bandwidth and the spectral sampling interval. However, in the case of radiation sources emitting line spectra there is a real danger that discrete spectral sampling may lead to appreciable errors in spectroradiometry. For this reason a method of continuous scanning is generally preferred in which a stepper motor drive system is used to scan the monochromator across the spectrum at a constant speed with respect to wavelength, while the signal from the detector is integrated continuously.[101]

At regular wavelength intervals throughout the scan, electronic gating pulses are used to divide the spectrum into a series of contiguous bands. This method of continuous scanning has the advantages of great flexibility and an ability to cope with all types of spectral power distribution, no matter how complex. The method is well suited to automatic control and data handling procedures, for example, correcting for the spectral responsivity of the spectroradiometer, or convoluting the spectral power distribution with some biological action spectrum. For this reason many commercial spectroradiometer systems incorporate microcomputers with associated software.

6.4. An Application of Spectroradiometry

In recent years the use of artificial sources of UV-A radiation has increased rapidly in both the medical and consumer fields. This is due largely to the widespread treatment of psoriasis by oral psoralen photochemotherapy (see Section 8), and to the increasing cosmetic use of sun-lamps, solaria, and sunbeds.[105]

The most common source of UV-A employed in both medical whole-body UV irradiation units and cosmetic sunbeds is the so-called "UV-A fluorescent lamp." These lamps emit small quantities of visible, infrared, and UV-B radiation in addition to the predominant UV-A emission. Since the quantity of UV-B radiation needed to produce erythema in normal human skin is some thousandfold less than that of UV-A radiation,[15] concern has been expressed about the quantity of biologically effective UV-B which may be emitted by "UV-A fluorescent lamps." It is therefore important to have data on the UV-B emission of these lamps so that the erythemal effectiveness of the UV-B component can be assessed.

It is not possible to estimate reliably this UV-B component using currently available radiometers, which incorporate a UV detector and one or more optical filters, since the spectral transmission curves of optical filters are not sufficiently discriminating to block the incident UV-A and visible radiation completely while at the same time transmitting the UV-B component without spectral distortion. Consequently the only satisfactory method is by spectroradiometry.

In a recent study,[51] the spectral irradiances at 1 m from the midpoints of single UV-A fluorescent tubes (1.8 m in length), each from a different manufacturer, were measured using a portable spectroradiometer. A schematic diagram of this instrument, which was developed by the U.K. National Radiological Protection Board (NRPB) is shown in Fig. 22 and its performance characteristics are listed in Table III.

The input optics of the spectroradiometer consist of an integrating diffusing sphere coated internally with Eastman Kodak Diffuse Reflectance Standard Coating 6080. This provides the instrument with a cosine and wavelength independent spatial efficiency of response over the range of wavelengths 200–600 nm. Wavelength scanning is provided by a

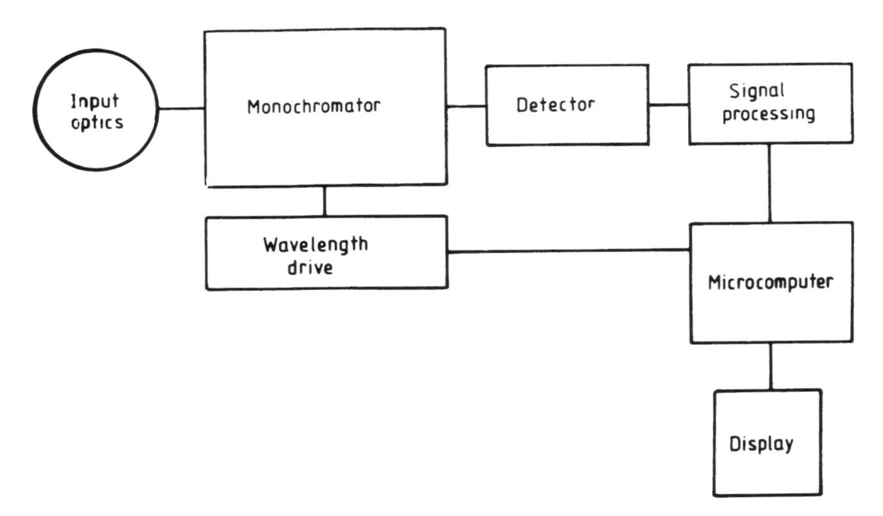

Figure 22. A simplified block diagram of the NRPB spectroradiometer.

TABLE III
Characteristics of NRPB Scanning Spectroradiometer

Input optics	Variable aperture (0.5–16 mm) integrating sphere
Coating	Eastman Kodak reflectance standard 6080
Monochromator	Yvon Jobin double grating, two reflecting surfaces only
Useful wavelength range	190–600 nm
Bandwidth range	2–10 nm
Minimum wavelength stepping increment	0.1 nm (100 control pulses per nm)
Detector	EMI Photomultiplier 9824Qb
Typical spectral sensitivity	400 nm 2.76×10^4 V per W m^{-2} nm^{-1} 350 nm 1.65×10^4 300 nm 8.4×10^3 250 nm 2.36×10^3 200 nm 38
Typical background (optical + electrical) (variation in parentheses)	$\sim 5 \times 10^{-3}$ V (2×10^{-3})

stepping motor coupled to the diffraction grating rotation mechanism of the monochromator via a precision gearbox. Optically aligned anti-backlash gears provide precise wavelength setting. Wavelength calibration is obtained using a HeNe laser (632.8 nm), an argon ion laser (514.5 and 488 nm), and a low-pressure mercury discharge lamp (253.7 nm). The spectral sensitivity calibration of the instrument was obtained by reference to a National Physical Laboratories calibrated deuterium spectral irradiance secondary standard. The uncertainty associated with electronic and optical "noise" is $< \pm 2 \times 10^{-3}$ V. This is equivalent to an uncertainty in spectral irradiance of 2.4×10^{-7} W m^{-2} nm^{-1} at 300 nm. The spectral irradiances of the lamps at 300 nm varied between 6×10^{-6} and 60×10^{-6} W m^{-2} nm^{-1}. Therefore for the minimum spectral irradiances at 300 nm (i.e., worst case) the uncertainty associated with "noise" is approximately ±5%. The uncertainty associated with sensitivity drift of the radiometer is approximately 5%. Hence the overall uncertainty between measurements is approximately ±10% (arithmetically summed). The uncertainty associated with calibration against the spectral irradiance standard is approximately ±12%. The expected overall error in the absolute values of irradiance should not be greater than 22% (arithmetically summed).

A relative spectral irradiance plot representing the emission from an "average" tube is illustrated in Fig. 23 (curve C). This was obtained by

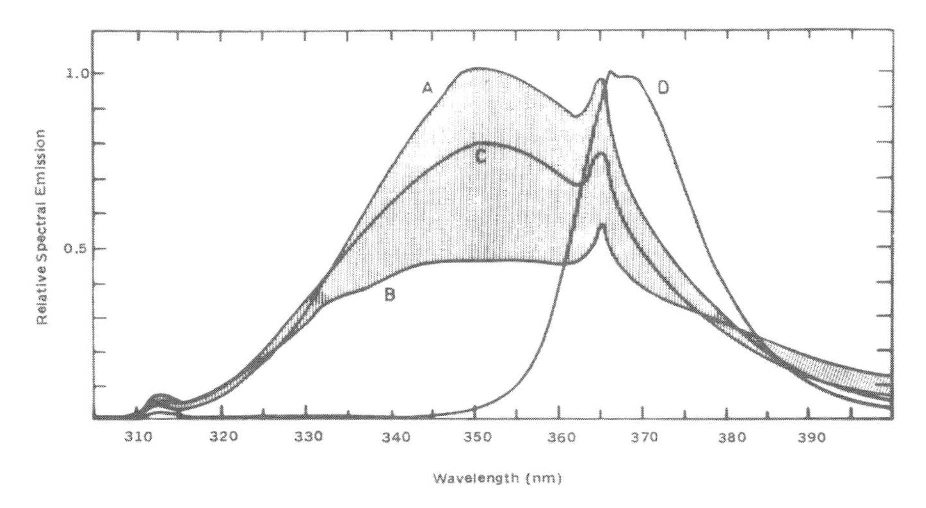

Figure 23. The relative spectral irradiance of the UV-A fluorescent lamps. A, lamp with greatest emission at 351 nm; B, lamp with smallest emission at 351 nm; C, average emission; D, lamp with fluorescent peak emission at 365–367 nm.

summing the measured values of spectral irradiance of all except one of the tubes at each wavelength and normalizing to the value corresponding to the fluorescent peak wavelength (351 nm). The overall variations in spectral irradiance which one would expect by using a tube type chosen at random are indicated by the shaded area. This area is bounded by plots of the spectral irradiance values obtained from the tubes displaying the highest (curve A) and lowest (curve B) irradiance. The absolute values of spectral irradiance of the average and lowest irradiance tubes are normalized against those of the highest irradiance tube (at 351 nm). The spectral irradiance characteristics of one of the tubes was markedly different from the rest. Its fluorescent peak emission coincided with the 365–367-nm characteristic line emissions of mercury vapor. Therefore it was excluded from the averaging procedure. The spectral irradiance of this tube is shown separately in Fig. 23 (curve D).

A summary of the results of the spectral and spectrally weighted summations for each tube, together with the values of an "average" tube, are presented in Table IV.

The variation in the overall UV-A irradiances of the 10 tubes is relatively low (standard deviation 16%) compared with the variation in UV-B (34%). Below 315 nm the spectral emissions from all of the tubes fall off very sharply and the characteristic mercury vapor lines at 313 and 297 nm comprise almost all of the radiation emitted, resulting in a mean

TABLE IV
Summary of Spectral Irradiance Data for 10 Types of UV-A Fluorescent Tubes

Lamp code	Fluorescence peak (nm)	UV-A ($W m^{-2}$)	UV-B ($W m^{-2} \times 10^2$)	$\frac{UV\text{-}B}{UV\text{-}A}$ (%)	$\frac{UV\text{-}B(EE)}{UV\text{-}A}$ (%)
1	384–354	1.18	1.82	1.55	0.093
2	366	1.35	1.06	0.79	0.180
3	351	1.55	0.86	0.56	0.030
4	350	1.84	2.03	1.10	0.084
5	351/2	1.51	1.71	1.14	0.086
6	351/2	1.86	0.81	0.44	0.020
7	351/2	1.83	1.27	0.69	0.035
8	351/2	1.68	0.96	0.57	0.027
9	351/2	1.54	2.19	1.42	0.135
10	350–357	1.17	1.29	1.10	0.072
Mean	—	1.55	1.40	0.94	0.076
Coefficient of variation (%)	—	16	34	38	64

UV-B emission of 0.94% of the UV-A. Similarly, the erythemally effective UV-B(EE), calculated according to Eq. (14) and expressed as a percentage of the UV-A emission and normalized to 297 nm (the peak of the CIE Standard Erythemal Curve) is low (0.076%). Because of the uncertainty in the erythema action spectrum in normal human skin shown by Diffey,[13] the tabulated erythemal spectral efficiency data published in the German DIN standard[106] has been chosen for the purpose of the present analysis. These data are based largely on the "standard erythemal curve" adopted by the International Commission on Illumination,[107] the main difference being that the DIN standard decreases more rapidly than the CIE standard at wavelengths greater than 297 nm.

6.5. Indirect Spectroradiometry

Although they are commercially available, spectroradiometers are expensive instruments. Nevertheless, estimates of spectral irradiance may still be possible using less expensive instrumentation. The accuracy of indirect estimates of spectral irradiance will not be as high as achieved spectroradiometrically, but may often be adequate for clinical purposes.

The total irradiance at the point of interest is simply equal to the integral of spectral irradiance over all wavelengths emitted by the source. The relative spectral intensity, or spectral power distribution, may therefore be reduced to absolute spectral irradiance by equating the integrated spectral power distribution to the total irradiance measured by a detector which has a uniform spectral response over the wavelength range of the source, and preferably a 180° field of view with a cosine-weighted response so that measurements can be performed on extended, linear sources. The spectral power distribution may be obtained either from the manufacturer's data or by means of a scanning monochromator and detector. It is not essential to incorporate input optics on the scanning monochromator since it is assumed that the spectral power distribution in the field of view of the entrance slit is representative of the spectral irradiance incident at the entrance slit. Although the response of the scanning monochromator/detector does not need to be absolutely determined, it is still important to estimate the spectral responsivity $R(\lambda)$ especially if measurements are to be carried out in a wavelength region where this function changes by a significant amount.

This technique of indirect spectroradiometry has been used by Diffey and Challoner[108] to determine the spectral irradiance from UV-A fluorescent tubes as follows.

The total irradiance, E ($\mathrm{W\,m^{-2}}$), was measured at the point of interest using an FT32 thermopile (Rank-Hilger). This device has a uniform wavelength response in the range 250–2200 nm, a wide field of view, and an approximately cosine-weighted angular response.

A Beckman DBG spectrophotometer, modified to perform as a scanning monochromator, was used to determine the spectral power distribution from the UV-A lamps.

The analog signal from the spectrophotometer was digitized by means of a voltage-to-frequency converter and the resultant pulses accumulated in a scalar. Every ten seconds the contents of the scalar were recorded on paper tape and the cycle repeated. During the 10-s data integration period, the wavelength change on the spectrophotometer was 1.70 nm. It was necessary to correct the recorded spectrum for the variation with wavelength of the sensitivity of the spectrophotometer. This was achieved by recording the spectrum of a 200-W quartz–iodine lamp of known spectral irradiance and deriving the spectral responsivity function $R(\lambda)$.

The photomultiplier tube employed in the Beckmann DBG spectrophotometer was such that the efficiency of detection fell rapidly at wavelengths greater than about 650 nm. Since the thermopile will respond to wavelengths up to 2500 nm, it was necessary to make an estimate of the fraction of the measured irradiance in the near-infrared region. This was achieved by using a Schott red glass filter (type RG630), 3 mm thick and 50 mm in diameter. This filter transmits uniformly from 650 nm to beyond 2500 nm with 50% internal transmittance at 630 nm and negligible transmittance at wavelengths below 600 nm. Measurements with this filter indicated that 7% of the irradiance measured by the thermopile was due to wavelengths greater than 600 nm, and so the total irradiance, E, which is equated with the integrated spectral power distribution, needs to be reduced by this amount.

Hence the spectral irradiance in the wavelength interval $\lambda_i - \Delta\lambda/2$ to $\lambda_i + \Delta\lambda/2$ is given as

$$E_s(\lambda_i, \Delta\lambda) = (1 - 0.07)E\left[P(i)\Big/\sum_{i=1}^{N} P(i)\Delta\lambda\right] \quad \mathrm{W\,m^{-2}\,nm^{-1}} \qquad (25)$$

where λ_i is the central wavelength in the ith timing period, $\Delta\lambda$ is the wavelength increment during each data integration period ($\Delta\lambda = 1.70$ nm), E is the total irradiance incident upon the detector, $P(i)$ is the number of pulses collected during the ith timing period corrected for nonlinearity of spectrophotometer response, and N is the number of timing periods ($N \sim 200$).

7. ULTRAVIOLET RADIATION PROTECTION DOSIMETRY

7.1. Occupational Ultraviolet Exposure Standards

At the present time there are no internationally formally agreed limits for occupational exposure to UVR. The most comprehensive occupational exposure standard for inchoerent (nonlaser) UV sources which has been published is that of the American Conference of Governmental Industrial Hygienists. This standard has been endorsed by the National Institute for Occupational Safety and Health (NIOSH)[109] in the United States and adopted as a voluntary standard in the United Kingdom by the Health and Safety Executive (HSE) and the National Radiological Protection Board (NRPB).[110]

The exposure standard is considered separately for the UV-A region, and for the UV-B and UV-C regions, since most acute biological effects are initiated by wavelengths less than 315 nm.

7.1.1. UV-A Exposure Standard

For the spectral region 400–315 nm (UV-A), the total irradiance incident on unprotected eyes and skin for periods of greater than 1000 s should not exceed 10 $\mathrm{W\,m^{-2}}$, and for exposure times of 1000 s or less than total radiant exposure on unprotected eyes and skin should not exceed $10^4\ \mathrm{J\,m^{-2}}$.

7.1.2. UV-B and UV-C Exposure Standard

The exposure standard for the spectral region 315–200 nm (UV-B and UV-C) is based on an envelope action spectrum which combines the photokeratitis and skin erythema action spectra and which is defined as a smooth curve somewhat below the energies required for the development of observable effects.[111] The standard applies to occupational exposure during an 8-h working day and shows maximum sensitivity at a wavelength of 270 nm and an exposure dose of $30\,\mathrm{J\,m^{-2}}$. Maximum permissible exposures (MPEs) are presented in Table V together with the spectral effectiveness of the radiation relative to a wavelength of 270 nm. The MPE for monochromatic UVR sources in the wavelength region 315–200 nm can be determined directly from Table V, but for broadband UVR sources, an effective irradiance, E_{eff}, is calculated by summing the contributions from all the spectral components of the source, each contribution being weighted by the relative spectral effectiveness, according to

$$E_{\mathrm{eff}} = \sum E_s(\lambda)S(\lambda)\Delta\lambda \tag{26}$$

where E_{eff} is effective irradiance relative to a monochromatic source at 270 nm ($\mathrm{W\,m^{-2}}$), $E_s(\lambda)$ is the spectral irradiance at wavelength λ ($\mathrm{W\,m^{-2}\,nm^{-1}}$), $S(\lambda)$ is the relative spectral effectiveness at wavelength λ,

TABLE V
Maximum Permissible Exposure (MPE) for 8-h Period

Wavelength (nm)	MPE ($\mathrm{J\,m^{-2}}$)	Relative spectral effectiveness, $S(\lambda)$
200	1000	0.03
210	400	0.075
220	250	0.12
230	160	0.19
240	100	0.30
250	70	0.43
254	60	0.5
260	46	0.65
270	30	1.0
280	34	0.88
290	47	0.64
300	100	0.30
305	500	0.06
310	2000	0.015
315	10,000	0.003

and $\Delta\lambda$ is the bandwidth employed in the measurement or calculation of $E_s(\lambda)$ (nm).

The maximum permissible exposure time t_{max}, is then calculated as

$$t_{max} = 30(\mathrm{J\,m^{-2}})/E_{eff}(\mathrm{W\,m^{-2}})\ \mathrm{s} \tag{27}$$

The method of summation described by Eq. (26) assumes that there are no synergistic or protective interactions between wavelengths, although it is unlikely that these assumptions are true.

For UVR sources which emit line spectra arranged in simple exposure geometries, it may be possible to assess the MPE by calculation. In most practical situations, however, recourse to measurement is probably necessary.

7.2. Instrumentation for Assessing UV Exposure Hazards

The experimental determination of a UV exposure hazard is necessary either when the spectral power distribution of the source is unknown, or when the geometry of the source makes calculation prohibitive.

The most fundamental method is to measure the spectral irradiance of the source, $E_s(\lambda)$, and to combine these data with the relative spectral data with the relative spectral effectiveness, $S(\lambda)$, in order to calculate an effective irradiance, E_{eff}, as given by Eq. (26). However, this technique requires a spectroradiometer with high spectral resolution coupled with extremely good rejection of stray radiation, and probably demands a double monochromator.

Because of the severe experimental difficulties associated with absolute spectral radiometry, an assessment of the potential hazard from a UVR source is usually made through the use of a direct reading instrument whose spectral response has been designed to match the NIOSH "hazard curve."[99]

A prototype ultraviolet hazard monitor was developed by Roach in 1973.[112] This device made use of a spectrally weighted UV filter, a silicon photodiode, and an optical chopper to reduce noise. Although the instrument was found to be quite reliable and followed the hazard action spectrum fairly accurately, it was never produced commercially since the optical filter was not available in quantity. There were several further attempts in the 1970s to develop an appropriate hazard monitor, but most failed because of the principal difficulty of combining sufficient sensitivity to UV-B and UV-C radiation with adequate rejection of longer wavelength radiation.[99]

Figure 24. The IL 730A actinic radiometer (courtesy of International Light Inc., Newburyport, Massachusetts, U.S.A.).

Possibly the best instrument that is commercially available at present is the IL730A actinic radiometer (see Fig. 24), manufactured by International Light, Inc.

This device incorporates a quartz wide-angle diffuser, an interference filter, a blocking filter, and a "solar blind" vacuum phototube as the detector. The published spectral response of a typical instrument is shown in Fig. 25. The spectral response of the instrument is an adequate match to the NIOSH curve in the wavelength range 250–300 nm, but the response at longer wavelengths may give cause for concern. The solar blind vacuum phototube can exhibit variations in its long-wavelength response due to changes in the photoemissivity of the cathode with time and temperature. This can give rise to severe errors when trying to estimate the actinic hazard associated with sources which have a high UV-A component but whose spectral distribution in the actinic region extends only a short way below 315 nm. The most common example of such a source in medicine is probably the UV-A fluorescent lamp (see Fig. 23).

A survey of the levels of irradiance around three common sources of ultraviolet radiation found in physiotherapy departments has been carried out by Diffey[113] using an IL730A actinic radiometer, and the results are summarized in Table VI. The levels of UVR are quoted in terms of an effective irradiance in $W\,m^{-2}$ and so the MPE time (in seconds) is calculated by dividing this irradiance into the MPE for 270-nm radiation ($30\,J\,m^{-2}$). It is evident that a severe occupational exposure hazard exists

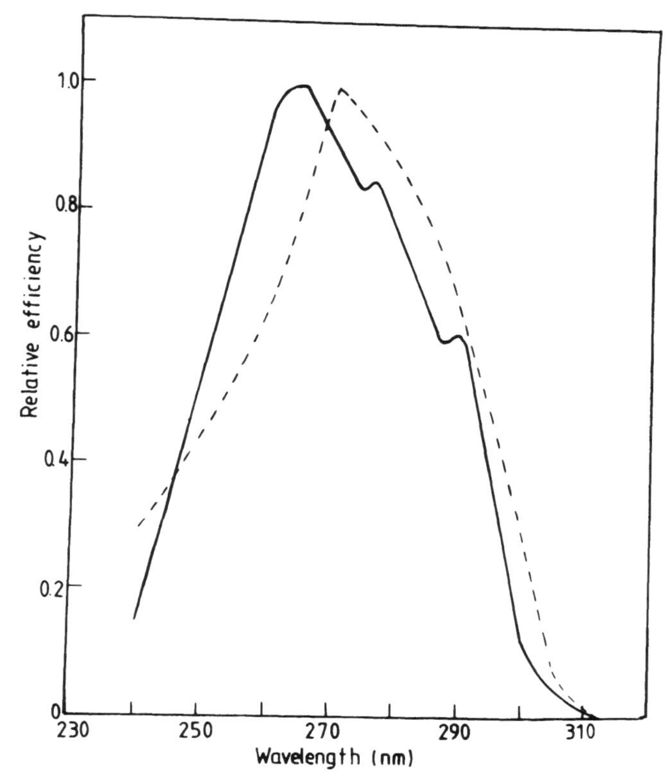

Figure 25. The published spectral response of the IL 730A actinic radiometer (——) used in conjunction with a SEE 240 solar blind vacuum phototube and an ACTS 270 filter compared with the NIOSH relative spectral efficiency curve (- - -).

in the direct beam of a Kromayer or Alpine lamp. Measurement of the effective irradiance in the vicinity of the beds used with the Alpine sunlamp and Theraktin UV bath indicated a MPE time of the order of 10 min. Consequently it would be inadvisable for staff to remain in the immediate vicinity of the patient during treatment with these two units due to the presence of scattered and reflected radiation from the patient's skin, sheets, and so on. In fact it is the custom in many physiotherapy departments for screens to be drawn around the patient and lamp during treatment and invariably the patient is issued UV opaque goggles.

8. PSORALEN PHOTOCHEMOTHERAPY

Probably the most exciting development of the use of ultraviolet radiation in clinical medicine in recent years has been psoralen photo-

TABLE VI
Exposure Hazard Associated with Three Sources of Ultraviolet Radiation Used in Physiotherapy Departments

Irradiation unit	Lamp	Typical treatment distance	Typical treatment time	Effective irradiance ($W\ m^{-2}$) at treatment distance (MPE time in parentheses)
Kromayer lamp	Medium-pressure mercury vapor arc in a quartz envelope	Contact	2 s	240 (0.1 s)
Alpine sun-lamp	Medium-pressure mercury vapor arc in an envelope designed not to emit ozone-producing radiation	0.5 m	3 min	7 (4 s)
Theraktin ultraviolet bath	Four Westinghouse FS40 fluorescent sunlamps (1.22 m long)	0.5 m	5–30 min	0.4 (75 s)

chemotherapy.[114] This form of treatment, known as PUVA, involves the combination of the photoactive drugs, psoralens (P), with long-wave ultraviolet radiation (UV-A) to produce a beneficial effect. Psoralen photochemotherapy has been used to treat many skin diseases in the past decade, although its principal success has been in the management of psoriasis, a disorder characterised by an accelerated cell cycle and rate of DNA synthesis. The mechanism of the treatment is thought to be that psoralens bind to DNA in the presence of UV-A, resulting in a subsequent transient inhibition of DNA synthesis and cell division.[115] The psoralens may be applied to the skin either topically or systematically; the latter route is generally preferred and the psoralens are administered as 8-methoxypsoralen (8-MOP).

Proper UV-A radiation dosimetry is important in PUVA therapy, not only to prevent severe erythema, which can result from over-exposure, but also to determine the lowest effective radiant exposure to minimize potential long-term damage.[116] As yet, PUVA therapy has not been shown to produce a higher than expected incidence of skin cancer, although its mutagenic potential has been indicated by the suspected mode of action of the psoralen group of drugs[117,118] and observations made on cultured cells[119] and in animals.[120–122] The only clinical study to date reporting an increased risk of cutaneous carcinoma and the reversal of the normal ratio of basal cell to squamous cell carcinoma[123] is open to criticism over (1) the use of inappropriate controls, (2) the inclusion of patients with previous exposure to carcinogenic agents, and (3) the failure to record the cumulative UV-A radiant exposure.

8.1. Treatment Regimen

The patient ingests the 8-MOP tablets and, two hours later, when the photosensitivity of the skin is at a maximum, he is exposed to UV-A radiation. If the psoriasis is generalized, whole-body exposure is given in the type of irradiation cabinet shown in Fig. 26. This unit incorporates 48 high-intensity UV-A fluorescent lamps with the spectral power distribution shown in Fig. 23. The initial UV-A radiant exposure (in $J\,m^{-2}$) is determined either from the criteria in Table VII or by evaluating the minimal photoxicity dose; small areas of psoralen-sensitized skin are irradiated with increasing exposures of UV-A, and that dose which produces a barely perceptible, well-defined erythema 72 h later is taken as the initial dose.[124] The radiant exposure is increased at regular intervals throughout the course of treatment to keep pace with increasing melanization. The exposure time may vary from a few minutes to up to an hour, depending upon the degree of pigmentation and also on the irradiance at the patient's skin, which reflects the design of the treatment cubicle. Values of UV-A irradiance in clinical treatment cubicles have been found to range from 16 to $140\,W\,m^{-2}$, although an irradiance of $50\,W\,m^{-2}$ is probably typical.

Treatment is given two or three times weekly until the psoriasis clears. The total time taken for this to occur will obviously vary considerably from one patient to another, and in some cases complete clearing of the lesions is never achieved. However, it would be fair to say that something like 25 treatments are required for clearing of the psoriatic lesions in most patients over a period of around 10 weeks. PUVA therapy is not a cure for psoriasis, and maintenance therapy is often needed at intervals of, say, once a week to once a month to prevent relapse. Since the encouraging results of PUVA therapy for psoriasis were first reported by Parrish and his colleagues in 1974, several other investigators have confirmed the efficacy of the treatment.[125–127]

8.2. Aspects of UV-A Dosimetry in PUVA Therapy

The ideal UV-A detector in photochemotherapy would be an instrument whose spectral response matched the therapeutic action spectrum of the treatment, and which had an angular response weighted according to a cosine function with a wide angle of view suitable for use

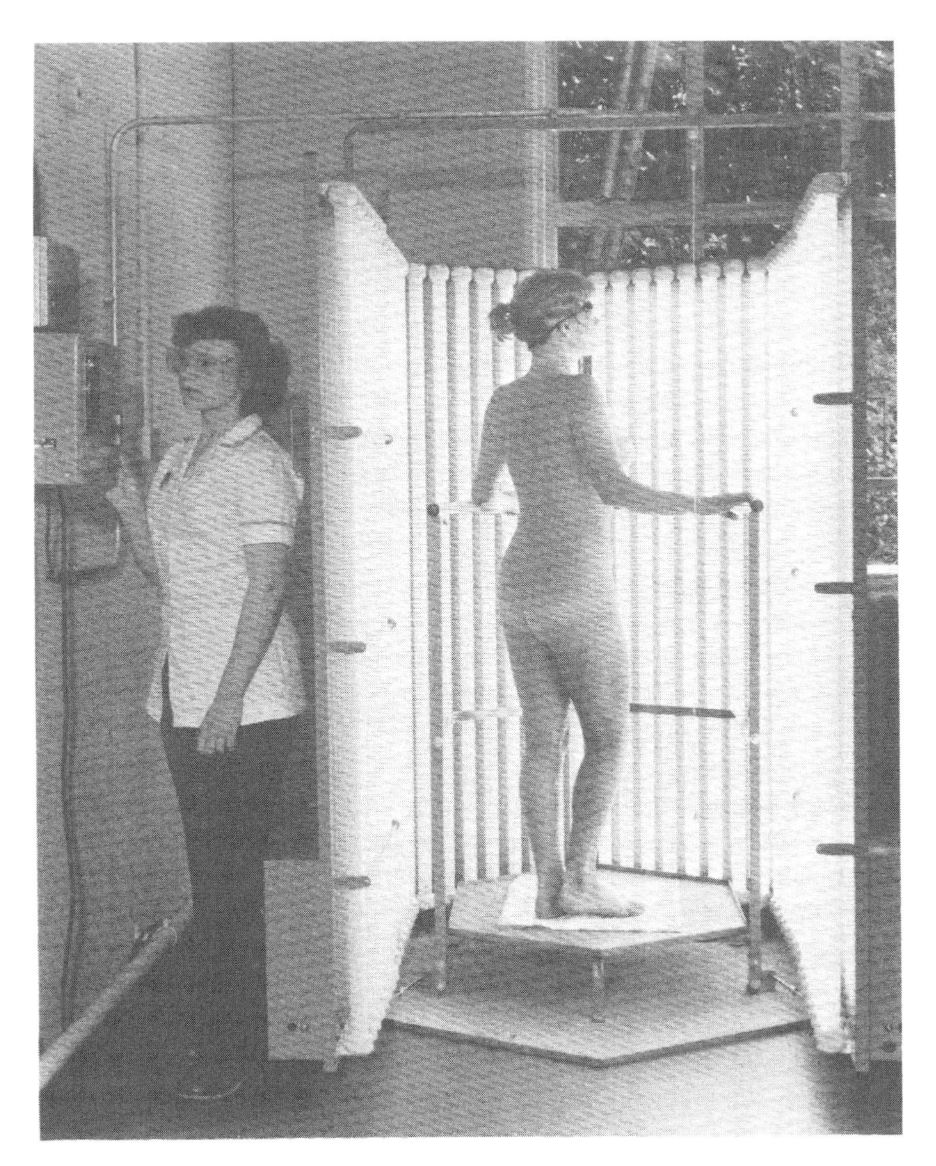

Figure 26. A whole-body UV-A irradiation cubicle (courtesy of Rank Stanley Cox, Ware, U.K.).

TABLE VII
Initial UV-A Radiant Exposures for Patients Embarking on PUVA Therapy

Skin type	Criteria	J m^{-2}
I	Always burn, never tan	1.5×10^4
II	Always burn, sometimes tan	2.5×10^4
III	Sometimes burn, always tan	3.5×10^4
IV	Never burn, always tan	4.5×10^4
V	Moderately pigmented subjects, e.g., Arabs	5.5×10^4
VI	Heavily pigmented subjects, e.g., Negroes	6.5×10^4

at short distances from extended arrays of fluorescent lamps. It is not yet possible to design such a detector, since the data relating to the therapeutic effectiveness of each wavelength in causing regression of the disease still remain uncertain. One end-point which has been considered is the erythema action spectrum of 8-MOP, but even in this field there is only limited agreement between workers. Nakayama *et al.*[43] found a broad spectrum for erythema in guinea-pig skin after 8-MOP in the range 320–380 nm with a maximum at 330 nm. Similar results were found by Owens *et al.*[128] whereas both Buck *et al.*[129] and Pathak[130] found the same range for humans, but with a maximum at 360 nm. The latter of the two papers also showed essentially the same results for guinea-pig skin. It should also be pointed out that the spectra obtained by both Buck *et al.*[129] and Pathak[130] were based on only small samples, i.e., three subjects in each study; and that, furthermore, these action spectra relate to the ability of different wavelengths to evoke an erythema response, which may bear no relationship to the ability to cause clearing of psoriatic lesions. More recently Young and Magnus[131] have shown that the action spectrum for 8-MOP induced sunburn cells in mammalian epidermis has maximal activity in the spectral interval 320–335 nm. These authors suggest that this spectral interval may also be the most active in PUVA therapy.

In practice, UV-A dosimetry is generally achieved by means of an optical radiation detector incorporating a solid state photodiode, one or more optical filters, some simple electronic circuitry, and a display.[132] One such detector is shown in Fig. 27.

The spectral responses of three different models of UV-A detector commonly found in clinical practice are shown in Fig. 28. The curves have been normalized to unity at a wavelength of 365 nm for each UV-A detector. The fractional deviation of the angular response of each of

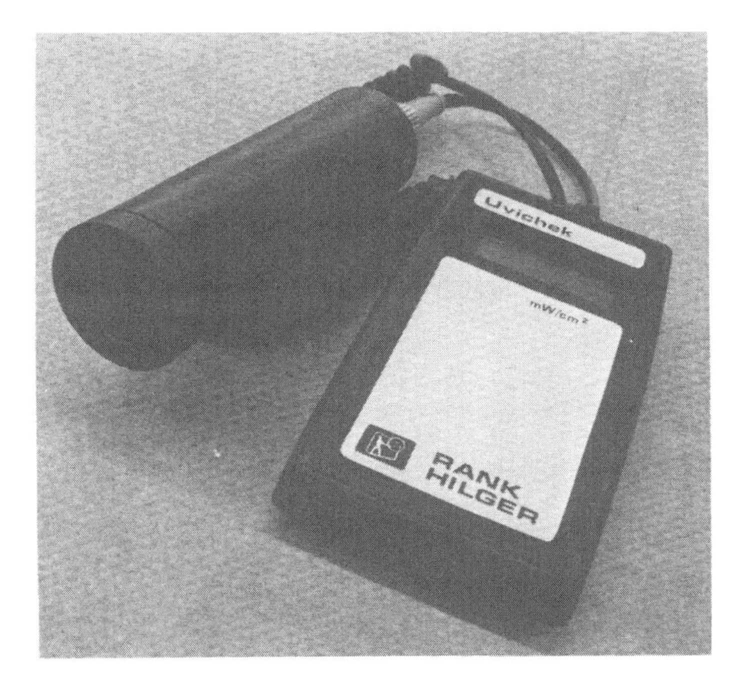

Figure 27. The "Uvichek" UV-A detector (courtesy of Rank-Hilger, Margate, England).

these UV-A detectors from a cosine-weighted response is shown as a function of angle of incidence in Fig. 29. The fractional deviation at angle θ from the normal is defined as

$$1 - \left[\frac{\text{meter reading at } \theta°}{\text{meter reading at } 0°} \Big/ \cos\theta\right] \tag{28}$$

The meter reading in $\mathrm{W\,m^{-2}}$ generally referes to monochromatic irradiation at 365 nm incident normally on the sensor since calibration can be conveniently carried out using this strong mercury line. In fact, with a medium pressure mercury vapor lamp it is possible to isolate radiation at wavelengths between 350 and 380 nm, with over 96% of the radiation coming from the group of mercury emission lines between 365 and 367 nm.

The UV-A irradiance at a particular point exposed to the radiation from the lamps in a treatment cubicle is simple equal to the integral of the spectral irradiance at that point over the UV-A range of wavelengths,

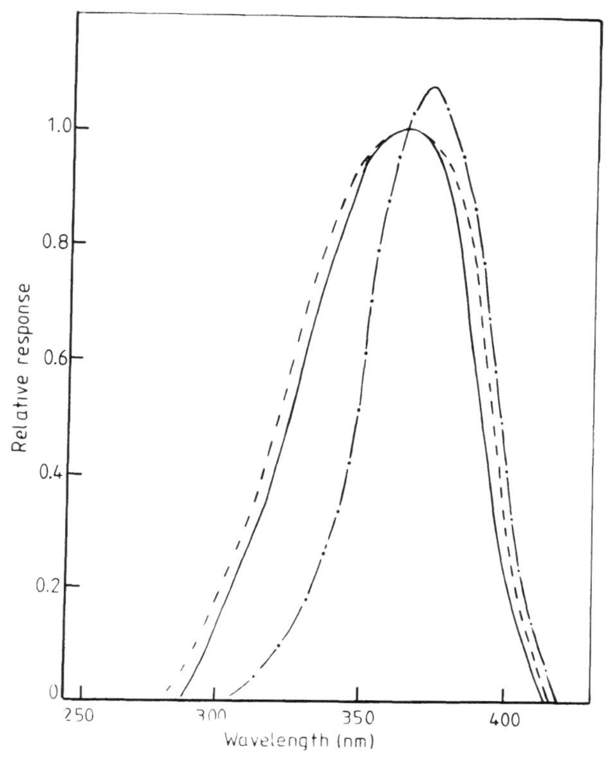

Figure 28. The spectral response of three commercially available UV-A detectors.[132] ——, Uvichek (Rank-Hilger, Margate, England); – – –, Blak-Ray J221 (Ultraviolet Products Inc., San Gabriel, California, U.S.A.); –·–·, PUVA Meter (Waldmann GmbH & Co., Villingen-Schwenningen, West Germany).

that is, 315–400 nm. This may be expressed mathematically as

$$\text{EV-A irradiance} = \int_{315}^{400} E_s(\lambda)\, d\lambda \ \text{W m}^{-2} \tag{29}$$

where $E_s(\lambda)$ is the spectral irradiance ($\text{W m}^{-2}\,\text{nm}^{-1}$) at a wavelength λ nm.

Ideally it is desirable to use a detector with a spectral response which matches the action spectrum for the regression of lesions in psoriasis. However, since this action spectrum still remains uncertain, perhaps the next best thing would be to use a detector which combined a cosine-weighted angular response with a flat spectral response from 315 to 400 nm and a zero response at all other wavelengths. This detector would measure true UV-A irradiance irrespective of the spectral power dis-

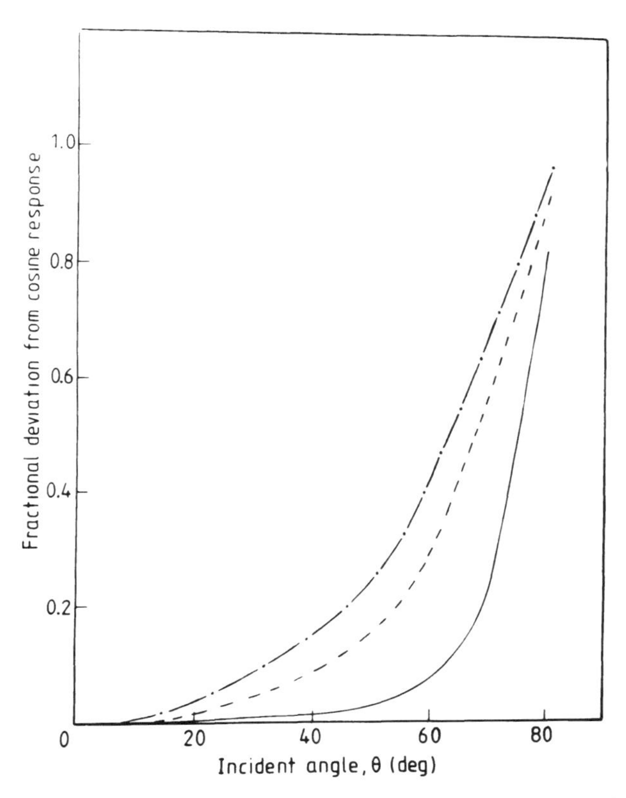

Figure 29. The angular response of three commercially available UV-A detectors.[132] ——, Uvichek (Rank-Hilger, Margate, England); – – –, Blak-Ray J221 (Ultraviolet Products Inc., San Gabriel, California, U.S.A.); –·–·, PUVA Meter (Waldmann GmbH & Co., Villingen-Schwenningen, West Germany).

tribution of the source. Unfortunately such a detector does not exist for various physical reasons.

Nevertheless, presently used UV-A detectors can still enable computation of the true UV-A irradiance or even some biological irradiance (assuming a therapeutic action spectrum is postulated) if there is access to the spectral power distribution of the radiation source, and the spectral response and angular response of the UV-A detector.

8.2.1. UV-A Dosimetry in Clinical Practice

A review[133] of published papers on PUVA therapy for psoriasis showed large differences in the total UV-A radiant exposure needed to clear psoriatic lesions. It was concluded that these large variations could be attributed partly to an uncertain approach to the measurement of ultraviolet radiation combined with poorly calibrated UV-A detectors. This finding prompted a survey[134] of UV-A dosimetry as practised in nine centers in England engaged in PUVA therapy.

At each treatment center a "standard" UV-A detector (IL442A phototherapy radiometer, International Light Inc., Newburyport, Massachusetts, U.S.A.) and the center's own UV-A detector were placed in turn in a specially constructed jig at approximately the position of a patient's skin 1 m above the podium on which the patient stands.

The results, shown in Table VIII, indicate a wide variation in the accuracy of the UV-A meters encountered. The extremes are a detector reading 63% of the UV-A irradiance as determined by the "standard" IL442A to a detector reading a factor of 2 high, a difference of over 300% in relative sensitivity.

The implication of this variability in detector response is that a patient attending center 1, for example, and receiving a UV dose of $5 \times 10^4\ \mathrm{J\,m^{-2}}$ on the basis of that center's UV-A meter would actually receive $3.2 \times 10^4\ \mathrm{J\,m^{-2}}$. On the other hand, a patient attending center 4 and again expected to receive $5 \times 10^4\ \mathrm{J\,m^{-2}}$ would in fact receive $5.8 \times 10^4\ \mathrm{J\,m^{-2}}$. Although each center would soon be able to assess the clinical efficacy of their own PUVA unit and would probably use their UV-A detector simply to monitor changes in lamp output, it is clear that comparison of UV dosimetry from the literature on PUVA treatment might be subject to some uncertainty.

TABLE VIII
Results of the Survey of UV-A Dosimetry in PUVA Treatment Centers[a]

Center	UV-A irradiance ($\mathrm{W\,m^{-2}}$) measured by "standard" IL442A	UV-A detector reading ($\mathrm{W\,m^{-2}}$) at treatment centers		
		International Light IL442A	Blak-Ray J-221	Waldmann PUVA meter
1	17.6		27.5	
2	9.3			
3	40.3	56.5	53.0	
4	53.1		46.0	
5	19.7		30.5	
6	44.1	27.7		
7	23.1		37.5	
8	41.3		47.5	82.0
9	79.3	100.0		
	27.6			29.2

[a] Reference 134.

8.2.2. Anatomical Uniformity of Irradiation

Since the topology of a human subject does not permit uniform irradiation, it is strictly incorrect to speak of a single treatment radiant exposure. The relative distribution of UV-A exposure on the surface of a manikin resulting from whole-body irradiation in a step-in cubicle containing 62 UV-A fluorescent lamps is illustrated in Fig. 30. The measurements were made by attaching 65 UV-A film badges incorporating phenothiazine as the chromophore to the surface of an unclothed manikin.[135] A large fraction of the body surface area receives more than 70% of the maximum exposure, although areas such as the axillae and groin receive a smaller fraction, as expected.

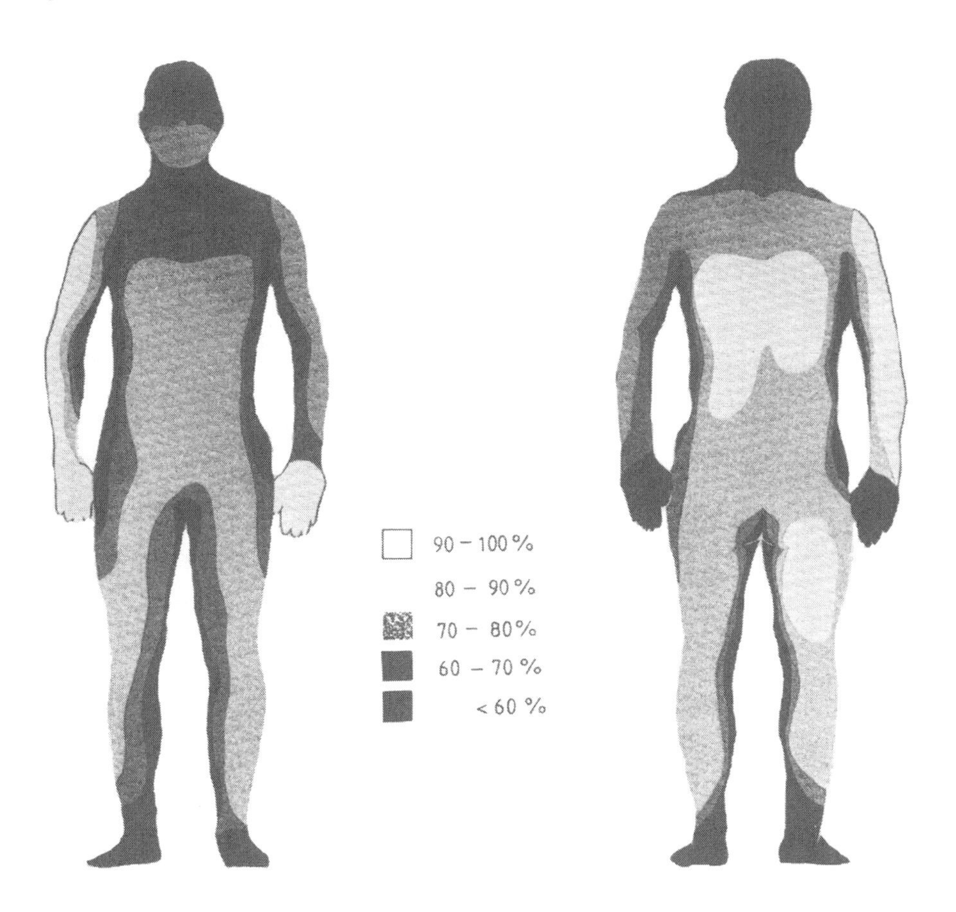

Figure 30. The relative distribution of UV-A exposure on the surface of a manikin resulting from whole-body irradiation in a step-in cubicle.[135]

8.3. Solar UV-A Received by Patients Undergoing PUVA Therapy

A psoriatic patient treated with oral psoralen photochemotherapy is exposed to two sources of UV-A resulting in a known dose from the artificial sources used in the PUVA treatment cubicle, and an unknown dose, generally to the hands and face, from the UV-A present in sunlight. Both of these sources of exposure will contribute to the photocarcinogenic risk associated with the treatment. Some estimate of the latter exposure would be valuable so that the solar UV-A exposures received can be judged in perspective to the dose of artificial UV-A when assessing the overall photocarcinogeneity of the treatment. It is generally recognized that the patient may be exposed to solar UV-A during a period when the drug is photoactive and, for this reason, the patient is advised to wear suitable UV-A protective glasses for the remainder of the day after ingestion of the psoralen to minimize the risk of cataract induction. Also some workers advise patients to avoid exposure to strong sunshine during the 8-h period after the administration of the 8-MOP tablets,[126] while others recommend protection of the skin by suitable clothing, which may include a brimmed hat and gloves.[136]

Even in the UK, the level of UV-A irradiance from the sun can be as high as that in some PUVA irradiation cubicles, yet there is little quantitative information available on the solar UV-A dose received by patients after they have ingested the drug 8-MOP and while it is still photoactive. The only experimental study of personal solar UV-A dosimetry reported to date[137] was carried out on patients undergoing PUVA therapy at two centers in the UK. Solar UV-A exposure was measured in this study using a thin plastic film which incorporates 8-MOP as the chromophore.[76]

Fifty-six patients took part in the study, 40 in Newcastle (latitude 55°N) and 16 in Glasgow (latitude 56°N). Patients were instructed to attach the film badge to their lapel at the same time as they took their psoralen tablets and to wear the badge for the remainder of that day. Their clothing is removed during treatment and so the film badge was not exposed to the UV-A lamps used in the treatment cubicle. At the end of that day the film badge was removed, put in an envelope and handed in to the nurse when they next attended for treatment.

A frequency histogram of the recorded solar UV-A doses is shown in Fig. 31. The results have been grouped in intervals of $2 \times 10^3\ \mathrm{J\,m^{-2}}$, which is of the order of the limit of resolution. Exposure levels range from less than $2 \times 10^3\ \mathrm{J\,m^{-2}}$ to almost $7 \times 10^4\ \mathrm{J\,m^{-2}}$, 7% of patients receiving doses in excess of $4 \times 10^4\ \mathrm{J\,m^{-2}}$. The distribution appears approximately log normal with a median radiant exposure of around $10^4\ \mathrm{J\,m^{-2}}$.

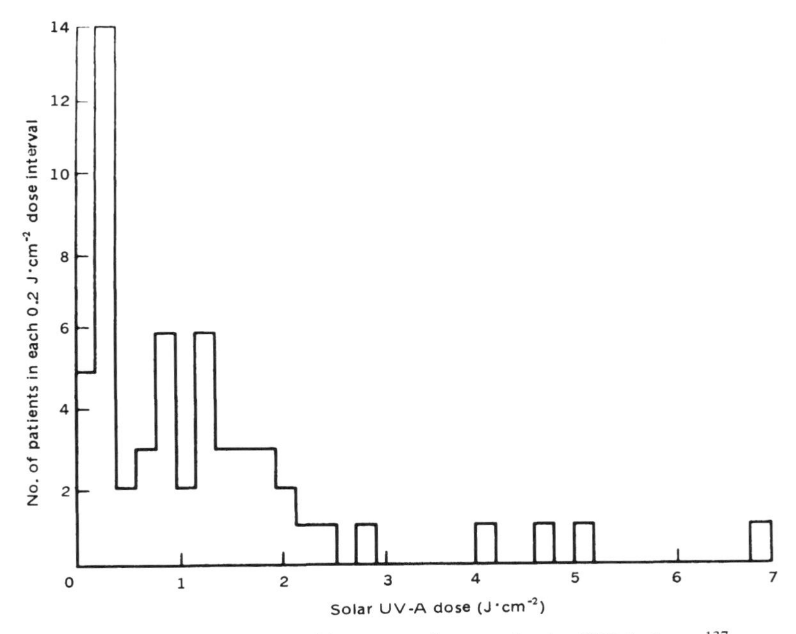

Figure 31. The frequency histogram of personal solar UV-A doses.[137]

In order to extend the usefulness of these experimental results, a theoretical model was developed by Diffey[138] for calculating the solar UV-A dose so that the effect of variables, such as geographical latitude and season, could be studied. The model considered four factors: the sunlight exposure habits of patients on PUVA therapy, the variation of solar UV-A irradiance from the time the patient ingests the psoralen until it is cleared, the effectiveness of different wavelengths in the solar UV-A spectrum in initiating psoralen photocarcinogenisis, and the pharmacology of parenteral 8-MOP.

The application of this model to the temporal and geographical conditions pertaining to the experimental study of Moseley *et al.*[137] yielded a calculated solar UV-A radiant exposure of $1.4 \times 10^4\, J\, m^{-2}$, a value which is in good agreement with the observed median dose of $10^4\, J\, m^{-2}$. Further application of the model showed that patients treated during the summer months in the UK may well receive a UV radiant exposure which is comparable with, or even in excess of, the dose of artificial UV-A received in the treatment center and consequently should be advised about limiting their exposure to sunlight on the treatment days. In the winter months in the UK the UV-A intensity in sunlight is such that no unduly restrictive precaution need apply. However, at locations between about 40°N and 40°S, the solar UV-A intensity in midwinter is still sufficiently high that unnecessary sunlight exposure on treatment days should be avoided.

9. PERSONAL UVR MONITORING

It is well established that exposure to ultraviolet radiation can have both beneficial and detrimental effects on humans.[139,140]

Although only relatively low doses of UVR are required to form measurable levels of vitamin D in the skin,[141] people deprived of UVR are more liable to develop osteomalacia, particularly when there is impaired utilization or dietary insufficiency of vitamin D. Groups which have been found to show low levels of vitamin D associated with chronic underexposure to UVR are the elderly, submariners, and Asiatic immigrants.[142–144]

On the other hand, excessive repeated exposure to UVR is well known to induce both skin cancer and aging effects.[12] Also the increasing use of drugs, which have photosensitivity side effects (e.g., phenothiazines, nalidixic acid, and certain tetracyclines), have made a greater percentage of the population more liable to show abnormal sensitivity to UVR.[145]

At present, environmental conservationists are becoming increasingly concerned that changes in the atmosphere ozone mantle may be induced both by high-flying supersonic aircraft and by the buildup of freons and related compounds. Should significant changes occur in the depth of the ozone layer, then it can be expected that the amount of UVR reaching the Earth's surface will be altered.[146,147] Apart from a possible rise in the incidence of nonmelanoma skin cancer following a reduction of stratospheric ozone,[148] there is also concern that depletion of the ozone layer could have a more serious effect on marine ecosystems such as primary producers, zooplankton, fish eggs, and larvae.[149]

Coupled with this is the steady increase in the number of artificial ultraviolet sources available for medical, industrial, military, and cosmetic use, which has necessitated the introduction of safety standards for UV exposure.

A recent report by the World Health Organization[3] on environmental health criteria for ultraviolet radiation has recommended that certain studies be carried out to obtain the information required both for the adequate evaluation of health risks and for the establishment of appropriate protective measures and guidelines. The report stresses that population studies using personal monitoring devices for UV-B radiation are needed to determine the fraction of the daily natural UV dose received by persons at risk either from UVR deficiency or excess, or from occupational exposure, and concludes that the development of personal UVR monitoring devices is of the utmost priority.

Natural UVR is generally measured with solid state detectors, often used in conjunction with optical filters. In particular the Robertson–Berger meter, which measures those wavelengths in the global spectrum less than 320 nm, has been used to monitor natural UVR continuously at several sites throughout the world.[150] A different, yet complimentary, approach is the use of various photosensitive films as UVR dosimeters. The principle is to relate the degree of deterioration of the films, usually in terms of changes in their optical properties, to the incident UVR dose. The principal advantages of the film dosimeter are that it provides a simple means of integrating UVR exposure continuously and that it allows numerous sites, inaccessible to bulky and expensive instrumentation, to be compared simultaneously. The most widely used photosensitive film is the polymer, polysulfone (see Section 5.4), and examples of its use as a personal UVR dosimeter will be described in this section.

9.1. Environmental Exposure to UVR

It is evident that the natural UVR exposure received by different individuals will depend not only upon the quality and quantity of the UV environment but also on the behavior of the individuals concerned. It would be expected that outdoor workers, for example, would by and large receive much greater personal UVR doses than indoor workers. Nevertheless, it is difficult to estimate from the recordings of stationary UV-B detectors the typical doses received by people under a variety of situations. It is then in the area of personal UVR dosimetry that polymer film badges have been most widely used.

In all of the studies reported to date the polysulfone film badges have been worn on the lapel site as this site is judged to receive approximately the same solar UVR exposure as the hands and face. The results of various field studies[65,66,68,69] have been combined to yield representative annual erythemally effective UV-B doses [termed UV-B(EE)] for groups of people in different occupations and pursuing different leisure activities. These UV-B(EE) doses, which are summarized in Table IX, are equivalent to that hypothetical radiant exposure of monochromatic 297-nm radiation which would produce the same erythemal effect as the solar radiation received in the various situations. A minimal erythema in unacclimatized Caucasian skin can be produced with a UV-B(EE) dose of around 150 J m^{-2}. The "sunburn units," given in column 3 of Table IX, are simply defined here as the UV-B(EE) dose divided by 150, and can be thought of as the number of minimal erythema doses received. Also included in Table IX are the annual

TABLE IX
Representative Annual UV-B(EE) Doses Received by People in Northern Europe (Latitude 50°–60°N)

Type of exposure	Annual UV-B(EE) dose (J m^{-2})	Number of "sunburn units"
Occupational		
Outdoor, e.g., farmers	6×10^4	400
Mixed, e.g., policemen	2×10^4	133
Indoor, e.g., office staff	10^4	67
Vacational (during summertime)		
Sunbathing for 3 weeks in Mediterranean area (40°N)	1.5×10^4	100
Sightseeing for 3 weeks in Central Europe (45°N)	4×10^3	27
Ambient		
Calculated under clear day conditions at 52°N	2.4×10^5	1610
Measured at Belsk-Duzy, Poland (52°N)	—	1521

ambient UV-B(EE) radiant exposure on an unshaded, horizontal surface under clear sky conditions at a latitude of 52°N calculated using a computer model,[151] and the measured annual sunburn units at Belsk-Duzy, Poland (52°N) averaged over the four years 1976–1979.[150] These data were obtained with the Robertson–Berger meter (see Section 5.3.9) and are less than one would expect for clear day conditions due principally to meteorological factors. It should also be remembered that the "sunburn units" recorded by the Robertson–Berger meter are not directly comparable with the calculated values because of differences between the spectral sensitivity of the meter and the CIE erythema action spectrum which was used in the calculations.

Inspection of Table IX shows that indoor workers may receive more erythemally effective (and presumably carcinogenic) solar UV-B in a 3-week period during the summer sunbathing along the shores of the Mediterranean Sea than in the rest of the year going about their normal duties in Northern Europe. This factor may be why malignant melanoma, unlike basal cell and squamous cell carcinomas, tend to be most common at a relatively young age, as it has been suggested that intense intermittent exposure to ultraviolet radiation may play a role in the induction of malignant melanoma.[24]

9.2. UVR and Drug Photosensitivy of the Skin

There are many oral drugs that appear to cause skin photosensitivity. The diagnosis of photosensitivity is often made clinically from the history presented by the patient. The dose of a therapeutic drug is normally known where a patient suffers an adverse photosensitive effect, but the radiant exposure of sunlight received is unknown. This factor is probably just as important as the drug and its dose in assessing diagnosis and in planning treatment and prevention. Personal UVR dosimetry may then be useful in this respect, and the findings of one such investigation[67] are summarized below.

The study was carried out in groups of 5–6 institutionalized psychiatric patients on phenothiazine therapy. Chlorpromazine was the most commonly prescribed drug and commonly causes skin photosensitivity as an adverse side effect. The project was done in five separate hospitals, four in England, one in Eire. The medical and nursing staff recorded daily the hours that each of the patients under study spent out of doors in bright sunlight over July and August, 1976. They also recorded symptoms of photosensitivity on a special form provided; from this a simple numerical scoring system for the severity of photosensitive symptoms was derived which could be compared with UVR dose recorded by the polysulfone film badge.

On analysis of the results it was found that badge dose and the recorded hours as spent in direct sunlight had a strong positive correlation ($p < 0.001$). What was especially interesting was that the badge dose also correlated positively with the symptom score of the patients ($p < 0.001$), but that symptom score and hours outdoors did not correlate; this is plausible, for the time outdoors in our climate will include many periods of overcast weather. These results suggest that, given suitable situations, this method is a valid approach for collecting objective data where previously results were subjective and conclusions had to be tentative.

9.3. UVR and Long-Stay Geriatric Patients

One of the features of long-stay geriatric patients is the high incidence of osteomalacia coupled with very low levels of plasma-25-hydroxy-vitamin D [25(OH)D]. Such levels are due to a combination of negligible exposure to sunlight, and poor absorption and dietary intake of vitamin D. The efficacy of artificial UVR in restoring plasma-25-(OH)D to normal in-patients in long-stay geriatric wards has been studied by Corless *et al.*[70] and Toss *et al.*[72] In both studies the radiation sources

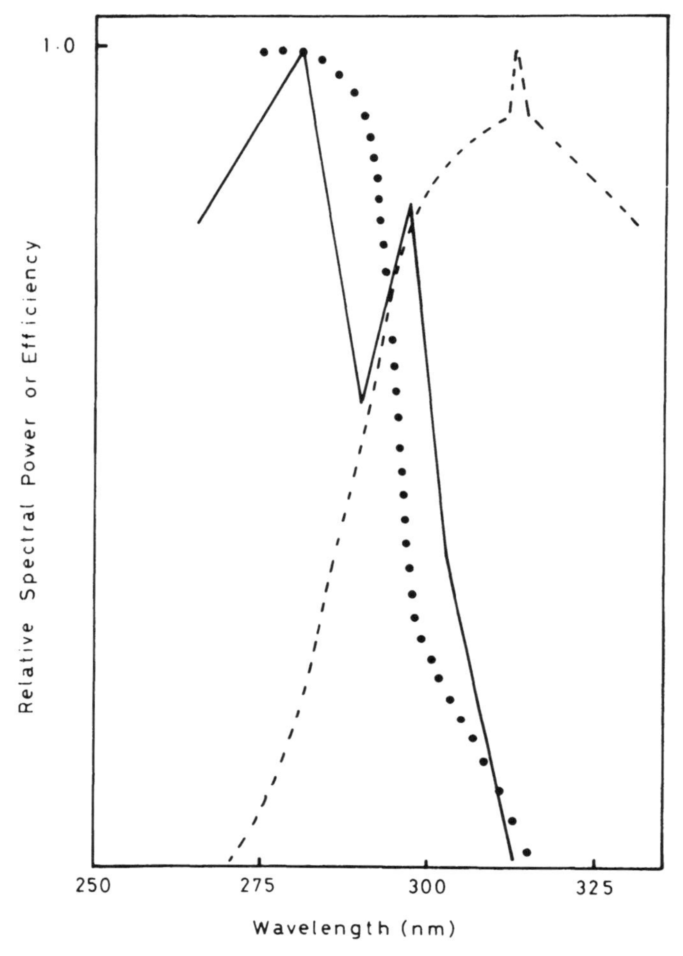

Figure 32. The action spectrum for antirachitic activity (——), the spectral response of 1-μm-thick polysulfone film (· · ·), and the spectral power distribution of Westinghouse FS fluorescent sunlamps (- - -).

consisted of Westinghouse FS fluorescent sunlamps, which emit a spectrum extending from 270 to 380 nm peaking at 313 nm, and the radiation doses received by the patients was monitored using polysulfone film. The UV dose recorded by the film badge was expressed in terms of equivalent radiant exposure at the most effective wavelength for healing of rickets, namely, 280 nm.[152] In order to express the dose in this manner, allowance was made for the differences between the spectral power distribution of the Westinghouse sunlamp, the spectral response of polysulfone film, and the action spectrum for antirachitic activity (see Fig. 32).

In both studies the results showed that supplementation of the ward lighting by UVR increased plasma-25(OH)D in the patients by an amount sufficient to bring the level in depleted subjects into the normal range.

ACKNOWLEDGMENT. I am grateful to Dr. A. F. McKinlay for his helpful comments on the text.

REFERENCES

1. W. W. Coblentz, Report on the Copenhagen meeting of the second international congress on light, *Science* **76,** 412–415 (1932).
2. J. A. Parrish, R. R. Anderson, F. Urbach, and D. Pitts, *UV-A: Biological Effects of Ultraviolet Radiation with Emphasis on Human Responses to Longwave Ultraviolet,* Wiley, Chichester (1978).
3. World Health Organisation, *Environmental Health Criteria 14: Ultraviolet Radiation,* WHO, Geneva (1979).
4. A. Stimson, *Photometry and Radiometry for Engineers,* Wiley, New York (1974).
5. C. L. Wyatt, *Radiometric Calibration: Theory and Methods,* Academic, New York (1978).
6. J. W. T. Walsh, *Photometry,* Dover, New York (1965).
7. J. Jagger, in *The Science of Photobiology* (K. C. Smith, ed.), pp. 1–26, Plenum, New York (1977).
8. W. Harm, *Biological Effects of Ultraviolet Radiation,* Cambridge University Press, Cambridge (1980).
9. H. J. Morowitz, Absorption effects in volume irradiation of microorganisms *Science* **111,** 229–230 (1950).
10. H. E. Johns, Dosimetry in photochemistry, *Photochem. Photobiol.* **8,** 547–563 (1968).
11. J. Jagger, T. Fossum, and S. McCaul, Ultraviolet irradiation of suspensions of microorganisms: Possible errors involved in the estimation of average fluence per cell, *Photochem. Photobiol.* **21,** 379–382 (1975).
12. B. E. Johnson, F. Daniels Jr., and I. A. Magnus, in *Photophysiology* (A. C. Giese, ed.), Vol. IV, pp. 139–202, Academic, London (1968).
13. B. L. Diffey, The consistency of studies of ultraviolet erythema in normal human skin, *Phys. Med. Biol.* **27,** 715–720 (1982).
14. J. A. Parrish, C. Y. Ying, M. A. Pathak, and T. B. Fitzpatrick, in *Sunlight and Man* (T. B. Fitzpatrick, ed.), pp. 131–142, Universtity of Tokyo Press, Tokyo (1974).
15. A. P. Warin, The ultraviolet erythemas in man, *Br. J. Dermatol.* **98,** 473–477 (1978).
16. M. F. Holick, J. A. MacLaughlin, J. A. Parrish, and R. R. Anderson, in *The Science of Photomedicine* (J. D. Regan and J. A. Parrish, eds.), pp. 195–218, Plenum, New York (1982).
17. H. F. DeLuca and H. K. Schnoes, Metabolism and mechanism of action of vitamin D, *Ann. Rev. Biochem.* **45,** 631–642 (1976).
18. A. C. Giese, *Living with Our Sun's Ultraviolet Rays,* Plenum, New York (1976).
19. P. Unna, *Histopathologie der Hautkrankheiten,* August Hirschwald, Berlin (1894).

20. W. Dubreuilh, Des hyperkeratoses circonscriptes, *Ann. Dermatol. Syphiligr.* (*Paris*) **7** (Ser. 3), 1158–1204 (1896).
21. H. F. Blum, *Carcinogenesis by Ultraviolet Light,* Princeton Univ. Press, Princeton (1959).
22. G. Swanbeck and L. Hillström, Analysis of etiological factors of squamous cell skin cancer of different locations: 4. Concluding remarks, *Acta Derm. Venereal.* (*Stockholm*) **51,** 151–156 (1971).
23. F. Urbach, J. H. Epstein, and P. D. Forbes, in *Sunlight and Man* (T. B. Fitzpatrick, ed.), pp. 259–284, University of Tokyo Press, Tokyo (1974).
24. F. Urbach, in *The Science of Photomedicine* (J. D. Regan and J. A. Parrish, eds.), pp. 261–292, Plenum, New York (1982).
25. D. G. Pitts and A. P. Cullen, Ocular ultraviolet effects from 295 to 335 nm in the rabbit eye. A preliminary report, DHEW (NIOSH) Publication No. 77–130. National Institute of Occupational Safety and Health, Division of Biomedical and Behavioral Science, Cincinnati, Ohio (1977).
26. S. Zigman, G. Griess, T. Yulo, and J. Schultz, Ocular protein alterations by near UV light, *Exp. Eye Res.* **15,** 255–265 (1973).
27. S. Duke-Elder, *System of Ophthalmology,* Vol. 14: *Injuries,* Part 2: *Nonmechanical injuries,* C. V. Mosby, St. Louis (1972).
28. S. Zigman, T. Yulo, T. Paxhia, S. Salceda, and M. Datiles, Comparative studies of human cataracts, Abstracts of the Association for Research in Vision and Ophthalmology, Sarasota, Florida (1977).
29. D. J. Rees, *Health Physics,* Butterworths, London (1967).
30. K. W. Hausser and W. Vahle, Die Abhängigkeit des Lichterythems und der Pigmentbildung von der Schwingungszahl (Wellenlänge) der erregenden Strahlung, *Strahlentherapie* **13,** 41–71 (1922).
31. M. Luckiesh, L. L. Holladay, and A. H. Taylor, Reactions of untanned skin to ultraviolet radiation, *J. Opt. Soc. Am.* **20,** 423–432 (1930).
32. K. W. Hausser and W. Vahle, Sonnenbrand und Sonnenbraünung, *Wiss. Veröff. Siemens Konzern* **6,** 101–120 (1927).
33. W. W. Coblentz, R. Stair, and J. M. Hogue, The spectral erythemic reaction of the human skin to ultraviolet radiation, *Proc. Natl. Acad. Sci.* **17,** 401–405 (1931).
34. W. W. Coblentz and R. Stair, Data on the spectral erythemic reaction of the untanned human skin to ultraviolet radiation, *Bur. Stand. J. Res.* **12,** 13–14 (1934).
35. I. A. Magnus, Studies with a monochromator in the common idiopathic photodermatoses, *Br. J. Dematol.* **76,** 245–264 (1964).
36. M. A. Everett, R. L. Olson, and R. M. Sayre, Ultraviolet erythema, *Arch. Dermatol.* **92,** 713–719 (1965).
37. R. G. Freeman, D. W. Owens, J. M. Knox, and H. T. Hudson, Relative energy requirements for an erythemal response of skin to monochromatic wavelengths of ultraviolet present in the solar spectrum, *J. Invest. Dermatol.* **47,** 586–592 (1966).
38. D. Berger, F. Urbach, and R. E. Davies, The action spectrum of erythema induced by ultraviolet radiation: Preliminary report, in *Proc. XIII Congressus Internationalis Dermatologiae* (W. Jadassohn and C. G. Schirren, eds.), Vol. 2, pp. 1112–1117, Springer-Verlag, Berlin (1968).
39. R. L. Olson, R. M. Sayre, and M. A. Everett, Effect of anatomic location and time on ultraviolet erythema, *Arch. Dermatol.* **93,** 211–215 (1966).
40. R. M. Sayre, R. L. Olson, and M. A. Everett, Quantitative studies on erythema, *J. Invest. Dermatol.* **46,** 240–244 (1966).
41. D. J. Cripps and C. A. Ramsay, Ultraviolet action spectrum with a prism-grating monochromator, *Br. J. Dermatol.* **82,** 584–592 (1970).

42. L. A. Mackenzie and W. Frain-Bell, The construction and development of a grating monochromator and its application to the study of the reaction of the skin to light, *Br. J. Dermatol.* **89,** 251–264 (1973).
43. Y. Nakayama, F. Morikawa, M. Fukuda, M. Hamano, K. Toda, and M. A. Pathak, in *Sunlight and Man* (T. B. Fitzpatrick, ed.) pp. 591–611. University of Tokyo Press, Tokyo (1974).
44. R. L. Olson, R. M. Sayre, and M. A. Everett, Effect of field size on ultraviolet minimal erythema dose, *J. Invest. Dermatol.* **45,** 516–519 (1965).
45. H. Brodthagen, in *The Biologic Effects of Ultraviolet Radiation* (*with Emphasis on the Skin*) (F. Urbach, ed.), pp. 459–467, Pergamon, Oxford (1969).
46. W. H. Goldsmith and F. F. Hellier, in *Recent Advances in Dermatology,* 2nd edition, p. 364, Blakiston, New York (1954).
47. B. L. Diffey, The variation of erythema with monochromator bandwidth, *Arch. Dermatol.* **111,** 1070–1071 (1975).
48. J. C. van der Leun, in *Research Progress in Organic, Biological and Medicinal Chemistry* (U. Gallo and L. Santamaria, eds.), Vol. 3, pp. 711–736, North-Holland, Amsterdam (1972).
49. I. Willis, A. Kligman, and J. Epstein, Effects of long ultraviolet rays on human skin: Photoprotective or photoaugmentative?, *J. Invest. Dermatol.* **59,** 416–420 (1972).
50. J. C. van der Leun and T. Stoop, in *The Biologic Effects of Ultraviolet Radiation* (*with Emphasis on the Skin*) (F. Urbach, ed.), pp. 251–254, Pergamon, Oxford (1969).
51. B. L. Diffey and A. F. McKinlay, The UVB content of "UVA fluorescent lamps" and its erythemal effectiveness in human skin, *Phys. Med. Biol.* **28,** 351–358 (1983).
52. W. Harm, *Biological Effects of Ultraviolet Radiation,* Cambridge University Press, Cambridge (1980).
53. J. G. Calvert and J. N. Pitts, Jr., *Photochemistry,* Wiley, New York (1966).
54. C. G. Hatchard and C. A. Parker, A new sensitive chemical actinometer. II. Potassium ferrioxalate as a standard chemical actinometer, *Proc. R. Soc. London* **A325,** 518–536 (1956).
55. A. H. Carter and J. Weiss, The transfer of excitation energy from uranium ions in solution, *Proc. R. Soc. London* **A174,** 351–370 (1940).
56. G. R. Seely, in *Photophysiology* (A. C. Giese, ed.), Vol. III, pp. 1–32, Academic, New York (1968).
57. R. E. Davies, in *The Biologic Effects of Ultraviolet Radiation* (*with Emphasis on the Skin*) (F. Urbach, ed.), pp. 437–443, Pergamon, Oxford (1969).
58. F. Urbach, in *The Biologic Effects of Ultraviolet Radiation* (*with Emphasis on the Skin*) (F. Urbach, ed.), pp. 635–650, Pergamon, Oxford (1969).
59. K. I. Jacobson and R. E. Jacobson, *Imaging Systems,* Focal Press, London (1976).
60. J. Kosar, *Light Sensitive Systems,* Wiley, New York (1965).
61. S. A. Jackson, A film badge dosimeter for UVA radiation, *J. Biomed. Eng.* **2,** 63–64 (1980).
62. J. B. Ali and R. E. Jacobson, The use of diazo film as a film badge dosimeter, *J. Photographic Sci.* **28,** 172–176 (1980).
63. J. B. A. Card, An investigation of diazo materials as dosimeters for ultraviolet and visible radiation, Ph.D. thesis, CNAA (1982).
64. A. Davis, G. H. W. Deane, and B. L. Diffey, Possible dosimeter for ultraviolet radiation, *Nature* **261,** 169–170 (1976).
65. A. V. J. Challoner, D. Corless, A. Davis, G. H. W. Deane, B. L. Diffey, S. P. Gupta, and I. A. Magnus, Personnel monitoring of exposure to ultraviolet radiation, *Clin. Exp. Dermatol.* **1,** 175–179 (1976).
66. J. F. Leach, V. E. McLeod, A. R. Pingstone, A. Davis, and G. H. W. Deane,

Measurement of the ultraviolet doses received by office workers, *Clin. Exp. Dermatol.* **3,** 77–79 (1978).
67. M. F. Corbett, A. Davis, and I. A. Magnus, Personnel radiation dosimetry in drug photosensitivity: Field study of patients on phenothiazine therapy, *Br. J. Dermatol.* **98,** 39–46 (1978).
68. B. L. Diffey, O. Larkö, and G. Swanbeck, UV-B doses received during different outdoor activities and UV-B treatment of psoriasis, *Br. J. Dematol.* **106,** 33–41 (1982).
69. O. Larkö and B. L. Diffey, Natural UV-B radiation received by people with outdoor, indoor and mixed occupations and UV-B treatment of psoriasis, *Clin. Exp. Dermatol.* **8,** 279–285 (1983).
70. D. Corless, S. P. Gupta, S. Switala, J. M. Barragry, B. J. Boucher, R. D. Cohen, and B. L. Diffey, Response of plasma 25-hydroxyvitamin D to ultraviolet irradiation in long-stay geriatric patients, *Lancet* **ii,** 649–651 (1978).
71. A. R. Young, A. V. J. Challoner, I. A. Magnus, and A. Davis, UVR radiometry of solar simulated radiation in experimental photocarcinogenesis studies, *Br. J. Dermatol.* **106,** 43–52 (1982).
72. G. Toss, R. Andersson, B. L. Diffey, P. A. Fall, O. Larkö, and L. Larsson, Oral vitamin D and ultraviolet radation for the prevention of vitamin D deficiency in the elderly, *Acta. Med. Scand.* **212,** 157–162 (1982).
73. M. G. Pepper and B. L. Diffey, Automatic read-out device for ultraviolet radiation polymer film dosimeters, *Med. Biol. Eng. Comp.* **18,** 467–473 (1980).
74. A. Davis, B. L. Diffey, and T. J. Tate, A personal dosimeter for biologically effective solar UV-B radiation, *Photochem. Photobiol.* **34,** 283–286 (1981).
75. B. L. Diffey, A. Davis, M. Johnson, and T. R. Harrington, A dosimeter for long wave ultraviolet radiation, *Br. J. Dermatol.* **97,** 127–130 (1977).
76. B. L. Diffey and A. Davis, A new dosimeter for the measurement of natural ultraviolet radiation in the study of photodermatoses and drug photosensitivity, *Phys. Med. Biol.* **23.** 318–323 (1978).
77. T. J. Tate, B. L. Diffey, and A. Davis, An ultraviolet radiation dosimeter based on the photosensitsing drug nalidixic acid, *Photochem. Photobiol.* **31,** 27–30 (1980).
78. B. L. Diffey, I. Oliver, and A. Davis, A personal dosimeter for quantifying the biologically-effective sunlight exposure of patients receiving benoxaprofen, *Phys. Med. Biol.* **27,** 1507–1513 (1982).
79. A. Zweig and W. A. Henderson, Jr., A photochemical mid-ultraviolet dosimeter for practical use as a sunburn dosimeter, *Photochem. Photobiol.* **24,** 543–549 (1976).
80. M. V. Mayneord and T. J. Tulley, The measurement of non-ionizing radiations for medical purposes, *Proc. R. Soc. Med.* **36,** 411–422 (1943).
81. E. J. Gillham, *Radiometric Standards and Measurements,* Notes on Applied Science No. 23, HMSO, London (1961).
82. E. Schwarz, Semi-conductor thermopiles, *Research* **5,** 407–411 (1952).
83. M. Slater and G. S. Melville, A method for improving the precision of inverse square measurements, *Br. J. Radiol.* **31,** 392–394 (1958).
84. W. M. Doyle, B. C. McIntosh, and J. Geist, Implementation of a system of optical calibration based on pyroelectric radiometry, *Opt. Eng.* **15,** 541–548 (1976).
85. P. O. Byrne and F. T. Farmer, A self-calibrating black-body radiometer for use in the UV, IR and visible spectrum, *J. Phys. E: Sci. Instr.* **5,** 590–591 (1972).
86. Hamamatsu catalog of silicon photocells, Hamamatsu TV Co. Ltd, Japan (1981).
87. 87. B. L. Diffey and R. J. Oliver, An ultraviolet radiation monitor for routine use in physiotherapy, *Physiotherapy* **67,** 64–66 (1981).
88. P. J. Mountford and M. G. Pepper, A wide-band ultraviolet radiation monitor for measuring the output of monochromators used in dermatology, *Phys. Med. Biol.* **26,** 925–930 (1981).

89. Mullard, *Cadmium Sulphide Photoconductive Cells,* Mullard Ltd., London (1969).
90. Hamamatsu catalog of CdS, CdSe photoconductive cells, Hamamatsu TV Co. Ltd., Japan.
91. B. L. Diffey, T. R. Elkerton, and M. F. Diprose, A simple light meter for use in dermatology, *Med. Biol. Eng.* **14,** 101–102 (1976).
92. B. L. Diffey and A. Miller, A detector for monitoring the output of ultraviolet radiation sources used in the photochemotherapy of psoriasis, *Phys. Med. Biol.* **23,** 514–517 (1978).
93. J. Jagger, A small and inexpensive ultraviolet dose-rate meter useful in biological experiments, *Rad. Res.* **14,** 394–403 (1961).
94. D. F. Robertson, Solar ultraviolet radiation in relation to human sunburn and skin cancer, Ph.D. thesis, University of Queensland, Australia (1972).
95. D. F. Robertson, in *The Biologic Effects of Ultraviolet Radiation* (*with Emphasis on the Skin*) (F. Urbach, ed.), pp. 433–436, Pergamon, Oxford (1969).
96. D. S. Berger, The sunburning ultraviolet meter: Design and performance, *Photochem. Photobiol.* **24,** 587–593 (1976).
97. D. S. Berger, in *The Role of Solar Ultraviolet Radiation in Marine Ecosystems,* (J. Calkins, ed.), pp. 181–192, Plenum, New York (1982).
98. G. Busuoli, UV-dosimetry with TLD materials, Proceedings of ISPRA course on Applied Thermoluminescence Dosimetry, Commission of the European Communities (1982).
99. D. H. Sliney and M. Wołbarsht, *Safety with Lasers and Other Optical Sources: A Comprehensive Handbook,* Plenum, New York (1980).
100. E. W. Palmer, M. C. Hutley, A. Franks, J. R. Verrill, and B. Gale, Diffraction Gratings, Rep. Prog. Phys. **38,** 975–1048 (1975).
101. J. R. Moore, Sources of error in spectroradiometry, *Lighting Res. Tech.* **12,** 213–220 (1980).
102. P. J. Key and T. H. Ward, The establishment of ultraviolet spectral emission scales using synchrotron radiation, *Metrologia* **14,** 17–29 (1978).
103. H. E. Johns and A. M. Rauth, Theory and design of high intensity UV monochromators for photobiology and photochemistry, *Photochem. Photobiol.* **4,** 673–692 (1965).
104. R. D. Saunders and H. J. Kostkowski, Accurate solar spectroradiometry in the UV-B, Optical Radiation News No. 24, U.S. Department of Commerce, NBS, Washington (1978).
105. B. L. Diffey, Sunlamps, sunbeds and solaria, *Dermatol. Practice* **1,** 13–18 (1982).
106. DIN 5031, Deutsche Normen, Berlin (1978).
107. CIE, Compte Rendu 9, Tagung, Berlin/Karlsruhe, S 596-625 (1935).
108. B. L. Diffey and A. V. J. Challoner, Absolute radiation dosimetry in photochemotherapy, *Phys. Med. Biol.* **23,** 1124–1129 (1978).
109. National Institute for Occupational Safety and Health, Criteria for a recommended standard . . . occupational exposure to ultraviolet radiation, DHEW, Washington DC (1972).
110. National Radiological Protection Board, Protection against ultraviolet radiation in the workplace, HMSO, London (1977).
111. D. H. Sliney, The merits of an envelope action spectrum for ultraviolet radiation exposure criteria, *Am. Ind. Hyg. Assn. J.* **33,** 644–653 (1972).
112. T. Roach, Final report on a method for field evaluation of UV radiation hazards, prepared by CBS Laboratories for the National Institute for Occupational Safety and Health (NIOSH), Contract No. HSM-99-72-144, NIOSH, Cincinnati (1973).
113. B. L. Diffey, *Ultraviolet Radiation in Medicine,* Adam Hilger, Bristol (1982).

114. J. A. Parrish, T. B. Fitzpatrick, L. Tannehbaum, and M. A. Pathak, Photochemotherapy of psoriasis with oral methoxsalen and longwave ultraviolet light, *New Engl. J. Med.* **291,** 1207–1211 (1974).
115. J. F. Walter, J. J. Voorhees, W. H. Kelsey, and E. A. Duell, Psoralen plus blacklight inhibits epidermal DNA synthesis, *Arch. Dermatol.* **107,** 861–865 (1973).
116. Task Forces on Psoriasis and Photobiology of the American Academy of Dermatology, PUVA Statement, *Arch. Dermatol.* **113,** 1195 (1977).
117. Lancet Editorial, Ultraviolet Radiation and Cancer of the Skin, *Lancet* **i,** 537–538 (1978).
118. J. H. Eptein, Risks and benefits of the treatment of psoriasis, *New England J. Med.* **300,** 852–853 (1979).
119. C. F. Arlett, Mutagenesis in cultured mammalian cells, *Stud. Biophys.* **36/37,** 139–147 (1973).
120. F. Urbach, Modification of ultraviolet carcinogenesis by photoactive agents, *J. Invest. Dermatol.* **32,** 373–387 (1959).
121. M. A. Pathak, F. Daniels, C. E. Hopkins, and T. B. Fitzpatrick, Ultraviolet carcinogenesis in albino and pigmented mice receiving furocoumarins: Psoralen and 8-methoxypsoralen, *Nature* **183,** 728–730 (1959).
122. A. C. Griffin, R. E. Hakim, and J. Knox, The wavelength effect upon erythemal and carcinogenic response in psoralen treated mice, *J. Invest. Dermatol.* **31,** 289–294 (1959).
123. R. S. Stern, L. A. Thibodeau, R. A. Kleinerman, J. A. Parrish, and T. B. Fitzpatrick, Risk of cutaneous carcinoma in patients treated with oral methoxsalen photochemotherapy for psoriasis, *New England J. Med.* **300,** 809–813 (1979).
124. K. Wolff, F. Gschnait, H. Hönigsmann, K. Konrad, J. A. Parrish, and T. B. Fitzpatrick, Phototesting and dosimetry for photochemotherapy, *Br. J. Dermatol.* **96,** 1–10 (1977).
125. K. Wolff, T. B. Fitzpatrick, J. A. Parrish, F. Gschnait, B. Gilchrest, H. Honigsmann, M. A. Pathak, and L. Tannenbaum, Photochemotherapy for psoriasis with orally administered methoxsalen, *Arch. Dermatol.* **112,** 943–950 (1976).
126. T. Lakshmipathi, P. W. Gould, L. A. Mackenzie, B. E. Johnson, and W. Frain-Bell, Photochemotherapy in the treatment of psoriasis, *Br. J. Dermatol.* **96,** 587–594 (1977).
127. J. W. Melski, L. Tanenbaum, J. A. Parrish, T. B. Fitzpatrick, and H. L. Bleich, Oral methoxsalen photochemotherapy for the treatment of psoriasis: A co-operative clinical trial, *J. Invest. Dermatol.* **68,** 328–335 (1977).
128. D. W. Owens, J. M. Glicksman, R. G. Freeman, and R. Carnes, Biologic action spectra of 8-methoxypsoralen determined by monochromatic light, *J. Invest. Dermatol.* **51,** 435–440 (1968).
129. H. W. Buck, I. A. Magnus, and A. D. Porter, The action spectrum of 8-methoxypsoralen for erythema in human skin: Preliminary studies with a monochromator, *Br. J. Dermatol.* **72,** 249–255 (1960).
130. M. A. Pathak, Mechanism of psoralen photosensitization and in vivo biological action spectrum of 8-methoxypsoralen, *J. Invest. Dermatol.* **37,** 397–407 (1961).
131. A. R. Young and I. A. Magnus, An action spectrum for 8-MOP induced sunburn cells in mammalian epidermis, *Br. J. Dermatol.* **104,** 541–548 (1981).
132. D. Stobbart and B. L. Diffey, A comparison of some commercially available UVA meters used in photochemotherapy, *Clin. Phys. Physiol. Meas.* **1,** 267–273 (1980).
133. B. L. Diffey, PUVA: A review of ultraviolet dosimetry, *Br. J. Dermatol.* **98,** 703–706 (1978).
134. B. L. Diffey, A. V. J. Challoner, and P. J. Key, A survey of the ultraviolet radiation emissions of photochemotherapy units, *Br. J. Dermatol.* **102,** 301–306 (1980).

135. B. L. Diffey, T. R. Harrington, and A. V. J. Challoner, A comparison of the anatomical uniformity of irradiation in two different photochemotherapy units, *Br. J. Dermatol.* **99,** 361–363 (1978).
136. S. Rogers, J. Marks, S. Shuster, D. Vella Briffa, A. Warin, and M. Greaves, Comparison of photochemotherapy and dithranol in the treatment of chronic plaque psoriasis, *Lancet* **i,** 455–458 (1979).
137. H. Moseley, B. L. Diffey, J. M. Marks, and R. M. Mackie, Personal solar UV-A doses received by patients undergoing oral psoralen photochemotherapy for psoriasis, *Br. J. Dermatol.* **105,** 573–577 (1981).
138. B. L. Diffey, A mathematical model of the biologically effective dose of solar UVA received by patients undergoing oral psoralen photochemotherapy for psoriasis, *Phys. Med. Biol.* **26,** 1129–1135 (1981).
139. F. Ellinger, *Medical Radiation Biology,* Charles C. Thomas, Springfield, Illinois (1957).
140. R. M. Neer, T. R. A. Davis, A. Walcott, S. Koski, P. Schepis, I. Taylor, L. Thorington, and R. J. Wurtman, Stimulation by artificial lighting of calcium absorption in elderly human subjects, *Nature* **229,** 244–257 (1971).
141. E. Gorter, On rickets, *J. Paediatr.* **4,** 1 (1934).
142. A. N. Exton-Smith, B. R. Stanton, and A. C. M. Windsor, *Nutrition of Housebound Old People,* King Edward's Hospital Fund for London (1972).
143. D. Corless, M. Beer, B. J. Boucher, S. P. Gupta, and R. D. Cohen, Vitamin-D status in long-stay geriatric patients, *Lancet* **i,** 1404 (1975).
144. M. A. Preece, S. Tomlinson, C. A. Ribot, J. Pietrek, H. T. Korn, D. M. Davies, J. A. Ford, M. G. Dunnigan, and J. L. H. O'Riordan, Studies of Vitamin-D deficiency in Man, *Q. J. Med. New Ser.* **XLIV,** 575–589 (1975).
145. Register of Adverse Reactions, Committee on Safety of Medicines (1964–71).
146. P. Goldsmith, A. F. Tuck, J. S. Foot, E. L. Simmons, and R. L. Newson, Nitrogen oxides, nuclear weapon testing, Concorde and stratospheric ozone, *Nature* **244,** 545–551 (1973).
147. R. D. Hudson and E. I. Reed (eds), *The Stratosphere: Present and Future,* NASA Reference Publication 1049 (1979).
148. F. R. de Gruijl and J. C. van der Leun, A dose–response model for skin cancer induction by chronic UV exposure of a human population, *J. Theor. Biol.* **83,** 487–504 (1980).
149. J. Calkins (ed.), *The Role of Solar Ultraviolet Radiation in Marine Ecosystems,* Plenum, New York (1982).
150. D. S. Berger and F. Urbach, A climatology of sunburning ultraviolet radiation, *Photochem. Photobiol.* **35,** 187–192 (1982).
151. B. L. Diffey, The calculation of the spectral distribution of natural ultraviolet radiation under clear day conditions, *Phys. Med. Biol.* **22,** 309–316 (1977).
152. A. Knudson and F. Benford, Quantitative studies of the effectiveness of UV-radiation of various wavelength in rickets, *J. Biol. Chem.* **124,** 287–290 (1938).

Index